BREAST CANCER AND THE ENVIRONMENT

A LIFE COURSE APPROACH

Committee on Breast Cancer and the Environment:
The Scientific Evidence, Research Methodology, and Future Directions

Board on Health Care Services

Board on Health Sciences Policy

INSTITUTE OF MEDICINE
OF THE NATIONAL ACADEMIES

THE NATIONAL ACADEMIES PRESS
Washington, D.C.
www.nap.edu

THE NATIONAL ACADEMIES PRESS   500 Fifth Street, NW   Washington, DC 20001

NOTICE: The project that is the subject of this report was approved by the Governing Board of the National Research Council, whose members are drawn from the councils of the National Academy of Sciences, the National Academy of Engineering, and the Institute of Medicine. The members of the committee responsible for the report were chosen for their special competences and with regard for appropriate balance.

This study was supported by a contract between the National Academy of Sciences and Susan G. Komen for the Cure®. Any opinions, findings, conclusions, or recommendations expressed in this publication are those of the author(s) and do not necessarily reflect the view of the organizations or agencies that provided support for this project.

Library of Congress Cataloging-in-Publication Data

Institute of Medicine (U.S.). Committee on Breast Cancer and the Environment: The Scientific Evidence, Research Methodology, and Future Directions.
   Breast cancer and the environment : a life course approach / Committee on Breast Cancer and the Environment: The Scientific Evidence, Research Methodology, and Future Directions, Board on Health Care Services, Board on Health Sciences Policy.
      p. ; cm.
   Includes bibliographical references.
   ISBN 978-0-309-22069-9 (pbk.) — ISBN 978-0-309-22070-5 (PDF)
   I. Title.
   [DNLM: 1. Breast Neoplasms—etiology. 2. Environmental Exposure—adverse effects. 3. Breast Neoplasms—prevention & control. 4. Risk Factors. WP 870]

   362.19699'449—dc23
                           2012007110

Additional copies of this report are available from the National Academies Press, 500 Fifth Street, NW, Keck 360, Washington, DC 20001; (800) 624-6242 or (202) 334-3313; http://www.nap.edu.

For more information about the Institute of Medicine, visit the IOM home page at: **www.iom.edu.**

Printed in the United States of America

Cover credit: Illustration by Diana Ong/Getty Images.

The serpent has been a symbol of long life, healing, and knowledge among almost all cultures and religions since the beginning of recorded history. The serpent adopted as a logotype by the Institute of Medicine is a relief carving from ancient Greece, now held by the Staatliche Museen in Berlin.

Suggested citation: IOM (Institute of Medicine). 2012. *Breast cancer and the environment: A life course approach*. Washington, DC: The National Academies Press.

"Knowing is not enough; we must apply.
Willing is not enough; we must do."
—Goethe

INSTITUTE OF MEDICINE
OF THE NATIONAL ACADEMIES

Advising the Nation. Improving Health.

THE NATIONAL ACADEMIES
Advisers to the Nation on Science, Engineering, and Medicine

The **National Academy of Sciences** is a private, nonprofit, self-perpetuating society of distinguished scholars engaged in scientific and engineering research, dedicated to the furtherance of science and technology and to their use for the general welfare. Upon the authority of the charter granted to it by the Congress in 1863, the Academy has a mandate that requires it to advise the federal government on scientific and technical matters. Dr. Ralph J. Cicerone is president of the National Academy of Sciences.

The **National Academy of Engineering** was established in 1964, under the charter of the National Academy of Sciences, as a parallel organization of outstanding engineers. It is autonomous in its administration and in the selection of its members, sharing with the National Academy of Sciences the responsibility for advising the federal government. The National Academy of Engineering also sponsors engineering programs aimed at meeting national needs, encourages education and research, and recognizes the superior achievements of engineers. Dr. Charles M. Vest is president of the National Academy of Engineering.

The **Institute of Medicine** was established in 1970 by the National Academy of Sciences to secure the services of eminent members of appropriate professions in the examination of policy matters pertaining to the health of the public. The Institute acts under the responsibility given to the National Academy of Sciences by its congressional charter to be an adviser to the federal government and, upon its own initiative, to identify issues of medical care, research, and education. Dr. Harvey V. Fineberg is president of the Institute of Medicine.

The **National Research Council** was organized by the National Academy of Sciences in 1916 to associate the broad community of science and technology with the Academy's purposes of furthering knowledge and advising the federal government. Functioning in accordance with general policies determined by the Academy, the Council has become the principal operating agency of both the National Academy of Sciences and the National Academy of Engineering in providing services to the government, the public, and the scientific and engineering communities. The Council is administered jointly by both Academies and the Institute of Medicine. Dr. Ralph J. Cicerone and Dr. Charles M. Vest are chair and vice chair, respectively, of the National Research Council.

www.national-academies.org

LAUREN ZEISE, Chief, Reproductive and Cancer Hazard Assessment Branch, Office of Environmental Health Hazard Assessment, California Environmental Protection Agency

Study Staff

LOIS JOELLENBECK, Study Director
JANE DURCH, Senior Program Officer
SHARYL NASS, Senior Program Officer
NIHARIKA SATHE, Research Assistant (from November 2010)
CASSANDRA CACACE, Research Assistant (until December 2010)
ASHLEY McWILLIAMS, Senior Program Assistant (until December 2010)
PATRICK BURKE, Financial Associate
ROGER HERDMAN, Director, Board on Health Care Services
ANDREW POPE, Director, Board on Health Sciences Policy

Commissioned Paper Authors

REBECCA SMITH-BINDMAN, University of California, San Francisco School of Medicine
LINDA DIX-COOPER, University of California, Berkeley

Reviewers

This report has been reviewed in draft form by individuals chosen for their diverse perspectives and technical expertise, in accordance with procedures approved by the National Research Council's Report Review Committee. The purpose of this independent review is to provide candid and critical comments that will assist the institution in making its published report as sound as possible and to ensure that the report meets institutional standards for objectivity, evidence, and responsiveness to the study charge. The review comments and draft manuscript remain confidential to protect the integrity of the deliberative process. We wish to thank the following individuals for their review of this report:

Mary Helen Barcellos-Hoff, NYU Langone Medical Center
Julia G. Brody, Silent Spring Institute
Diana Chingos, USC Norris Comprehensive Cancer Center
Kathryn Guyton, U.S. Environmental Protection Agency
Leena Hilakivi-Clarke, Georgetown Lombardi Comprehensive Cancer Center
William A. Knaus, The University of Virginia Health Sciences Center
Ruth M. Parker, Emory University School of Medicine
Lorenz Rhomberg, Gradient
Stephen H. Safe, Texas A&M University
Michael Thun, American Cancer Society

David M. Umbach, National Institute of Environmental Health Sciences, National Institutes of Health
Sandy Walsh, California Breast Cancer Organizations
Noel S. Weiss, University of Washington

Although the reviewers listed above have provided many constructive comments and suggestions, they were not asked to endorse the conclusions or recommendations, nor did they see the final draft of the report before its release. The review of this report was overseen by **David G. Hoel,** Medical University of South Carolina, and **David A. Savitz,** Brown University. Appointed by the National Research Council and the Institute of Medicine, they were responsible for making certain that an independent examination of this report was carried out in accordance with institutional procedures and that all review comments were carefully considered. Responsibility for the final content of this report rests entirely with the authoring committee and the institution.

Acknowledgments

The committee is grateful for the contribution of participants at its three public meetings. The presentations and discussions at these meetings were valuable in informing the committee about relevant research findings, issues of interest in the research community, the perspectives of advocacy organizations, and the concerns of individuals with breast cancer and their families. The agendas for these meetings appear in Appendix A.

The committee was also ably assisted by Linda Dix-Cooper from the University of California, Berkeley, and Dr. Rebecca Smith-Bindman from the University of California, San Francisco, from whom the committee commissioned papers.

The study was conducted with the generous support of Susan G. Komen for the Cure®. The Institute of Medicine staff worked closely with Dr. Amelie Ramirez, who is a member of the Susan G. Komen for the Cure Scientific Advisory Board and Director of the Institute for Health Promotion Research at the University of Texas Health Science Center at San Antonio. Dr. Ramirez was helpful and supportive as the committee's point of contact with the Scientific Advisory Board, which originally requested this study. In addition, Kendall Bergman graciously assisted the staff in the administrative coordination with Komen for the Cure.

The committee and project staff appreciate the work of copy editor Laura Penny and gratefully acknowledge valuable assistance within the National Academies from Laura Harbold DeStefano, Greta Gorman, Diedtra Henderson, Jillian Laffrey, William McLeod, Janice Mehler, Abbey Meltzer, Michael Park, Christine Stencel, Vilija Teel, and Lauren Tobias.

Contents

APPENDIXES

*Appendix F is available online at http://www.nap.edu/catalog.php?record_id=13263.

Tables, Figures, and Boxes

TABLES

xv

FIGURES

BOXES

Acronyms and Abbreviations

ACS	American Cancer Society
ADH	alcohol dehydrogenase
ADI	acceptable daily intake
AFFF	aqueous fire fighting foams
AFP	alpha fetoprotein
AhR	aryl hydrocarbon receptor
AHRQ	Agency for Healthcare Research and Quality
AICR	American Institute for Cancer Research
ALDH	aldehyde dehydrogenase
AMPK	AMP-activated protein kinase
AR	attributable risk
ASTDR	Agency for Toxic Substances and Disease Registry
ATH	adipose tissue hypoxia
BaP	benzo[a]pyrene
BBD	benign breast disease
BCAC	Breast Cancer Association Consortium
BMI	body mass index
BPA	bisphenol A
BPC3	Breast and Prostate Cancer Cohort Consortium
CAR	constitutive androstate receptor
CDC	Centers for Disease Control and Prevention
CFR	Code of Federal Regulations
CGEMS	Cancer Genetic Markers of Susceptibility project

CI	confidence interval
CISNET	Cancer Intervention and Surveillance Modeling Network
CT	computed tomography
CTS	California Teachers Study
CYP	cytochrome P-450
DAG	directed acyclic graphs
DBP	disinfection by-products
DCIS	ductal carcinoma in situ
DDE	dichlorodiphenyldichloroethylene
DDT	dichlorodiphenyltrichloroethane
DEHP	di(2-ethylhexyl) phthalate; *also* bis(2-ethylhexyl) phthalate
DEP	diethyl phthalate
DES	diethylstilbestrol
DHEA	dehydroepiandrosterone
DINP	diisononyl phthalate
DMBA	7,12-dimethylbenz[a]anthracene
DNA	deoxyribonucleic acid
DSHEA	Dietary Supplement Health and Education Act
ECA	European Chemical Agency
ECM	extracellular matrix
EDC	endocrine disrupting compound
EFSA	European Food Safety Authority
EGFR	epidermal growth factor receptor
ELF-EMF	extremely low frequency electromagnetic field
EPA	Environmental Protection Agency
ER	estrogen receptor
ER–	estrogen receptor negative
ER+	estrogen receptor positive
ERE	estrogen response element
FAO	Food and Agriculture Organization of the United Nations
FDA	Food and Drug Administration
FSH	follicle stimulating hormone
GAO	Government Accountability Office
GST	glutathione S-transferase
GWAS	genome-wide association studies
HER	human epidermal growth factor receptor
HERS	Heart and Estrogen/Progestin Replacement Study

HHS	Department of Health and Human Services
HPA axis	hypothalamic–pituitary–adrenal axis
HPG axis	hypothalamic–pituitary–gonadal axis
HPV program	High Production Volume Chemical program
HR	hazard ratio
HT	hormone therapy
IARC	International Agency for Research on Cancer
IGF	insulin-like growth factor
IGFBP	insulin-like growth factor binding protein
IOM	Institute of Medicine
IPCS	(WHO) International Program on Chemical Safety
LH	leutenizing hormone
MAPK	mitogen-activated protein kinase
MBzP	monobenzyl phthalate
MCPP	mono-(3-carboxylpropyl) phthalate
MEP	monoethyl phthalate
MMTV	mouse mammary tumor virus
MNU	N-methyl-N-nitrosourea; *also* N-nitroso-N-methylurea
MPA	medroxyprogesterone acetate
MQSA	Mammography Quality Standards Act
MRFIT	Multiple Risk Factor Intervention Trial
mRNA	messenger RNA
NAS	National Academy of Sciences
NAT	N-acetyltransferase
NCI	National Cancer Institute
NHANES	National Health and Nutrition Examination Survey
NHS	Nurses' Health Study
NIAAA	National Institute on Alcohol Abuse and Alcoholism
NIEHS	National Institute of Environmental Health Sciences
NIH	National Institutes of Health
NNH	number needed to harm
NNT	number needed to treat
NPCR	National Program of Cancer Registries
NRC	National Research Council
NTP	National Toxicology Program
OEHHA	Office of Environmental Health Hazard Assessment
OR	odds ratio
OSHA	Occupational Safety and Health Administration

OVX	ovariectomized
PAH	polycyclic aromatic hydrocarbon
PAR	population attributable risk
PBDE	polybrominated diphenyl ether
PCB	polychlorinated biphenyl
PDQ	Physician Data Query
PFOA	perfluorooctanoic acid
PFOS	perfluorooctanesulfonic acid
PI3K	phosphoinositide 3-kinase
PND	postnatal day
PPAR	peroxisome proliferator-activated receptor
ppm	parts per million
PR	progesterone receptor
PR–	progesterone receptor negative
PR+	progesterone receptor positive
PSA	prostate specific antigen
REACH	Registration, Evaluation, Authorisation and Restriction of Chemical Substances
RNA	ribonucleic acid
ROC	receiver operating characteristic
ROS	reactive oxygen species
RR	relative risk
SEER	Surveillance, Epidemiology, and End Results Program
SERD	selective estrogen receptor down-regulator
SERM	selective estrogen receptor modulator
SHBG	sex hormone binding globulin
SNP	single nucleotide polymorphism
STAR	Study of Tamoxifen and Raloxifene
TCDD	2,3,7,8-tetrachlorodibenzo-*p*-dioxin
TDLU	terminal duct lobular unit
TNBC	triple negative breast cancer
TNF	tumor necrosis factor
TSCA	Toxic Substances Control Act
USPSTF	U.S. Preventive Services Task Force
UV	ultraviolet radiation
VOC	volatile organic compound

WCRF	World Cancer Research Fund
WECARE	Women's Environment, Cancer, and Radiation Epidemiology Study
WHI	Women's Health Initiative
WHO	World Health Organization
XOR	xanthine oxidoreductase

Summary

Abstract: *Breast cancer accounts for substantial morbidity among women in the United States, with an estimated 230,480 new cases of invasive disease in 2011. Susan G. Komen for the Cure® and its Scientific Advisory Board commissioned a study by the Institute of Medicine (IOM) to assess the current evidence on the contribution of environmental exposures, alone or in combination with genetic factors, to the risk of developing breast cancer; review the challenges in investigating potential environmental contributions; explore evidence-based actions that women might take to reduce the risk of breast cancer; and recommend research directions.*

"Environment" was broadly defined to encompass all factors that are not directly inherited through DNA, and a qualitative review examined current evidence on selected factors that illustrate various environmental agents and conditions that may be more amenable to modification. For some of these factors, epidemiologic studies consistently support associations with increased risk for breast cancer (e.g., ionizing radiation, combination hormone therapy, greater postmenopausal weight) or reduced risk (e.g., more physical activity). For many other factors, however, the epidemiologic evidence is more limited, contradictory, or absent. Evidence from animal or mechanistic studies sometimes adds support to the epidemiologic evidence or suggests biologic plausibility when human evidence is lacking for a particular factor.

Knowledge about the complexity of breast cancer and its relation to environmental exposures continues to grow, but researchers face many challenges. To move toward greater opportunities for prevention, more needs to be learned about the biologic significance of the life stages at which

1

environmental risk factors are encountered; optimal approaches to assessing exposures, designing and analyzing epidemiologic studies, and integrating analysis of genetic and environmental influences; the possible combined effects of a multitude of low-level chemical exposures; and interpretation of findings from studies in animals and in vitro systems.

Although many questions remain regarding the contributions of environmental factors to breast cancer risk, evidence suggests that women may have some opportunities to reduce their risk of breast cancer through personal actions: avoiding unnecessary medical radiation throughout life, avoiding use of some forms of postmenopausal hormone therapy, avoiding smoking, limiting alcohol consumption, increasing physical activity, and, for postmenopausal breast cancer, minimizing weight gain. The potential risk reductions for any individual woman will vary and may be modest, but the impact of these actions could be important at a population level. In many cases, however, lack of robust data on environmental agents' effects on human breast cancer risk, especially during different life stages, and some sense of the trade-offs involved, are major challenges for identifying evidence-based actions that could be taken at the individual or societal level to reduce breast cancer risk.

Recommendations for research include applying a life course perspective and a transdisciplinary approach to studies of breast cancer, developing new and better tools for epidemiologic research and carcinogenicity testing of chemicals and other substances, developing effective preventive interventions, developing better approaches to modeling breast cancer risks, and improving communication about breast cancer risks to health care providers, policy makers, and the public.

Breast cancer has long been the most common invasive noncutaneous cancer among women in the United States, accounting for an estimated 230,480 new cases in 2011.[1] After lung cancer, it is the second most common cause of women's cancer mortality, with about 39,520 deaths expected in 2011. In 2011, there were also approximately 2,140 new cases of breast cancer and 450 breast cancer deaths among men in the United States.

Knowledge about the complexity of breast cancer continues to grow: the characterization of multiple tumor subtypes; the likelihood that critical events in the origins of breast cancer can occur very early in life; the variety of pathways through which breast cancer risks may be shaped; the likely contribution to breast cancer of some fundamental biological processes;

---

[1] Approximately 57,650 noninvasive (in situ) breast tumors will also have been diagnosed in 2011.

and the potential significance of the timing and combinations of environmental exposures in determining their effect on risks for different types of breast cancer. This growing knowledge is helping to stimulate a transition in breast cancer research, with new ideas influencing the design and analysis of epidemiologic studies, experimental studies in animals, and mechanistic studies of breast cancer biology. As this work elucidates how endogenous and exogenous factors may influence the development of breast cancer, new opportunities for prevention may emerge.[2]

Susan G. Komen for the Cure and its Scientific Advisory Board requested that the Institute of Medicine (IOM) review the current evidence on environmental risk factors for breast cancer, consider gene–environment interactions in breast cancer, explore evidence-based actions that might reduce the risk of breast cancer, and recommend research in these areas. The Statement of Task for the study appears in Box S-1.

The committee interpreted "environment" broadly, to encompass all factors that are not directly inherited through DNA. As a result, this definition includes elements that range from the cellular to the societal: the physiologic and developmental course of an individual, by-products of innate metabolic processes that can be modulated by external stressors, diet and other ingested substances, physical activity, microbial agents, physical and chemical agents encountered in any setting, medical treatments and interventions, social factors, and cultural practices. With the potentially vast scope of the study task, the committee focused on areas that it considered to be the most significant and the most pertinent to its charge. In particular, the study focused primarily on breast cancer in women and on the initial occurrence of a tumor, not recurrence. The committee took into account the changes in the breast over a woman's life and the potential for the timing of exposures to influence risks they may pose for breast cancer. The committee did not address practices in the diagnosis and treatment of breast cancer or policies or practices for breast cancer screening.

REVIEWING EVIDENCE ON CERTAIN ENVIRONMENTAL FACTORS

The committee explored the available evidence concerning breast cancer risks associated with a necessarily limited selection of specific factors that illustrate a variety of environmental agents and conditions (see Box S-2 and Chapter 3). The committee drew on evidence reviews by authoritative bodies, especially the International Agency for Research on Cancer (IARC) and the World Cancer Research Fund (WCRF) International, supplemented

---

[2]The term "breast cancer" is used in this report to refer to disease in humans, and "mammary cancer" or "mammary tumor" to refer to disease in animals.

BOX S-1
Study Charge

In response to a request from Susan G. Komen for the Cure®, the Institute of Medicine will assemble a committee to:

1. Review the evidentiary standards for identifying and measuring cancer risk factors;
2. Review and assess the strength of the science base regarding the relationship between breast cancer and the environment;
3. Consider the potential interaction between genetic and environmental risk factors;
4. Consider potential evidence-based actions that women could take to reduce their risk of breast cancer;
5. Review the methodological challenges involved in conducting research on breast cancer and the environment; and
6. Develop recommendations for future research in this area.

In addition to reviewing the published literature, the committee will seek input from stakeholders, in part by organizing and conducting a public workshop to examine issues related to the current status of evidentiary standards and the science base, research methods, and promising areas of research. The workshop will focus on the challenges involved in the design, conduct, and interpretation of research on breast cancer and the environment. The committee will generate a technical report with conclusions and recommendations, as well as a summary report for the lay public.

---

by reviews and original research reports in the peer-reviewed literature. The committee qualitatively reviewed relevant literature, without a formal systematic review or quantitative analysis (e.g., meta-analysis) or the intensive weighing of evidence undertaken by IARC or WCRF. Several familiar topics, such as diet and most dietary components, received less attention because of ongoing systematic review by other groups. Providing a review of a complete set of environmental agents and conditions was not feasible. Of the large number of environmental factors with potential but uncertain impact on breast cancer, the committee reviewed only a selected number that illustrated particular types of challenges in assessment.

The aim was to characterize the available evidence on whether the selected environmental factors are associated with breast cancer, and to identify areas of substantial uncertainty. Evidence from epidemiologic studies carried the greatest weight in identifying risk factors. Evidence from experimental studies in animals or in vitro systems, especially in the absence

BOX S-2
Environmental Factors Included in the
Committee's Evidence Review[a]

Exogenous hormones
- Hormone therapy: androgens, estrogens, combined estrogen–progestin
- Oral contraceptives

Body fatness and abdominal fat

Adult weight gain

Physical activity

Dietary factors
- Alcohol consumption
- Dietary supplements and vitamins
- Zeranol and zearalenone

Tobacco smoke
- Active smoking
- Passive smoking

Radiation
- Ionizing (including X-rays and gamma rays)
- Non-ionizing (extremely low frequency electric and magnetic fields [ELF-EMF])

Shift work

Metals
- Aluminum
- Arsenic
- Cadmium
- Iron
- Lead
- Mercury

Consumer products and constituents
- Alkylphenols
- Bisphenol A (BPA)
- Nail products
- Hair dyes
- Parabens
- Perfluorinated compounds (PFOA, PFOS)
- Phthalates
- Polybrominated diphenyl ethers (PBDEs; flame retardants)

Industrial chemicals
- Benzene
- 1,3-Butadiene
- PCBs
- Ethylene oxide
- Vinyl chloride

Pesticides
- DDT/DDE
- Dieldrin and aldrin
- Atrazine and S-chloro triazine herbicides (atrazine)

Polycyclic aromatic hydrocarbons (PAHs)

Dioxins

---

[a]The committee reviewed a selected set of factors for illustration; the chemicals were not chosen to be representative of any class. Some epidemiologic, mechanistic, or animal data relevant to mammary tumorigenesis or breast cancer are available for numerous other chemicals.

of human data, was the basis for noting that some factors may present a hazard, and thus potentially contribute to breast cancer risk, alone or in combination with other factors, depending on the nature of an exposure (e.g., amount, timing). A hazard has the potential to cause an adverse effect under certain conditions of exposure; a risk is the probability that the adverse effect will occur in a person or a population as a result of an exposure to a hazard.

Among the environmental factors reviewed, those most clearly associated with increased breast cancer risk in epidemiologic studies are use of combination hormone therapy products, current use of oral contraceptives, exposure to ionizing radiation, overweight and obesity among postmenopausal women, and alcohol consumption. Greater physical activity is associated with decreased risk. Some major reviews have concluded that the evidence on active smoking is consistent with a causal association with breast cancer, and other large-scale reviews describe the evidence as limited. For several other factors reviewed by the committee, the available epidemiologic evidence is less strong but suggests a possible association with increased risk: passive smoking, shift work involving night work, benzene, 1,3-butadiene, and ethylene oxide. For some of the reviewed factors (e.g., bisphenol A or BPA), animal or mechanistic data suggest biological plausibility as a hazard. A few factors, such as non-ionizing radiation and personal use of hair dyes, have not been associated with breast cancer risk in multiple, well-designed human studies. For several other factors, evidence was too limited or inconsistent to reach a conclusion (e.g., nail products, phthalates). In all cases, these conclusions are based on assessments of the currently available evidence; it is always possible for new evidence to point to different conclusions.

As the committee considered the current state of knowledge, it sees a need for research on the etiology of breast cancer to do more to incorporate new understanding of breast development over the life course, recent advances in elucidating the molecular biology of tumorigenesis, and the challenges of assessing the potential impact of a multitude of low-level chemical exposures. A more integrative approach to breast cancer research may accelerate progress in understanding the role that environmental factors may have in breast cancer.

CHALLENGES IN STUDYING BREAST CANCER AND THE ENVIRONMENT

Trying to determine which environmental exposures may influence rates of breast cancer poses substantial challenges. The biology of breast development and the origins and progression of breast cancer are not fully understood, and much research in the past lacked tools to differentiate among types of breast cancer. Also, a focus primarily on exposures during adulthood, as in past research, may miss critical windows during early life in which some environmental exposures may influence risk for breast cancer later in life.

Tracing multiple and potentially interacting causes of breast cancer will be difficult. Some risk factors may have very weak effects or effects in only a small portion of the population, making their contribution to risk hard to detect. People are exposed to a complex and changing mix of environmental agents over the course of a lifetime; discerning the effects of an individual agent, or knowing whether the components of the mixture may interact to influence the development of disease, is not straightforward. Moreover, many of these agents have never been studied in ways that could indicate whether they might be relevant to breast cancer. Several challenges appear especially formidable.

Assessing Human Exposure

It can be difficult to identify and measure exposures because few tools and opportunities are available for doing so directly, especially if relevant exposures occurred well in the past or the timing of such exposures is unclear. Many studies must base estimates of exposure on error-prone indicators such as self-reports of past product use or proxies for exposure, such as holding a particular type of job or living in a particular location at a particular time. Even when it is possible to detect evidence of exposure from biological samples (e.g., blood or urine), single measurements are rarely sufficient to establish the duration and levels of past exposure, and few studies have the benefit of multiple samples from the same study participant for comparisons over time. Determining the number of samples needed and interpreting comparisons among them requires a good understanding of the biological processes that influence variation in the production and retention of these biomarkers of exposure.

Conducting Epidemiologic Studies

Experimental studies in humans (i.e., controlled clinical trials), in which host factors and exposures can be carefully controlled, would provide the strongest evidence of causal associations, but they are rarely an option in studying causes of breast cancer because study participants should not be exposed to substances suspected of causing harm. As a result, researchers must generally rely on observational studies that depend on either collecting retrospective information about critical exposures and life events or conducting large prospective studies of extended duration. A few large cohorts of adult women have provided a valuable base for investigating breast cancer risk factors. But because it is likely that breast cancer diagnosed in older adult women is influenced by exposures at various stages of life, ideal prospective investigations would follow a study population throughout life. Such studies are very costly and logistically difficult to implement. Reliable predictors of increased risk for breast cancer that could be assessed at much younger ages (e.g., during adolescence) would greatly aid investigation of the influence of early-life exposures, but current understanding is limited to risk factors such as age at menarche and at first full-term pregnancy. An additional complication is that for some environmental pollutants, low-level exposures are so widespread and co-occur with low levels of numerous other possible contributors that it is difficult to identify an unexposed comparison group or adequately control for other exposures.

Identifying Genetic Influences

Only a few genetic markers of substantially increased risk are well established (e.g., *BRCA1* and *BRCA2* mutations), and these are rare in the general population. Studies suggest that other, more common mutations and polymorphisms may also be associated with breast cancer, but have a much smaller influence on risk. The multitude of potential associations and the relatively small differences in risk mean that studies must be very large to detect statistically significant effects, and efforts to replicate findings are often not successful because false positive rates are high in small studies. Gene–environment interactions for breast cancer risk have been shown in several epidemiologic studies for high alcohol intake combined with polymorphisms in enzymes involved in alcohol metabolism. For most chemicals, however, exposures are generally low, and efforts to study interactions between genetics and environmental factors are also hampered by lack of data on environmental exposures of interest in most datasets currently used for genomic studies.

Interpreting Findings from Studies in Animals and In Vitro Systems

Experimental studies in whole animals and in vitro systems are an essential component of research on breast cancer and of regulatory risk assessment to limit exposure to carcinogens, but the results remain approximations of human experience. In vitro systems are used to explore mechanisms by which environmental agents alter cellular and tissue behavior and to identify chemicals that cause genetic damage (genotoxic substances) in regulatory safety testing. Such systems currently do not fully account for the multiplicity of biological processes (e.g., pharmacokinetics, cell interactions) that occur in response to an exposure in a whole organism, and the degree to which they detect nongenotoxic carcinogens is uncertain. Even studies in human cell lines, though they may provide useful mechanistic insights, are ill equipped to capture the full complexity of intact humans. In testing with whole animals (i.e., in vivo animal models), the small numbers of test animals make it statistically impossible to detect small increases in risk. It may also be difficult to interpret results from studies that use doses or routes of exposure that do not correspond to typical human exposures. Adding to the complexity of interpreting in vivo animal studies are differences in responses among the commonly used rat and mouse strains and assessing the significance of underlying differences in anatomy and physiology between humans and rodents.

EMPHASIZING THE LIFE COURSE IN STUDYING RISK FACTORS AND BREAST CANCER MECHANISMS

As in most types of adult cancer, breast cancer is thought to develop as a result of accumulated damage induced by both internal and external triggers resulting in initial carcinogenic events. The affected cells and tissues then progress through multiple stages, with accompanying alterations in surrounding tissue likely playing a role in permitting or potentiating the cancer process. These events contributing to subsequent cancers may occur spontaneously as a by-product of errors in normal processes, such as DNA replication, or through effects of environmental exposures, such as damage from exposure to sunlight or tobacco carcinogens; or they can be sustained and furthered by physiologic conditions, such as obesity.

The breast undergoes substantial changes from the time it begins developing in the fetus through old age, especially in response to hormonal changes during puberty, pregnancy, lactation, and menopause. The timing of a variety of environmental exposures may be important in directly increasing or reducing breast cancer risks or in acting indirectly by influencing the developmental events. There may be critical windows of susceptibility (e.g., periods of rapid cell proliferation or maturation) when specific mechanisms that increase the likelihood of a breast cancer developing may be more likely to come into play.

Research is continuing on many fronts to increase understanding of the mechanisms that contribute to breast cancer and the ways they relate to or may be modulated by exposure to environmental factors. Some exposures act principally at early stages of carcinogenesis (activating oncogenes or inactivating tumor suppressor genes within affected cells) whereas others act later (stimulating cell division and proliferation), so that mutations are less likely to be repaired and more likely to have detrimental consequences. Others may act to alter susceptibility to exposures later in life. Estrogen produced in the body is critical to normal breast development, but it also appears to play a major role in breast carcinogenesis. It may do this by promoting proliferation of cells (mitogenesis) and possibly via mutagenic activity of its metabolites. Some environmental factors can have estrogenic properties, but the implications for breast cancer are not entirely clear. Environmental exposures might cause damage (mutations) to DNA; they may also act through epigenetic reprogramming, which alters gene expression without altering DNA. Factors that modify the functioning of the immune system may also contribute to carcinogenic processes. Also important may be disruption of the stromal component of the breast that normally functions to maintain the structural and functional integrity of the breast tissues through regulatory and homeostatic mechanisms.

OPPORTUNITIES FOR EVIDENCE-BASED ACTION TO REDUCE RISK OF BREAST CANCER

On average, girls born in the United States today have approximately a 12 percent risk of developing invasive breast cancer that will be diagnosed at some point in their lifetime. Among 50-year-olds, 2.4 percent of white women (or 24 out of 1,000) are likely to be diagnosed with invasive breast cancer over the next 10 years, compared with 2.2 percent of black women, 2.0 percent of Asian women, and 1.7 percent of Hispanic women. Within average values such as these, groups of women have characteristics that give them a higher or lower 10-year risk, and of course, larger risks if followed through the remainder of their lives.

Research findings that certain factors are associated with increased or decreased risk of breast cancer are typically reported in terms of measures that compare the risk in exposed and unexposed populations (i.e., relative risks, odds ratios, hazard ratios, or risk differences). In general, the environmental factors reviewed by the committee were associated with less than a doubling of risk. These findings become more meaningful when they are linked back to the actual rates of illness. Thus, a doubling of risk might mean that the 10-year risk of breast cancer is 5 percent for a group of women who have a risk factor rather than 2.5 percent for those who do not.

Finding ways to reduce risk and avert cases of breast cancer is a high priority, but at present, the evidence-based options are limited (see Chapter 6). Many of the well-known risk factors for breast cancer—older age, being female, and older age at menopause—appear to offer little or no opportunity to intervene. For a limited set of other risk factors, women have a greater opportunity to act in ways that may have the potential to reduce risk for breast cancer while carrying limited risks of increasing other adverse health outcomes (Table S-1). Some of these actions may have health benefits beyond any contribution they may make to reducing risk of breast cancer.

The potentially risk-reducing—but not necessarily easily accomplished—actions identified by the committee include eliminating exposure to unnecessary medical radiation throughout life; avoiding use of combination estrogen–progestin menopausal hormone therapy, unless it is considered medically appropriate and the benefits are expected to outweigh the risks; avoiding active and passive smoking; limiting alcohol consumption; increasing physical activity; and minimizing overweight and weight gain to reduce risk of postmenopausal breast cancer.

Chemoprevention using tamoxifen or raloxifene may be an appropriate choice for some women at high risk of breast cancer, but use of these medications also raises the risk of serious adverse events such as stroke and endometrial cancer. Women who qualify for use of chemoprevention should receive appropriate counseling on its benefits and risks to be able to make an informed choice.

For some of the chemicals reviewed by the committee, it may be prudent to avoid or minimize exposure because the available evidence suggests biological plausibility for exposure to be associated with an increased risk of breast cancer, or there is suggestive evidence from epidemiology, or both. The evidence is clearest for benzene, 1,3-butadiene, and ethylene oxide because there is suggestive evidence from both epidemiologic and nonhuman data. Occupational exposures to these chemicals can occur in industrial settings, and the general public is exposed through transportation-related air pollution, industrial emissions, and tobacco smoke. For cosmetics and dietary supplements, the Food and Drug Administration (FDA) can remove from the market products found to be hazardous or adulterated, but it generally lacks authority to test the safety of these products before they are sold. The committee urges efforts to better inform consumers and health professionals about the limits of FDA's role, to encourage manufacturers to identify hormonally active ingredients in cosmetics and dietary supplements, and to ensure that FDA has effective tools to identify contaminants or ingredients that are potential contributors to increased risk of breast cancer. Similarly for chemicals in consumer products, interested organizations can help inform the public about the current provisions for testing chemicals and encourage manufacturers to improve testing and make existing information on their products more readily available.

The limited set of opportunities for individual action noted by the committee reflects the scientific community's still incomplete understanding of which exposures might best be avoided and when, of the actions following exposure that might have a long-term benefit in reducing risk for breast cancer, and, in some cases, of the potential for unintended consequences of interventions. Few intervention studies have investigated whether factors associated with increased postmenopausal risk, such as overweight or alcohol consumption, should be avoided completely, or whether reducing or

TABLE S-1 Summary of Committee Assessment of Opportunities for Actions by Women That May Reduce Risk of Breast Cancer

Opportunity for Action	Modification of Exposure	
	Personal Action Possible	Requires Action by Others
Avoid inappropriate medical radiation exposure[c]	Yes	Yes
Avoid combination menopausal hormone therapy, unless medically appropriate[d]	Yes	Confer with physician
Avoid or end active smoking	Yes	Others can facilitate
Avoid passive smoking	Varies	Yes
Limit or eliminate alcohol consumption	Yes	Others can facilitate
Maintain or increase physical activity	Yes	Others can facilitate
Maintain healthy weight or reduce overweight or obesity to reduce postmenopausal risk	Yes	Others can facilitate
Limit or eliminate workplace, consumer, and environmental exposure to chemicals that are plausible contributors to breast cancer risk while considering risks of substitutes[e]	Varies by chemical	Varies
If at high risk for breast cancer, consider use of chemoprevention	Yes	Confer with physician

[a]Actions to address risk factors can take various forms, some of which may be more effective than others, and some of which may have to be taken at a specific time in life to be effective. For example, increasing physical activity might be based on the amount of time spent in any one exercise opportunity, on increasing specific types of exercise, or on increasing the frequency of exercise, or perhaps some combination of any of these. Studies have not been done that provide evidence that a specific form of physical activity is optimal for reducing breast cancer risk.

[b]The committee's comments on other benefits or risks highlight major considerations, but they are not intended to be exhaustive.

[c]While recognizing the risks of ionizing radiation exposure, particularly for certain higher dose methods (e.g., computed tomography [CT] scans), it is not the committee's intent to dissuade women from routine mammography screening, which aids in detecting early-stage tumors.

Action		
Target Population Defined	Effective Form and Timing Established[a]	Other Prominent Known Risks or Benefits from Taking Action[b]
All ages	Yes, especially at younger ages	May result in loss of clinically useful information in some instances Likely to decrease risk for other cancers
Postmenopausal women	Yes	May experience moderate to severe menopausal symptoms without hormone therapy
All ages, especially before first pregnancy	Yes (form) No (timing)	Likely to reduce risk for other cancers, heart disease, stroke
All ages	Yes	Likely to reduce risk for other cancers, heart disease
All women	Yes (form) No (timing)	May increase risk for cardiovascular disease No known benefit of high alcohol consumption
All ages	No	Likely to reduce risk for cardiovascular disease, diabetes May increase risk for injury
Unclear	No	Likely to reduce risk for cardiovascular disease, diabetes, other cancers
Varies	No	May reduce risk for other forms of cancer or other health problems May result in replacement with products that have health or other risks not yet identified
High-risk women	Yes	Depending on the agent, increased risk of endometrial cancer, stroke, deep-vein thrombosis, among others

[a]Combination hormone therapy with estrogen and progestin increases the risk of breast cancer, and the associated risk is reduced upon stopping therapy. Oral contraceptives are also associated with an increased risk of breast cancer while they are being used. This risk is superimposed on a low background risk for younger women, who are most likely to use oral contraceptives. Use of oral contraceptives is associated with long-term risk reduction for ovarian and endometrial cancer.

[b]Plausibility may be indicated by epidemiologic evidence, animal bioassays, or mechanistic studies.

eliminating the exposure later in adulthood will reduce the risk that might have accrued from exposure at younger ages. It is also difficult to judge what any *individual* woman's change in risk might be. Moreover, much of the evidence on breast cancer risk factors has come from studies of postmenopausal breast cancer in white women, and it has pointed to a greater potential to reduce risk for estrogen receptor–positive (ER+) cancers than other types. A much better basis is needed for guidance for risk reduction for younger women and women of other races and ethnicities. Nevertheless, many of the suggested actions are likely to not only reduce breast cancer risk, but also reduce risks for other major health conditions.

DIRECTIONS FOR FUTURE RESEARCH

The research needed to better understand the relation between breast cancer risks and environmental factors ranges from further examination of elements of the biology of breast development and carcinogenesis to tests of potential interventions to reduce risk. Important components of the work recommended here are support for the research necessary to develop better tools for assessing the carcinogenicity of chemicals and pharmaceuticals as well as tools needed to strengthen epidemiologic research. The importance of a life course perspective runs throughout the recommendations.

Applying a Life Course Perspective to Studies of Breast Cancer

Progress has been made in understanding the biology of breast development and many aspects of breast cancer, but important gaps remain in understanding its causes and the extent of environmental influences on its development. Future research should increasingly focus on the influence of exposure to a variety of environmental factors during potential windows of susceptibility over the full life course, from the prenatal experience throughout adult life.

Recommendation 1: Breast cancer researchers and research funders should pursue integrated and transdisciplinary studies that provide evidence on etiologic factors and the determinants of breast cancer across the life course, with the goal of developing innovative prevention strategies that can be applied at various times in life.

These studies should seek to integrate animal models that capture the whole life course and human epidemiologic cohort studies that follow individuals over long periods of time and allow for investigation of windows of susceptibility. Long-term follow-up of cohorts is critical because new, unexpected evidence frequently arises with longer follow-up.

Topics warranting attention include (but are not limited to) the biology of breast development; the mechanisms of carcinogenesis early in life, including the role of the tissue microenvironment in tumor suppression and development, and differences that may be related to tumor type; differences in effects of exposures by tumor type; the potential contribution of timing of exposure to variation in risk; and analytical tools for investigating the potential for interactions among exposures and the impact of mixtures of environmental agents on biological processes.

Other work to aid investigation of environmental influences on breast cancer risk includes

- identifying cellular, biochemical, or molecular biomarkers of early events leading to breast cancer and validating their predictive value for future risk for breast cancer;
- determining whether intermediate endpoints (e.g., indicators of breast development, peak height growth velocity) are valid and predictive biomarkers of differences in risk for breast cancer;
- investigating the role that environmental factors may have in the origins of the different types of breast cancer to better understand disparities in incidence among racial and ethnic groups;
- exploring the value of linking information across cohort studies focused on different stages of life as a way to overcome the challenges of mounting single long-term follow-up studies; and
- ensuring that cohorts established primarily to study genetic determinants of cancer and other diseases improve the capacity of these cohorts to capture information about environmental exposures over the life course.

Targeting Specific Concerns

From its examination of evidence on a selection of environmental factors, the committee sees particular benefit in further research to clarify the mechanisms underlying breast cancer.

Recommendation 2: Breast cancer researchers and research funders should pursue research to increase knowledge of mechanisms of action of environmental factors for which there is provocative, but as yet

inconclusive, mechanistic, animal, life course, or human health evidence of a possible association with breast cancer risk.

High-priority topics include the following:

- *Shift work*: The biological processes and pathways through which shift work and circadian rhythm disruption relate to breast cancer; more detailed and standardized approaches to exposure assessment.
- *Endocrine activity*: Interactions between chemicals, such as BPA, polybrominated diphenyl ethers (PBDEs), zearalenone, and certain dioxins and dioxin-like compounds, and timing of exposure, diet, and other factors that may influence the relationship of these types of compounds to breast cancer risk.
- *Genotoxicity*: The degree to which mutagenic chemicals, such as polycyclic aromatic hydrocarbons (PAHs), benzene, and ethylene oxide, acting alone or in combination with other exposures at specific life stages, contribute to breast cancer risk at current levels of exposure.
- *Epigenetic activity*: Fundamental research on the role of epigenetic modifications in breast cancer risk, and the potential importance of epigenetic modifications by environmental chemicals such as BPA.
- *Gene–environment interactions*: Continued research to identify genes relevant to breast cancer that modify risk from discrete environmental exposures.

Epidemiologic Research

Studies of Occupational Cohorts and Other Highly Exposed Populations

Many known human carcinogens were first identified through studies in occupational settings where workers had chemical and physical exposures that were higher than those of the general population. With many more women in the workforce, occupational studies may now be a means to identify some exposures that increase risk for breast cancer. Other identifiable groups of women with long-term or event-related high-dose exposures may also be promising study populations.

Recommendation 3: Breast cancer researchers and research funders should pursue studies of populations with higher exposures, such as occupational cohorts, persons with event-related high exposures, or patient groups given high-dose or long-term medical treatments. These studies should include collection of information on the prevalence of known breast cancer risk factors among the study population. Support for these studies should include resources for the development of improved exposure assessment methods to quantify chemical and other environmental exposures potentially associated with the development of breast cancer.

New Exposure Assessment Tools

A life course perspective on breast cancer suggests that critical periods of vulnerability may exist during in utero development, in childhood, adolescence, and early adulthood, and at older ages. Exposure assessment becomes particularly challenging over such extended intervals.

> **Recommendation 4: Breast cancer and exposure assessment researchers and research funders should pursue research to improve methodologies for measuring, across the life course, personal exposure to and biologically effective doses of environmental factors that may alter risk for or susceptibility to breast cancer.**

Such research should encompass

- improving measurements in the environment and assessing variation over time and space;
- determining routes of exposures and how they vary over time and over the life course;
- evaluating how products are used and the extent to which actual usage deviates from label instructions (e.g., home pesticide applications) as a critical component of exposure assessment, focusing on the impact on personal exposures;
- incorporating use of advanced environmental dispersion modeling techniques with accurate emissions and air monitoring data to characterize specific population exposures;
- measuring compounds and their metabolites in biospecimens, including specimens obtained by noninvasive means;
- understanding pharmacodynamics and pharmacokinetics and how they vary by life stage, body weight, nutrition, comorbidity, or other factors;
- developing other biomarkers of exposure through early biologic effects (DNA adducts, methylation, tissue changes, gene expression, etc.);
- using existing and yet-to-be-established human exposure biomonitoring programs (e.g., breast milk repositories) by geographic areas; and
- validating exposure questionnaires through various strategies.

Research to Advance Preventive Actions

Minimizing Exposure to Ionizing Radiation

Some of the strongest evidence reviewed by the committee supports a causal association between breast cancer and exposure to ionizing radiation. However, population exposures to ionizing radiation in medical imaging are increasing. Standards exist to ensure that mammography minimizes radiation exposures, but more needs to be learned to determine how to minimize exposures from other medical procedures.

> **Recommendation 5: The National Institutes of Health, the Food and Drug Administration, and the Agency for Healthcare Research and Quality should support comparative effectiveness research to assess the relative benefits and harms of imaging procedures and diagnostic/follow-up algorithms in common practice. This research effort should also assess the most effective ways to fill knowledge gaps among patients, health care providers, hospitals and medical practices, industry, and regulatory authorities regarding practices to minimize exposure to ionizing radiation incurred through medical diagnostic procedures.**

Developing and Validating Interventions to Prevent Breast Cancer

Some breast cancer risk factors appear to be modifiable, but it is important to determine what modifications can be most effective in reducing risk and when during the life course these changes need to occur. For example, overweight and obesity are recognized as increasing risk for postmenopausal breast cancer, but the contribution of weight loss to reducing risk is much less clear.

> **Recommendation 6: Breast cancer researchers and research funders should pursue prevention research in humans and animal models to develop strategies to alter modifiable risk factors, and to test the effectiveness of these strategies in reducing breast cancer risk, including timing considerations and population subgroups likely to benefit most.**

Particular aspects of prevention that require attention include

- when weight loss is most likely to be beneficial in reducing risk for postmenopausal breast cancer;
- effective strategies for achieving and maintaining weight loss in different risk groups;
- effective and sustainable methods to prevent obesity;

- the feasibility of interventions in early life and development that may influence breast cancer risk in adult life, such as preventing childhood obesity, increasing physical activity, and minimizing exposures to potentially harmful environmental carcinogens;
- approaches to prevention that respond to the differing breast cancer experience of various racial and ethnic groups; and
- dissemination and adoption of effective prevention strategies.

Chemoprevention—Medications to Reduce Breast Cancer Risk

Tamoxifen and raloxifene have been shown to substantially reduce risk of ER+ breast cancer in women who have not been diagnosed with the disease, and they are approved by the FDA for this use by women at increased risk of breast cancer. Other medications (e.g., aromatase inhibitors, bisphosphonates, metformin) are being studied to assess their effectiveness for reducing the risk of either ER+ or estrogen receptor–negative (ER–) breast cancer.

However, tamoxifen and raloxifene increase the risk of other potentially serious events (e.g., endometrial cancer [tamoxifen], stroke) and are not widely used. Additional research is needed to identify other drugs that can reduce risk of all forms of breast cancer with minimal risk of other adverse health effects.

Recommendation 7: Breast cancer researchers and research funders should pursue continued research into new breast cancer chemoprevention agents that have minimal risk for other adverse health effects. This work should include efforts to identify chemopreventive approaches for hormone receptor negative breast cancer.

Adequately sized primary prevention studies will be needed to allow for estimation of both benefits and risks. Research plans should also include long-term follow-up to identify any changes in risk patterns for types of breast cancer or other effects that only become evident beyond the time frame of current analyses.

Testing to Identify Potential Breast Carcinogens

In Vivo Testing for Carcinogenicity

Current whole-animal (rodent) protocols for carcinogenicity testing may not be ideally suited to screening for possible human breast carcinogens because they typically do not address changing sensitivity during the life course, such as during in utero and early postnatal periods, to carcinogens or to exposures that may alter susceptibility to later carcinogenic exposures. Because of study power constraints, these tests are not usually structured to assess the low-dose exposures to mixtures that are characteristic of human experience, and interpretation of findings (positive or negative) may be complicated by the test animal strains' characteristic susceptibility (or lack of susceptibility) to mammary tumors.

> Recommendation 8: The research and testing communities should pursue a concerted and collaborative effort across a range of relevant disciplines to determine optimal whole-animal bioassay protocols for detection and evaluation of chemicals that potentially increase the risk of human breast cancer.

The development of these protocols should include consideration of the appropriateness of the rodent species and strains used for testing; the utility of genetically engineered mouse models to address specific mechanisms; the frequency, magnitude, and route of dosing that may be most relevant for predicting human risk; potential differences in sensitivity in different life stages; and standard practices for conducting studies and reporting results.

New Approaches to Toxicity Testing

New toxicity testing approaches are being developed to more rapidly and accurately screen chemicals and minimize in vivo testing. Because breast cancer is a major contributor to women's morbidity, these tests should be relevant to the basic mechanisms of breast cancer—for example, mutagenesis, estrogen receptor signaling, epigenetic programming, modulation of immune functioning, and alterations at the whole-organ level—and to human exposures (low doses and mixtures).

> Recommendation 9:
> a. The research and testing communities should ensure that new testing approaches developed to serve as alternatives to long-term rodent carcinogenicity studies include components that are relevant for breast cancer. The tests should be able to account for changes in susceptibility through the life course and mechanisms characteristic of hormonally active agents. The test development should also include exploring the predictive value of in vitro and in vivo experimental testing for site-specific cancer risks for humans.
> b. A research initiative should assess the persistence and consequences for mammary carcinogenicity of abnormal mammary gland development and related intermediate outcomes observed in some

toxicological testing. As useful predictors of increased mammary cancer risk become available, intermediate outcomes may aid in identifying chemicals that may pose increased risk of human breast cancer when exposures occur early in life.

c. Research should be conducted to improve understanding of the potential cumulative effects of multiple, small environmental exposures on risk for breast cancer and the interaction of these exposures with other factors that influence risk for breast cancer.

New Approaches to Testing Hormonally Active Candidate Pharmaceuticals

Given the evidence for hormonal influences on the development of breast cancer, the committee is concerned that testing required to gain marketing approval for various hormonally active pharmaceuticals, including oral contraceptives and menopausal hormone therapies, does not adequately address the potential for increased risk for breast cancer.

Recommendation 10: The pharmaceutical industry and other sponsors of research on new hormonally active pharmaceutical products should support the development and validation of better preclinical screening tests that can be used before such products are brought to market to help evaluate their potential for increasing the risk of breast cancer.

A suite of in vitro and in vivo tests will likely be needed to address the different mechanisms of action that may be relevant over the life course. If such tests can be developed and validated, FDA should require submission of the results as part of the process for approving the introduction of new hormonal preparations for prescription or over-the-counter use. These tests may also prove useful in testing environmental chemicals.

Postmarketing Studies of Hormonally Active Drugs

With the demonstration that use of certain hormonally active prescription drugs is associated with an increased risk of breast cancer and other adverse health effects, it is important to investigate whether use of other hormonally active drugs is also associated with increased risk.

Recommendation 11: FDA should use its authority under the Food and Drug Administration Amendments Act of 2007 to engage the pharmaceutical industry and scientific community in postmarketing studies or clinical trials for hormonally active prescription drugs for which the potential impact on breast cancer risk has not been well characterized.

The studies should be adequately powered to quantitatively explore the possible contribution of the products to breast cancer risk. Products that represent a substantial change in pharmacologic composition or dosage schedule from products currently on the market should be a particular focus of attention.

Understanding Breast Cancer Risks

Researchers, health care providers, and the public all have an incomplete picture of the components of breast cancer risk. Further work is needed to clarify the contribution of recognized risk factors to differences and changes in the incidence of breast cancer and to determine the most effective ways to convey information about breast cancer risk.

Risk Modeling

Systematic modeling approaches are needed to refine estimates of the proportion of breast cancer in the United States and other countries that can be attributed to established risk factors (individually and in combination), especially those that can be modified. Additionally, better data are needed on the prevalence of these risk factors. Improved estimates of risk associated with established factors should help in determining the scale of residual risk, which may be associated with other environmental exposures. A collaborative approach, such as that used by the Cancer Intervention and Surveillance Modeling Network (CISNET) consortium, may be a cost-effective way to pursue this work.

> **Recommendation 12: Breast cancer researchers and research funders should pursue efforts to (1) develop statistical methodology for the estimation of risk of breast cancer for given sets of risk factors and that takes the life course perspective into account, (2) determine the proportion of the total temporal and geographic differences in breast cancer rates that can be plausibly attributed to established risk factors, and (3) develop modeling tools that allow for calculation of breast cancer risk, in both absolute and relative terms, with the goal of assessing potential risk reduction strategies at both personal and public health levels.**

Communicating About Breast Cancer Risks

Accurate and effective communication of breast cancer risks is challenging, and developing better approaches should be a research target. Uncertainty is inherent in risk prediction, but it is important to inform a broad range of stakeholders and constituencies on both those exposures that are associated with increased risk and those that have no evident association with breast cancer.

Recommendation 13: Breast cancer researchers and research funders should pursue research to identify the most effective ways of communicating accurate breast cancer risk information and statistics to the general public, health care professionals, and policy makers.

This work should include identifying ways to improve translation of research results into messages that can effectively convey the implications for women in different risk categories, women from diverse racial and ethnic groups, health care providers, and public health decision makers. It also should include ways to convey information about chemicals for which there is suggestive evidence of risk from experimental studies.

CONCLUDING OBSERVATIONS

Breast cancer is a leading cause of cancer morbidity among women in the United States and many other countries. Major advances have been made in understanding its biology and diversity, but more needs to be learned about the causes of breast cancer and how to prevent it. Familiar advice about healthful lifestyles appears relevant, but it remains difficult to discern what contribution a diverse array of other environmental factors may be making. Important targets for research are the biologic significance of life stages at which environmental risk factors are encountered, what steps may counter their effects, when preventive actions can be most effective, and whether opportunities for prevention can be found for the variety of forms of breast cancer.

1

Introduction

The prospect of developing breast cancer is a source of anxiety for many women. Breast cancer remains the most common invasive cancer among women (aside from nonmelanoma skin cancers), accounting in 2011 for an estimated 230,480 new cases among women in the United States and another 2,140 new cases among men (ACS, 2011). After lung cancer, it is the second most common cause of mortality from cancer for women, with about 39,520 deaths expected in the United States in 2011. Another 450 breast cancer deaths are expected among men in 2011 (ACS, 2011). Since the mid-1970s, when the National Cancer Institute (NCI) began compiling continuous cancer statistics, the annual incidence of invasive breast cancer rose from 105 cases per 100,000 women to 142 per 100,000 women in 1999 (NCI, 2011). Since then, however, the incidence has declined. In 2008, the incidence of breast cancer was 129 cases per 100,000 women.

Further reduction of the incidence of breast cancer is a high priority, but finding ways to achieve this is a challenge. As in most types of adult cancer, breast cancer is thought to develop as a result of accumulated damage induced by both internal and external triggers resulting in initial carcinogenic events. The affected cells and tissues then progress through multiple stages, with accompanying alterations in the surrounding tissue likely playing a role in whether the damage leads to a cancer. These events contributing to subsequent cancers may occur spontaneously as a by-product of errors in normal processes, such as DNA replication, or potentially through effects of environmental exposures. The early procarcinogenic events from endogenous and exogenous processes may be sustained and furthered by physiologic conditions such as obesity. It is likely that many such procarcinogenic events may never be entirely preventable because, although potentially modifiable, they are consequences of basic biologic processes, such as oxidative damage to DNA from endogenous metabolism, or stimulation of cell growth through normal hormonal processes.[1] Although such biological "background" mutagenesis is unavoidable, highly efficient protective pathways, such as DNA repair and immune surveillance, are effective at reducing the impacts of procarcinongenic events (Loeb and Nishimura, 2010; Bissell and Hines, 2011).

Although more needs to be learned about both the mechanisms by which breast cancers arise and the array of factors that influence risk for them, much has been established. Among the factors generally accepted as increasing women's risk are older age, having a first child at an older age or never having a child, exposure to ionizing radiation, and use of certain forms of postmenopausal hormone therapy (HT). Inherited mutations in the *BRCA1* and *BRCA2* genes also markedly increase risk for breast cancer (and other cancers as well), but these mutations are rare in the general population and account for only 5 to 10 percent of cases (ACS, 2011).

Even though aging, genetics, and patterns of childbearing account for some of the risk for breast cancer, they are not promising targets for preventive measures. More helpful would be identifying modifiable risk factors. For example, the publication of findings from the Women's Health Initiative (Writing Group for the Women's Health Initiative Investigators, 2002) confirming earlier indications that estrogen–progestin HT was contributing to an increase in the risk of postmenopausal breast cancer was followed by a rapid reduction in use of HT and in the incidence of invasive breast cancer. As reflected in NCI data, the incidence in 2002 was 136 cases per 100,000 women, compared with 127 in 2003 (NCI, 2011). A portion of the decline in breast cancer incidence since 1999 is attributed to this reduced use of HT (e.g., Ravdin et al., 2007; Farhat et al., 2010). But there are long-standing and still unresolved concerns that aspects of diet, ambient chemicals, or other potentially modifiable environmental exposures may be contributing to high rates of breast cancer.

At present, a large but incomplete body of evidence is available on the relationship between breast cancer and the wide variety of external factors that can be said to comprise the environment. Information on interactions between genetic susceptibility and environmental factors is particularly sparse. In contrast, knowledge of the complexity of breast cancer is growing, with the characterization of multiple tumor subtypes; the possibility

---

[1]Loeb and Nishimura (2010, p. 4270) note that each normal cell in a person's body may be exposed to as many as 50,000 DNA-damaging events each day, and that oxygen free radicals are a major source of DNA damage.

that critical events in the origins of breast cancer can occur very early in life; the variety of pathways through which breast cancer risks may be shaped; and the potential significance of both the timing of exposures and the way combinations of factors determine the effect on risks for different types of breast cancer. This growing knowledge has stimulated a transition in breast cancer research. The new perspectives on breast cancer highlight the limitations of the current understanding of the disease, and innovative ideas are beginning to influence the design and analysis of epidemiologic studies, experimental studies in animals, and mechanistic studies of breast cancer biology, all directed toward elucidating how external factors may influence the etiology of breast cancer.

This report presents the results of a study commissioned to review the current evidence on environmental risk factors for breast cancer, consider gene–environment interactions in breast cancer, explore evidence-based actions that might reduce the risk of breast cancer, and recommend research in these areas.

STUDY CHARGE AND COMMITTEE ACTIVITIES

This study resulted from a request to the Institute of Medicine (IOM) by Susan G. Komen for the Cure and its Scientific Advisory Board. Komen for the Cure funds research on prevention, diagnosis, and treatment of breast cancer, and also provides educational information and support services for the public and health care providers. The Statement of Task for the IOM study appears in Box 1-1.

The members of the study committee were selected to contribute expertise in epidemiology, toxicology, risk assessment, biostatistics, molecular carcinogenesis, gene–environment interactions, communication of health messages, environmental health science, exposure assessment, and health care. The committee includes a member from the patient advocacy community.

The committee met in person five times from April 2010 through February 2011 and conducted additional deliberations by conference call. During these meetings and calls, the committee reviewed and discussed the existing research literature on the topics central to its charge and developed and revised this report. At three of its meetings, the committee held public sessions during which it heard presentations by researchers, representatives of advocacy organizations, and members of the public.

The committee also commissioned work on two topics. One project was a review of data available to assess temporal changes in the potential for exposure to a selected set of chemicals and other environmental agents. The agents included in this paper have been discussed in the research literature and the popular press as possible contributors to increased risk for

BOX 1-1
Study Charge

In response to a request from Susan G. Komen for the Cure®, the Institute of Medicine will assemble a committee to:

1. Review the evidentiary standards for identifying and measuring cancer risk factors;
2. Review and assess the strength of the science base regarding the relationship between breast cancer and the environment;
3. Consider the potential interaction between genetic and environmental risk factors;
4. Consider potential evidence-based actions that women could take to reduce their risk of breast cancer;
5. Review the methodological challenges involved in conducting research on breast cancer and the environment; and
6. Develop recommendations for future research in this area.

In addition to reviewing the published literature, the committee will seek input from stakeholders, in part by organizing and conducting a public workshop to examine issues related to the current status of evidentiary standards and the science base, research methods, and promising areas of research. The workshop will focus on the challenges involved in the design, conduct, and interpretation of research on breast cancer and the environment. The committee will generate a technical report with conclusions and recommendations, as well as a summary report for the lay public.

breast cancer. This work served as an information resource for the committee and helped to identify some data presented in Chapter 4. The other project resulted in a paper examining temporal changes in the United States in exposure to ionizing radiation, with a particular focus on exposure from medical imaging (see Appendix F, available electronically at http://www.nap.edu/catalog.php?record_id=13263).

APPROACH TO THE STUDY

The committee began its work with recognition of the potentially vast scope of the study task and the need to develop a perspective and approach that could lead to a useful and timely report. The committee sought to focus its attention in areas that it considered to be the most significant and the most pertinent to the charge placed before it.

For purposes of this report, the committee interpreted "environment" broadly, to encompass all factors that are not directly inherited through

DNA. As a result, this definition includes elements that range from the cellular to the societal: the physiologic and developmental course of an individual, diet and other ingested substances, physical activity, microbial agents, physical and chemical agents encountered at home or at work, medical treatments and interventions, social factors, and cultural practices. This perspective was a foundation for the committee's work; application of it in its broadest sense is something that the committee hopes will expand the scope of future research. For some readers, this interpretation will differ from their association of the phrase "environmental risk factors" primarily with pollutants and other products of industrial processes (Baralt and McCormick, 2010). Furthermore, throughout the report the term "breast cancer" is used to refer to disease in humans and "mammary cancer" or "mammary tumor" to refer to disease in animals.

The committee explored the available evidence concerning breast cancer risks associated with a varied but limited collection of specific substances and factors (Chapter 3), and it also reviewed the many challenges that researchers have had to contend with in studying breast cancer, including those pertaining to gene–environment interactions (Chapter 4). But in its examination of the relation between breast cancer and the environment, the committee chose to highlight an approach that emphasizes the biologic mechanisms through which environmental factors may be operating and the importance of the changing picture over the life course (Chapter 5). This perspective played a major role in shaping the committee's conclusions and recommendations.

A Life Course Perspective

Breast cancer is primarily (but far from exclusively) a disease of adult women who are approaching or have reached menopause. In 2009, approximately 90 percent of new cases in U.S. women were diagnosed at age 45 or older (ACS, 2009). But the breast undergoes substantial changes from the time it begins developing in the fetus through old age, especially in response to hormonal changes during puberty, pregnancy, lactation, and menopause. With the timing of these developmental events related to risk for some types of breast cancer, there has been growing interest in exploring whether the timing of a variety of environmental exposures also is important in understanding what influences breast cancer risks. In Chapter 5, the committee has sought to link its examination of the mechanisms of carcinogenesis with a life course perspective on when and how those pathologic pathways may be particularly relevant in relation to when and how environmental exposures occur. Attention was paid to growing evidence for critical windows of susceptibility (e.g., periods with rapid cell proliferation or maturation) when specific mechanisms that increase the likelihood of a breast cancer developing may be more likely to be activated.

Identifying Environmental Risks for Breast Cancer

Trying to determine which environmental exposures may be influencing rates of breast cancer poses substantial challenges, many of which are discussed in Chapter 4. Cancer is a complex disease, and its "causes" are generally harder to trace than the bacteria and viruses that cause infectious diseases. People who are never exposed to the measles virus will never get measles. But the impact of removing a particular environmental exposure associated with breast cancer is less clear because many other factors can still contribute to the development of breast cancer. The role of underlying susceptibility from inherited genes appears to involve both rare variants and common ones, but it is still not well characterized. Moreover, people are exposed to a complex and changing mix of environmental agents over the course of a lifetime, so discerning the effects of an individual agent, or knowing which components of the mixture may influence the development of disease or how the mixture's components may interact with each other or with genes, is not straightforward.

Observational epidemiologic studies are a critical tool for learning about elevated risks, but they can be difficult to do well. They typically are the basis for demonstrating correlations between risk factors and outcomes, but establishing a causal inference is much more difficult. The challenges in establishing causality in such studies include difficulties with exposure measurement and accounting for undetected or poorly measured differences that may exist between the groups designated as exposed and unexposed. Furthermore, the timing and duration of observational studies may affect whether sufficient time has elapsed to detect differences in the incidence of a cancer that may not appear until many years after an exposure. Randomized controlled trials, which assign participants to a specific exposure or a comparison condition, are easier to interpret. However, for ethical and methodological reasons, such studies are rarely possible, especially when the goal is to determine whether the exposure is associated with an adverse event.

Experimental studies in animal models and in vitro systems offer an important opportunity to study the effects of well-defined exposures and to explore mechanisms of carcinogenicity in ways that are not possible in epidemiologic studies. They can signal potential hazards to human health that cannot be identified in other ways, but their results have to be interpreted with an understanding of differences across species and the comparability of an experimental exposure to the conditions encountered in the human population.

Reviewing Evidence on Specific Risk Factors

The literature on risk factors for cancer in general and breast cancer in particular is large and varied. In the United States, the Environmental Protection Agency (EPA) and the National Toxicology Program (NTP) in the National Institute of Environmental Health Sciences have programs to review the evidence on the carcinogenicity of various substances.[2] The International Agency for Research on Cancer (IARC), which is part of the World Health Organization, is a focal point for major international collaboration in such reviews.[3] In addition, a collaborative project between the World Cancer Research Fund International and the American Institute for Cancer Research has an ongoing program to review evidence on diet, physical activity, and cancer (WCRF/AICR, 2007).[4] All of these review programs consider evidence concerning breast cancer (or mammary cancers in animal studies) when it is available, but it is not their focus. Reviews specifically concerning breast cancer have also been conducted. These reviews include one conducted by the California Breast Cancer Research Program (2007) and a review sponsored by Komen for the Cure and conducted by the Silent Spring Institute (e.g., Brody et al., 2007; Rudel et al., 2007).

Assembling a comprehensive review of evidence on the relation between a complete set of environmental factors and breast cancer was not feasible for this study. Instead, the committee chose to focus on a limited selection of various types of environmental factors and potential routes of exposure. These factors are discussed in Chapter 3. The committee's aim was to characterize the available evidence and identify where substantial areas of uncertainty exist.

Observations About Risk

One component of the committee's task was to comment on actions that can be taken to reduce the risk of breast cancer. Opportunities for action are discussed in Chapter 6, but it is important to emphasize from the outset the challenge of interpreting evidence regarding risk and risk reduction. The widely quoted estimate that women in the United States have a 1-in-8 chance of being diagnosed with breast cancer during their lifetimes

---

[2]Information on the EPA and NTP review programs is available at http://www.epa.gov/ ebtpages/pollcarcinogens.html and http://ntp.niehs.nih.gov/?objectid=7 2016262-BDB7-CEBA-FA60E922B18C2540.

[3]Information on IARC reviews is available at http://www.iarc.fr/ and http://monographs. iarc.fr/index.php.

[4]Information on the review by the World Cancer Research Fund International and the American Institute for Cancer Research is available at http://www.wcrf.org/cancer_research/ expert_report/index.php.

can be restated as approximately a 12 percent lifetime risk of developing invasive breast cancer (NCI, 2010). The risk can also be presented for shorter, more comprehensible intervals. For example, among white women who are 50 years old, 2.4 percent are likely to be diagnosed with invasive breast cancer over the next 10 years (NCI, 2010). This 10-year risk is 2.2 percent for 50-year-old black women, 2.0 percent for Asian women, and 1.7 percent for Hispanic women. For 70-year-olds, the 10-year risks are 3.9 percent for white women, 3.2 percent for black women, and 2.4 percent for both Asian and Hispanic women. Estimates for longer follow-up periods (e.g., 20 or 30 years) will only increase those risks. Within average values such as these, there are always groups of women whose particular characteristics give them a higher or lower 10-year risk.

These estimates of risk are a critical reference point for understanding the implications of findings from epidemiologic studies on factors associated with increased or decreased risk of breast cancer. These findings are typically reported in terms of relative risk, which reflects a comparison between the risk in a population exposed to a particular factor and that in a similar population that is not exposed. Thus, a relative risk of 2.0 (a doubling of risk) might mean that for women with that risk factor, the 10-year risk of breast cancer is 5 percent rather than 2.5 percent. Similarly, a relative risk of 0.5 for a protective factor means that women with that characteristic may have a 10-year risk of 1.3 percent rather than 2.5 percent. These examples are offered to illustrate the scale of the change in risk implied by typical epidemiologic findings; they are not a formal analysis.

From a public health perspective, another important piece of information is the prevalence of the risk factor in the population. Finding that an environmental factor is associated with a large relative risk may still mean that it accounts for few cases of disease if the disease or the exposure is rare in that population. Alternatively, an environmental exposure that is associated with only a small increase in risk may be contributing to a large number of cases if the exposure is very common in the population. However, if the exposure is so common that there is little variability across the population (virtually everyone is exposed), it can be extremely difficult to identify the contribution from that exposure.

Virtually all of the epidemiologic evidence regarding breast cancer risk is drawn from population-level analyses. As a result, the conclusions reached on the basis of that evidence apply to an exposed *population*. With current knowledge, it is not possible to apply those conclusions to predict which *individuals* within that population are most likely to develop breast cancer. Nevertheless, an understanding of population-based estimates of risk can help people make personal choices that may lead to better health outcomes.

TOPICS BEYOND THE SCOPE OF THE STUDY

Several topics were defined as falling beyond the scope of the study. With the focus on environmental risk factors for breast cancer, the committee chose to devote little attention to the established associations between increased risk for breast cancer and reproductive events such as younger age at menarche, older age at first birth, lack of lactation, and older age at menopause. The committee also chose not to evaluate the established associations between breast cancer risk and higher birth weight and attained stature. Although some of them might fall under the committee's very broad definition of environmental factors, they were not the focus of its review. Background is provided on many of these other factors in Chapter 2, and the possibility that some environmental exposures may have an indirect influence on risk for breast cancer because they may affect the timing of these reproductive events is discussed in Chapter 5.

The committee also agreed that the nature and effectiveness of breast cancer screening, diagnosis, and treatment were generally beyond the scope of the study. It noted but did not analyze the impact of increased mammography and changes in screening practices since the 1970s on the observed incidence of breast cancer. The paper commissioned by the committee on medical sources of exposure to ionizing radiation took into account the contribution of mammography. The committee did not examine the appropriateness of screening recommendations or practices.

The committee decided as well that its charge called for a focus on risk for the initial occurrence of breast cancer and not on recurrence or factors that might be associated with the risk of recurrence. Although environmental exposures may well influence the risk of recurrence, that risk is also influenced by characteristics of tumors at the time of diagnosis and subsequent treatment and follow-up practices. Consideration of clinical practice in the treatment of women (and men) with diagnosed breast cancers is substantially different from the study's primary focus on prevention of breast cancer through improved understanding of and response to environmental risks. Similarly, the committee concluded that its charge called for a focus on the incidence of breast cancer and not mortality. Influences on breast cancer mortality patterns include factors that affect diagnosis and treatment that are separate from the effects of environmental exposures on the incidence of the disease.

The committee did not explicitly assess environmental risk factors for male breast cancer, beyond the general assumption that some of the risk factors identified through studies in women may also be relevant to the development of breast cancer in men.

THE COMMITTEE'S REPORT

This report reviews the current evidence on the biology of breast cancer, examines the challenges of studying environmental risk factors, and presents the committee's findings and research recommendations from its review of evidence on environmental risk factors. Specifically, Chapter 2 provides important background for evaluating factors influencing breast cancer risk with a brief review of the biology of breast cancer and trends in incidence in the United States, along with discussion of the kinds of studies used to investigate breast cancer and environmental exposures. Chapter 3 presents the committee's review of evidence on selected environmental risk factors. Chapter 4 discusses the variety of challenges that complicate the study of environmental risk factors for breast cancer, as well as gene–environment interactions. Chapter 5 examines mechanisms of carcinogenesis and links them to a life course perspective on breast development and the potential for environmental factors to influence risk for breast cancer. In Chapter 6, the committee examines opportunities for evidence-based action to reduce risks for breast cancer and also considers the challenges of avoiding the unintentional introduction of new risks. Chapter 7 concludes the report with the committee's recommendations for future research efforts. Included as appendixes are agendas for the committee's public sessions (Appendix A), biographical sketches of committee members (Appendix B), a summary of weight-of-evidence categories used by major organizations that evaluate cancer risks (Appendix C), a table summarizing reports of population attributable risks for breast cancer (Appendix D), a glossary (Appendix E), and the paper commissioned on exposure to ionizing radiation (Appendix F).

REFERENCES

ACS (American Cancer Society). 2009. *Breast cancer facts and figures 2009–2010*. Atlanta, GA: ACS. http://www.cancer.org/Research/CancerFactsFigures/BreastCancerFactsFigures/index (accessed November 17, 2010).

ACS. 2011. *Breast Cancer facts and figures 2011–2012*. Atlanta, GA: ACS. http://www.cancer.org/acs/groups/content/@epidemiologysurveilance/documents/document/acspc-030975.pdf (accessed November 15, 2011).

Baralt, L. B., and S. McCormick. 2010. A review of advocate–scientist collaboration in federally funded environmental breast cancer research centers. *Environ Health Perspect* 118(12):1668–1675.

Bissell, M. J., and W. C. Hines. 2011. Why don't we get more cancer? A proposed role of the microenvironment in restraining cancer progression. *Nat Med* 17(3):320–329.

Brody, J. G., K. B. Moysich, O. Humblet, K. R. Attfield, G. P. Beehler, and R. A. Rudel. 2007. Environmental pollutants and breast cancer: Epidemiologic studies. *Cancer* 109(12 Suppl):2667–2711.

California Breast Cancer Research Program. 2007. *Identifying gaps in breast cancer research: Addressing disparities and the roles of the physical and social environment.* http://cbcrp. org/sri/reports/ identifyingGaps/index.php (accessed October 25, 2011).

Farhat, G. N., R. Walker, D. S. Buist, T. Onega, and K. Kerlikowske. 2010. Changes in invasive breast cancer and ductal carcinoma in situ rates in relation to the decline in hormone therapy use. *J Clin Oncol* 28(35):5140–5146.

Loeb, L. A., and S. Nishimura. 2010. Princess Takamatsu Symposium on DNA repair and human cancers. *Cancer Res* 70(11):4269–4273.

NCI (National Cancer Institute). 2010. *SEER cancer statistics review, 1975–2007.* Edited by S. F. Altekruse, C. L. Kosary, M. Krapcho, N. Neyman, R. Aminou, W. Waldron, J. Ruhl, N. Howlader, Z. Tatalovich, H. Cho, A. Mariotto, M. P. Eisner, D. R. Lewis, K. Cronin, H. S. Chen, E. J. Feuer, D. G. Stinchcomb, and B. K. Edwards. Bethesda, MD: NCI. http://seer.cancer.gov/csr/1975_2007/ (accessed January 6, 2011).

NCI. 2011. *SEER cancer statistics review, 1975–2008.* Edited by N. Howlader, A. M. Noone, M. Krapcho, N. Neyman, R. Aminou, W. Waldron, S. F. Altekruse, C. L. Kosary, J. Ruhl, Z. Tatalovich, H. Cho, A. Mariotto, M. P. Eisner, D. R. Lewis, H. S. Chen, E. J. Feuer, K. A. Cronin, and B. K. Edwards. Bethesda, MD: NCI. (Based on November 2010 SEER data submission, posted to the SEER website, 2011.) http://seer.cancer.gov/ csr/1975_2008/ (accessed June 1, 2011).

Ravdin, P. M., K. A. Cronin, N. Howlader, C. D. Berg, R. T. Chlebowski, E. J. Feuer, B. K. Edwards, and D. A. Berry. 2007. The decrease in breast-cancer incidence in 2003 in the United States. *N Engl J Med* 356(16):1670–1674.

Rudel, R. A., K. R. Attfield, J. N. Schifano, and J. G. Brody. 2007. Chemicals causing mammary gland tumors in animals signal new directions for epidemiology, chemicals testing, and risk assessment for breast cancer prevention. *Cancer* 109(12 Suppl):2635–2666.

WCRF/AICR (World Cancer Research Fund/American Institute for Cancer Research). 2007. *Food, nutrition, physical activity, and the prevention of cancer: A global perspective.* Washington, DC: AICR.

Writing Group for the Women's Health Initiative Investigators. 2002. Risks and benefits of estrogen plus progestin in healthy postmenopausal women: Principal results from the Women's Health Initiative randomized controlled trial. *JAMA* 288(3):321–333.

2

Background, Definitions, Concepts

The committee's examination of breast cancer and the environment required considerations at the intersection of diverse fields, including the biology and epidemiology of breast cancer, the identification of carcinogens and cancer-promoting agents, exposure assessment, toxicity and carcinogenicity testing, and the design and interpretation of research studies. This chapter provides some brief, fundamental background on these topics as a basis for the discussions in subsequent chapters.

AN INTRODUCTION TO BREAST CANCER

The breast begins forming during the prenatal period and undergoes substantial changes during adolescence and adulthood. Breast cancer arises when abnormal cellular growth occurs in certain structures and types of cells within the breast.

Although breast cancer is often spoken of as if it were a single disease, evolving techniques of analysis of the molecular characteristics of tumors are pointing to a variety of types of potentially differing origins. Gaining a better understanding of the nature of the heterogeneity of breast cancer will be critical in helping researchers improve the design and interpretation of studies of possible risk factors, and it may influence approaches to prevention.

Described here are the basics of the anatomy of the breast and breast development, types of breast cancer, and levels and trends in the incidence of the disease, focusing primarily on experience in the United States. The mechanisms that appear to result in female breast cancers and the pathways

37

BOX 2-1
Breast Cancer in Men

Approximately 1 percent of breast cancer cases occur in men, and less than 1 percent of men's cancer diagnoses are for breast cancer (ACS, 2011b). Because it is rare, breast cancer in men has been difficult to study. Based on what is known, however, it is considered to resemble breast cancer in postmenopausal women (Korde et al., 2010).

As in women, men's breasts respond to changes in sex hormone concentrations (both estrogens and androgens), but under normal circumstances they do not undergo the differentiation and lobular development that women's breasts experience with puberty, pregnancy, and lactation (Johansen Taber et al., 2010). Either an excess of estrogens or deficit of androgens appears to increase risk of breast cancer in men (Korde et al., 2010). Beginning after age 20, rates rise steadily with age. Approximately 92 percent of male breast cancers are estrogen receptor positive, compared with approximately 78 percent of breast cancers in women (Anderson et al., 2010). As is the case for women, inherited mutations in *BRCA1* and especially *BRCA2*, as well as other mutations, are associated with an increased risk of male breast cancer, but the majority of cases are not associated with a family history of the disease (Korde et al., 2010).

along which they operate are one of the main topics in Chapter 5. A brief description of breast cancer in men is provided in Box 2-1.

The Breast, Breast Development, and Breast Cancer

The development of the human female breast begins during gestation but is not complete at the time of birth. Further development and differentiation of breast tissue occurs over time and especially in response to fluctuating estrogen and other hormonal signals beginning in puberty, continuing through the reproductive years, during pregnancy and lactation, and at menopause. Monthly ovulatory cycles are accompanied by cyclical changes in the form and behavior of cells and structures in the breast, including progressive differentiation. Pregnancy and lactation trigger maximal differentiation of the breast. When pregnancy and lactation end, as well as at menopause, breast tissue regresses to a less differentiated state.

Within the breast are adipose and connective tissues that surround multiple collections of lobules in which milk is produced during lactation. Milk moves to the nipple through ductal structures. The ducts are lined by luminal epithelial cells and have an outer layer of myoepithelial cells. Populations of stem cells that can give rise to either luminal or myoepithelial cells are also found in the ductal tissue. The ducts are anchored to a basement membrane, which contributes to both the structure and the function of the ductal tissue. Connective tissue within and between the lobules, known as the stroma, further contributes to the structure of the breast and plays an important role in regulating both normal and abnormal breast cell growth and function (Arendt et al., 2010). Cell types within the stroma include (but are not limited to) fibroblasts, adipocytes, macrophages, and lymphocytes (Johnson, 2010). These cells and structures in the breast generate and respond to a diverse mix of hormones, especially estrogen, and other regulatory factors.

Certain disruptions in the complex processes that govern the structure and function of breast tissue may set the stage for breast cancer. Some carcinogenic events occur spontaneously in the course of normal biological processes and others are triggered by external factors. Although the body has efficient protective responses, such as DNA repair and immune surveillance, that can reduce the effect of such events, these protective responses are not always successful. The interval between the earliest "event" and the detection of a cancer may span several decades.

Specific mechanisms that may play a role in breast cancer are noted here but discussed further in Chapter 5. The contribution of genetic mutations to cancer is well known. They may be inherited (e.g., germline mutations in the *BRCA1* or *BRCA2* genes, which normally have a role in DNA repair) or develop in some cells during a person's lifetime (somatic mutations) as a result of reactive by-products of normal biological processes, or from the effects of external exposures. Other mechanisms include epigenetic changes that can alter gene expression without changes to DNA, promotion of cell growth by estrogen and other hormones or cell-signaling proteins, and evasion of the immune system.

Types of Breast Cancer

Most commonly, breast cancers develop in the ducts, but cancers also develop in the lobules or take other forms. Several systems are used to characterize breast cancers, with the systems developed primarily to provide information on prognosis and treatment decisions. For example, breast tumors may be classified by tumor size, extent of spread beyond the tumor site (localized, regional, distant), the anatomical characteristics of the tumor cells (e.g., ductal or lobular histology), and the molecular features of the tumor cells, such as presence or absence of estrogen and progesterone receptors and human epidermal growth factor receptor 2 (HER2/neu).

The age at which a woman is diagnosed with breast cancer is associated with tumor characteristics, such as the likelihood that the breast cancer is estrogen receptor positive or negative (ER+ or ER–). In addition, age or menopausal status also guides treatment decisions. For example, aromatase inhibitors are part of treatment for postmenopausal women who have ER+ breast cancers, but tamoxifen is used among premenopausal women. Except for reference to menopausal status, breast cancers in men are characterized in similar ways. Differences in patterns of such features as tumor histology, grade, and receptor status may distinguish between a more aggressive form of breast cancer with a generally earlier onset and a more common and less aggressive form that tends to occur at older ages (see Anderson et al., 2006b, 2007; Kravchenko et al., 2011).

Another major distinction is between invasive and noninvasive (or in situ) tumors. As the terms suggest, invasive tumors spread beyond the site at which they arise, while in situ tumors remain within the tissue where they originate, such as the epithelial cells lining the breast ducts. About 20 percent of reported tumors are noninvasive (ACS, 2011a). Ductal carcinoma in situ (DCIS) is the most common form of abnormal but noninvasive growth in the breast. Although DCIS can, in some cases, progress to an invasive cancer, the natural history of these tumors is poorly understood, and it is not yet possible to identify which ones are likely to progress (Allred, 2010). As a result, most women with in situ tumors receive treatment that is similar to the treatment for early-stage invasive tumors.

Estrogen and Progesterone Receptor Status

The molecular and genetic characteristics of breast tumors are used to guide treatment and assess prognosis. A feature for which breast tumors are now commonly evaluated is whether the cells express estrogen or progesterone receptors. Tumors that express these receptors are designated ER+ or PR+, and those that do not as ER– or PR–. In the United States, approximately 75 percent of invasive tumors for which receptor status is reported are ER+ and 65 percent are PR+ (Ries and Eisner, 2007; Kravchenko et al., 2011). ER+ and PR+ tumors have a generally better prognosis than tumors that do not express these receptors. These receptor characteristics are correlated with other tumor markers related to regulation of cell growth and proliferation and appear to reflect important differences in tumor origin (Phipps et al., 2010). Researchers are also finding that they are associated with differences in response to risk factors (e.g., Althuis et al., 2004; Yang et al., 2011).

Triple Negative Breast Cancer

Tumors lacking not only ER and PR expression but also HER2 are called triple negative breast cancers (TNBCs), and they are considered closely related to basal-like breast cancers (Carey et al., 2006; Foulkes et al., 2010). Triple negative breast tumors are typically aggressive and are more likely to be diagnosed in women who are younger (below age 50) and are African American. These cancers in African American women tend to be more advanced and of higher grade at the time of diagnosis than tumors in other racial groups (Carey et al., 2006; Stead et al., 2009; Trivers et al., 2009). Triple negative tumors have been associated with *BRCA1* and *BRCA2* mutations (Armes et al., 1999; Foulkes et al., 2003; Turner et al., 2007; Atchley et al., 2008). Additionally, a large proportion of TNBCs have altered p53 levels (Carey et al., 2006; Kreike et al., 2007; Rakha et al., 2007).

Genetic Susceptibility to Breast Cancer

Genetic mutations may contribute to breast cancer by altering various critical processes such as those related to DNA repair, hormone synthesis, and metabolism of carcinogens. Two types of genetic mutations are possible. Germline mutations are genetic variants that are passed from parents to offspring and are present in all cells. Genetic changes can also occur in specific cells during a person's lifetime; these changes, which can persist as cells divide, are called somatic mutations. They can arise by chance, as a by-product of normal processes such as cellular respiration or DNA replication, or from external exposures. Such mutations may lead to that cell becoming a cancer cell.

Inherited genetic variation is found across the population. Many of these variations, called polymorphisms, may have little or no impact on the function of a gene, but some of them are associated with increased susceptibility to disease. Common genetic variants are found in 1 percent or more of the population.

Every breast cancer contains somatic genetic changes, but only a few inherited mutations are known to convey a high risk of breast cancer in the carrier. The strongest evidence of inherited genetic susceptibility is for germline mutations in the *BRCA1* and *BRCA2* genes. Research suggests that a larger number of lower-risk germline variants also exist.

Hereditary Syndromes

A family history of breast cancer is an established breast cancer risk factor. This risk factor represents both inherited genetic risks as well as environmental factors that may cluster in families. Overall an inherited susceptibility to breast cancer contributes to about 10 percent of breast cancer cases, and in about 5 percent of breast cancer cases this inherited susceptibility is attributed to mutation in the *BRCA1* or *BRCA2* genes.

Mutations in these two genes are associated with increased susceptibility not only for breast cancer, but also for other cancers such as ovarian cancer.

BRCA1/2 mutations are high-penetrance mutations, meaning that women with these mutations have a very high lifetime risk of developing breast cancer. This risk is estimated to be at least 40 percent and possibly as high as 85 percent (Oldenburg et al., 2007). However, these mutations are rare, with substantially less than 1 percent of women in most populations carrying them (Narod and Offit, 2005). In addition to increasing the risk of breast cancer for women, they also increase risk for male breast cancer. Families in which such mutations may be present may have multiple cases of breast cancer, occurring at younger ages and in multiple generations, and a family history of ovarian cancer (Narod and Offit, 2005). Other sources of increased familial genetic risk include the Li-Fraumeni syndrome[1] from germline mutations in the p53 gene (Malkin et al., 1990) and Cowden disease[2] from germline mutations in the PTEN gene (Liaw et al., 1997).

Genetic testing is available to identify BRCA1 and BRCA2 mutations. Identification of a familial mutation that carries an increased risk of breast cancer allows women, and men, who carry such a mutation to seek closer monitoring of their health and to consider primary and secondary preventive measures, such as increased screening, bilateral prophylactic mastectomy and, for women, bilateral salpingo-oophorectomy (Walsh et al., 2006). Use of medications that can reduce the risk of breast cancer (i.e., tamoxifen and raloxifene) may also be appropriate for some women (USPSTF, 2002).

Breast Cancers in Women Without a Strong Family History

Most women diagnosed with breast cancer do not have a strong family history of the disease and do not carry mutations in highly penetrant cancer-susceptibility genes. They may, however, have other more common genetic variants that affect gene function and that may be responsible for a proportion of the breast cancer cases that develop. These genetic variants are called low-penetrance variants because they are associated with only a small degree of risk for breast cancer. Yet because they are common, they may contribute to the burden of disease. In addition, these variants may interact with environmental exposures such that risk is only expressed in the presence of the environment exposure (gene–environment interaction).

Two approaches have been used to identify low-penetrance genetic variants: a candidate gene approach and genome-wide association studies.

---

[1]Li-Fraumeni syndrome is characterized by a predisposition to sarcomas, lung cancer, brain cancer, leukemia, lymphoma, adrenal-cortical carcinoma, and breast cancer.

[2]Cowden disease is a syndrome involving mucocutaneous and gastrointestinal lesions and breast cancer.

Studies initially relied on the candidate gene approach, in which poly-morphic variants of genes that plausibly influence breast cancer risk are assessed in epidemiologic studies (i.e., case–control or cohort studies) for their association with breast cancer. For example, the Breast and Prostate Cancer Cohort Consortium has conducted extensive analyses of genetic variation in large numbers of specific genes in biological pathways thought to be most relevant to breast cancer, such as the steroid hormone metabo-lism and insulin-like growth factor pathways (Canzian et al., 2010; Gu et al., 2010). These studies did not find an association with breast cancer risk. In general, the candidate gene approach has had limited success in consis-tently identifying specific variants associated with breast cancer.

Genome-wide association studies (GWAS) allow for a comprehen-sive and unbiased search for modest associations across the genome. The approach in these studies is to identify a relatively limited set of readily recognized single nucleotide polymorphisms (SNPs) that are highly cor-related with a larger block of genetic variants and to use the limited set of "tagSNPs" in the analysis (Manolio, 2010). These studies require very large sample sizes (thousands or tens of thousands of cases and controls) because these variants tend to be associated with a small degree of risk. Because these studies make use of large numbers of statistical tests, they require extreme levels of statistical significance to identify true positive results (Hunter et al., 2008).

Results from several GWAS of breast cancer in women of European ancestry have been published (Easton et al., 2007; Hunter et al., 2007; Stacey et al., 2007; Turnbull et al., 2010), and one of women of Asian ancestry (Zheng et al., 2009). Out of the many variants studied, approxi-mately 20 risk variants have been robustly associated with breast cancer risk, all having only modest influence on risk (relative risks in the range of 1.05–1.3 per allele). Stronger associations with common variants are unlikely to exist, but they may be possible for rarer variants (e.g., those with minor allele frequencies of <5 percent) that have not been tested with the technologies available to date. Even so, statistical modeling suggests that low-penetrance gene variants may do at least as well in predicting risk as using traditional risk factors such as age at first birth, family history of breast cancer, and history of breast biopsy(ies) (Wacholder et al., 2010). This is a rapidly evolving area of research.

BREAST CANCER INCIDENCE IN THE UNITED STATES

As noted in Chapter 1, an estimated 230,480 new cases of invasive breast cancer were diagnosed among women in the United States in 2011 and another 2,140 new cases among men (ACS, 2011a). In addition, approximately 57,650 in situ cases were diagnosed in women, of which

---

BOX 2-2
Data on Breast Cancer

For data on patterns and trends in incidence and mortality for all forms of cancer in the United States, researchers generally rely on data from the National Cancer Institute's Surveillance, Epidemiology, and End Results (SEER) Program. In 1973, SEER began systematic collection of data from cancer registries in sites selected to characterize the diversity of the U.S. population. The number of participating registries has increased, and as of 2005 covered approximately a quarter of the U.S. population (NCI, 2005). The SEER Program establishes standards for completeness and quality of the data provided to it, and it works with participating registries to achieve those standards. As practices change, new data elements may be collected. For breast cancer, for example, data on estrogen and progesterone receptor status of tumors were added in 1990 (Ries and Eisner, 2007). Annual reports present data and analysis on cancer incidence, mortality, survival, and trends since 1975. Datasets can also be made available to qualified researchers for independent analyses.

States also have cancer registries, but some of these registries are less than 20 years old (CDC, 2010). Through the National Program of Cancer Registries (NPCR), which was established by federal legislation in 1992 and is administered by the Centers for Disease Control and Prevention, states receive assistance to improve the quality and completeness of their cancer registries. The NPCR now produces an annual report that combines data from state registries with data from the SEER program.

---

about 85 percent were DCIS (ACS, 2011a). Sources of surveillance data on breast cancer are described in Box 2-2.

Age Patterns and Changes Over Time

Breast cancer can occur in women and men of any age, but it is predominantly a disease of middle and older ages. Rates of invasive cancer increase rapidly after age 35 and currently peak at approximately 432 cases per 100,000 women in the age group 75–79 years (NCI, 2011) (see Figure 2-1). Rates of in situ disease rise more slowly and increase as women reach ages at which mammographic screening becomes common. The peak rate is 99 cases per 100,000 women at ages 65–69 (NCI, 2011). Among men, cases of invasive breast cancer are found at young ages, but incidence peaks at ages 85 and older at a rate of approximately 10 cases per 100,000 men (NCI, 2011).

The incidence of breast cancer has increased since at least the mid-1970s but has dropped from its peak in 1999. Figure 2-2 shows the rates

Cases per 100,000 women

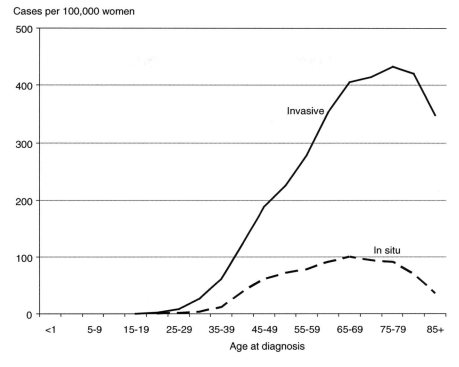

FIGURE 2-1 Age-specific incidence rates for invasive and in situ breast cancer among women in the United States, 2004–2008.
SOURCE: NCI (2011).

over time for both older (age 50 and older) and younger women (ages 20–49) and for invasive and in situ cases. Among older women, rates of invasive cancer rose during the 1980s and showed a slower increase during the 1990s. During the 1980s, use of menopausal hormone therapy had increased (Hersh et al., 2004; Glass et al., 2007). The 1980s and 1990s were also a period when use of screening mammography increased (Breen et al., 2001; Anderson et al., 2006a; Glass et al., 2007). In 1987, roughly 23 to 32 percent of women were screened, depending on their age, and by 1997, screening rates were as high as 74 percent among women ages 50–64 (Breen et al., 2001). Increased screening allowed for the earlier detection of tumors and for the detection of tumors that might never have progressed. When more tumors are detected at earlier stages, it will appear as if incidence rates are rising even if they are not, or are rising more rapidly than they actually are.

Cases per 100,000 women

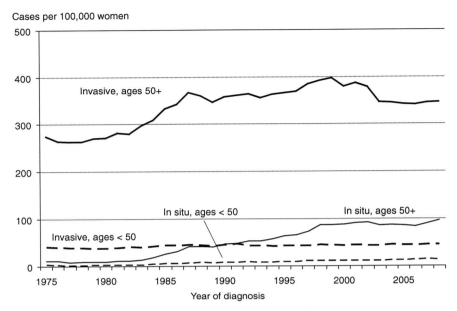

FIGURE 2-2 Age-adjusted incidence of invasive and in situ breast cancer in women, United States, 1975–2008.
SOURCE: NCI (2011).

A decline in breast cancer incidence occurred between 1999 and 2003 (Figure 2-2), principally in ER+ tumors in women ages 50–69 (Jemal et al., 2007). The decline is widely attributed to reductions in the use of hormone therapy (HT) (Clarke et al., 2006; Ravdin et al., 2007; Robbins and Clarke, 2007). In 1998, the Heart and Estrogen/Progestin Replacement Study (HERS) reported that use of combined estrogen–progestin HT failed to show an anticipated protective effect against coronary heart disease and was associated with an increase in risk for blood clots (Hulley et al., 1998). The subsequent publication of findings from the Women's Health Initiative confirmed the lack of benefit for heart disease and also showed an increased risk for breast cancer with use of combined estrogen–progestin therapy (Writing Group for the Women's Health Initiative Investigators, 2002). Reports from these studies were a major factor in the decline in use of HT.

As reflected in Figure 2-2, a recent analysis found that for 2003–2007 incidence rates of invasive cancer did not significantly change, although use of HT continued to decline (DeSantis et al., 2011). Use of screening mammography in 2008 remained similar to rates seen in 1997 (Breen et al., 2011). Rates of in situ cancer among older women also rose somewhat in the 1980s and into the 1990s, but they have remained relatively stable since the late 1990s.

Although the perception is widespread that breast cancer is becoming more common among young women, the best data available indicate that invasive breast cancer incidence rates have been almost unchanged since 1975 in women ages 20–49 (Figure 2-2). What has changed is the rate of in situ breast cancer, which has been rising since the introduction of mammography screening in the 1980s (Breen et al., 2001; Kerlikowske, 2010). The perception that breast cancer is increasing in younger women may come from several factors. First, any cancer diagnosis in a young woman in her prime working and reproductive years is notable, emotionally laden, and an event that will gain attention in many settings. An analysis of vignettes about breast cancer in popular magazines found that nearly half the stories were about women who were diagnosed before age 40 (Burke et al., 2001), a group that accounts for approximately 5 percent of cases (ACS, 2011a). Second, diagnosis of cases of "carcinoma in situ," especially DCIS, has increased, but its relation to invasive cancer can be unclear to women, at least in part because of the terminology and because of the aggressive treatment that may be recommended (De Morgan et al., 2002; Partridge et al., 2008; Liu et al., 2010). As noted, even within the research and medical communities, the natural history of DCIS is poorly understood, so the proportion of DCIS cases that would become invasive if untreated is unclear (Allred, 2010).

Race and Ethnicity

Differences can be seen in the age patterns and trends in breast cancer among the country's racial and ethnic groups. For 2004–2008, the overall incidence of breast cancer was 136 cases per 100,000 among non-Hispanic white women, 120 per 100,000 among African American women, 94 per 100,000 among Asian and Pacific Islander women, and 78 per 100,000 among Hispanic women (who can be of any race) (NCI, 2011).[3]

For African American women, the lower incidence rates compared with white women are most evident at older ages (Figure 2-3). However, incidence rates are higher among African American women under age 45. At ages 30–34, for example, African American women have an incidence of breast cancer of 31.8 cases per 100,000, compared with a rate of 25.8 for

---

[3]Throughout the report, incidence rates such as these are age-adjusted using the U.S. standard population for 2000. Age adjustment applies each group's incidence rates at specific ages to a single common population, the U.S. population for 2000 in this case. This process ensures that comparisons of rates are not affected by differences among the groups the age distribution of their populations.

Cases per 100,000 women

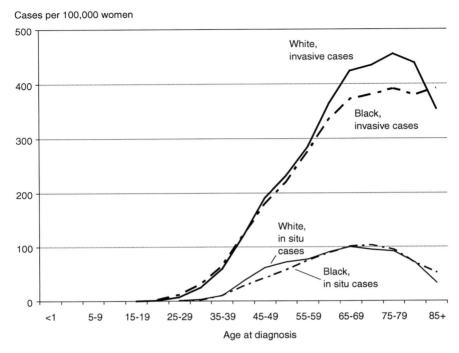

FIGURE 2-3 Age-specific incidence rates for invasive and in situ breast cancer among white and black women in the United States, 2004–2008. SOURCE: NCI (2011).

white women in that age group (NCI, 2011). At ages 40–44 the differences are smaller; the incidence rates are 123.6 for African American women and 122.4 for white women.

Despite ongoing efforts to improve detection and treatment of breast cancer for all women, African American women continue to experience greater mortality from breast cancer compared to women from other ethnic and racial groups. Surveillance, Epidemiology, and End Results (SEER) data from the National Cancer Institute show that the 5-year survival rate for women diagnosed with breast cancer during the period 2001–2007 was 77 percent among African American women and 91 percent among white women (NCI, 2011). These differences in breast cancer survival have been attributed in part to a higher proportion of African American women being diagnosed with advanced-stage disease; only 51 percent of breast cancers among African American women are localized at diagnosis compared with 61 percent of cancers among white women (NCI, 2011). Among women diagnosed with localized cancer, the 5-year survival rate for 2001–2007 was

93 percent for African American women and 99 percent for white women (NCI, 2011), reflecting a smaller but persistent difference in outcomes. Other factors contributing to poorer survival rates for African American women may include less access to early detection and treatment services as well as differences in tumor characteristics.

Among Hispanic women, the incidence of breast cancer is consistently lower than for non-Hispanic white women or African American women, with greater differences at older ages (NCI, 2006; Hines et al., 2010; Liu et al., 2011). Data from California show that the incidence of breast cancer for the period 1988–2004 was lower among the foreign-born Hispanic women: 68.2 per 100,000 for the foreign-born, 93.8 per 100,000 for U.S.-born Hispanic women, and 125.7 per 100,000 for non-Hispanic white women (Keegan et al., 2010). Approximately 40 percent of the Hispanic population living in the United States in 2007 was born in other countries (Grieco, 2010).

Analysis of the breast cancer experience of Hispanic women is still limited and based primarily on populations in specific areas of the United States, such as California (e.g., Keegan et al., 2010; Liu et al., 2011) or the Southwest (e.g., Hines et al., 2010). Additional research will be needed to assess whether the observations in these areas are representative of the experience of Hispanic women who live in other parts of the country and whose countries of origin and history of residence in the United States may differ from those of the women in the available studies.

The incidence of breast cancer has also traditionally been lower in Asian women, compared to white and black women, as reflected in both international and U.S. surveillance data (Stanford et al., 1995; Parkin et al., 1997, 2005; Jemal et al., 2005; Joslyn et al., 2005; Miller et al., 2008). Incidence rates commonly transition to higher levels as Asian women who migrate to the United States and their descendents experience greater acculturation. This pattern of increasing incidence among immigrants is often cited as evidence for the influence of social and environmental factors in disease risk because genetic factors are unlikely to be able to account for differences from the rates in their countries of origin (Buell, 1973; Thomas and Karagas, 1987; Ziegler et al., 1993; Kolonel and Wilkens, 2006).

Evaluating breast cancer incidence in the Asian and Pacific Islander population[4] is challenging because it is highly heterogeneous, with more than 60 distinct ethnicities. There is increasing evidence that the aggregate data on breast cancer incidence for these women tend to obscure large differences, including striking elevations in incidence for some subgroups (Deapen et al., 2002; Keegan et al., 2007; McCracken et al., 2007; Miller

---

[4]The Asian and Pacific Islander populations are combined as a standard reporting category for race and ethnicity for many federal data collection activities.

et al., 2008). Moreover, two studies that used different methods for assessing nativity suggest that young U.S.-born women from some Asian groups, especially women of Japanese and Filipina ancestry, are actually experiencing a higher risk for breast cancer than their white or African American contemporaries (Gomez et al., 2010; Reynolds et al., 2011).

Although Asian and Pacific Islanders, as a group, are less likely to receive an initial diagnosis of late-stage breast cancer than non-Hispanic white women (Hedeen et al., 1999; Morris and Kwong, 2004), foreign-born Asian women and some ethnic groups, including Hawaiians and South Asian Indians, are diagnosed with significantly more late-stage tumors than non-Hispanic white women (Li et al., 2003). Likewise, data from the 2001 California Health Interview Survey suggest that Asian women and Pacific Islander women have lower rates of mammography screening (67.2 percent and 63.4 percent, respectively) than non-Hispanic white women (78.1 percent) (Ponce et al., 2003a). The differences are further accentuated when disaggregated by ethnicity (53.1 percent among Korean women, 56.6 percent among Cambodian women) (Ponce et al., 2003b).

Racial and ethnic differences are also seen in terms of tumor types. The likelihood of having triple negative breast cancer, which is more difficult to treat, is significantly higher in African American women compared to women from other racial and ethnic groups (Bauer et al., 2007; Kwan et al., 2009; Stead et al., 2009). An analysis of SEER data for California found that African American women had a 1.98 percent lifetime risk of developing triple negative breast cancer, whereas Hispanic women had a 1.04 percent lifetime risk and white women had a 1.25 percent risk (Kurian et al., 2010). A high prevalence of triple negative tumors has also been reported in breast cancer cases from Nigeria and Senegal; of 507 cases, 27 percent were triple negative (Huo et al., 2009).

Reproductive Risk Factors

Several factors that are generally considered associated with increased risk for breast cancer include having a family history of the disease, particular reproductive characteristics (e.g., earlier age at menarche, later age at menopause, later age at first live birth), and certain forms of benign breast disease, as determined by breast biopsies (ACS, 2011a). Greater mammographic density, which reflects a higher proportion of connective and epithelial tissue in the breast, is a physiologic characteristic that is consistently associated with increased risk of breast cancer (Boyd et al., 2010). Studies in twins indicate that it is a heritable trait (e.g., Boyd et al., 2002; Ursin et al., 2009).

Differences in breast cancer incidence among population groups may reflect, in part, differences among them in the patterns of these types of risk factors. For example, data from the Third National Health and Nutrition Examination Survey (NHANES III) show that the median age at menarche for non-Hispanic black girls is 12.06 years compared to 12.25 years for Mexican American girls, and 12.55 years for non-Hispanic white girls (Chumlea et al., 2003).

In a review of epidemiologic studies, Bernstein and colleagues (2003) also found differences between African American and white women in reproductive risk factor profiles. For example, African American women have a higher birth rate than white women until age 30. This is important because while there may be a short-term increase in breast cancer risk immediately following pregnancy, earlier childbearing and higher numbers of births appear to be associated with a long-term reduction in risk. Lactation has been associated with a reduced risk of developing breast cancer; it induces additional differentiation in the breast and delays the re-initiation of ovulation. Studies included in the review conducted by Bernstein et al. (2003) found that, compared to African American women, white women are about twice as likely to breastfeed, and their cumulative time spent breastfeeding is longer.

Differences in breast cancer incidence and reproductive risk factor profiles have also been reported for Hispanic and non-Hispanic white women (e.g., Hines et al., 2010). Both premenopausal and postmenopausal Hispanic women had a higher prevalence of factors that have been associated with decreased breast cancer risk, including younger age at first birth and greater parity. But they were also more likely to have a younger age at menarche and to breastfeed less, characteristics associated with greater risk.

However, some of the associations between reproductive factors and breast cancer risk may be stronger for white non-Hispanic women than for women of other races and ethnicities. Hines and colleagues (2010) found that among premenopausal Hispanic women, only late age at first birth had a statistically significant association with increased risk of breast cancer. Reproductive factors were not associated with breast cancer risk among postmenopausal Hispanic women.

The contribution of differences in patterns of reproductive factors may also be influenced by racial and ethnic differences in risk for particular subtypes of breast cancer. Some reproductive factors appear to be more closely associated with ER+/PR+ tumors (Althuis et al., 2004; Ma et al., 2006) or lobular (versus ductal) tumors (Kotsopoulos et al., 2010; Newcomb et al., 2011). The risk for ER–/PR– and triple negative breast cancers is greater for African American women than for non-Hispanic white women, and reproductive factors have a more limited influence on risk for these forms of breast cancer.

A BROAD PERSPECTIVE ON THE ENVIRONMENT

As noted in Chapter 1, the committee adopted a broad interpretation of the environment that encompasses all factors that are not directly inherited through DNA. This definition allows for the consideration of a broad range of factors that may be encountered at any time in life and in any setting: the physiologic and developmental course of an individual, diet and other ingested substances, physical activity, microbial agents, physical and chemical agents encountered at home or work, medical treatments and interventions, social factors, and cultural practices. Figure 2-4 illustrates the multiple levels of biologic and social organization through which potential environmental exposures can influence breast cancer, and Figure 2-5 illustrates one approach to integrating this socio-ecologic perspective into investigation of potential contributions to breast cancer over the life course.

Many of these environmental influences overlap. For example, the physical environment encompasses medical interventions, dietary exposures to nutrients, energy and toxicants, ionizing radiation, and chemicals from industrial and agricultural processes and from consumer products. These in turn are influenced by the social environment, because cultural and economic factors influence diet at various stages of life, reproductive choices, energy balance, adult weight gain, body fatness, voluntary and involuntary physical activity, medical care, exposure to tobacco smoke and alcohol, and

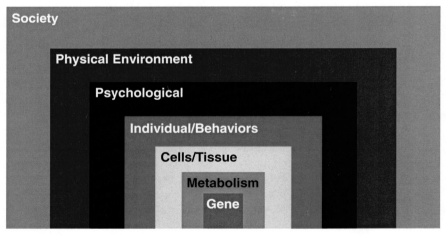

FIGURE 2-4 Multiple levels on which environmental exposures may act to influence breast cancer.
SOURCE: Personal communication, R. A. Hiatt, University of California, San Francisco, September 16, 2010.

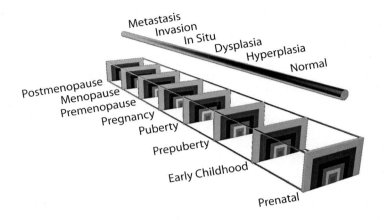

FIGURE 2-5 A schematic illustration of the potential for environmental exposures at various levels and times over the life course to influence the initiation and progression of breast cancer.
SOURCE: Personal communication, R. A. Hiatt, University of California, San Francisco, September 16, 2010.

occupational exposures, including shift work. Exposures at the tissue level are further influenced by metabolic and physiologic processes that modify the body's internal environment.

A full appreciation of environmental influences on breast cancer calls for an analysis at multiple levels (Anderson and May, 1995), from genetic and cellular mechanisms to the influence of societal factors. Applying this perspective to research requires a transdisciplinary approach. A previous Institute of Medicine committee advanced this socio-ecologic model as a way to understand the relationship of health and disease to complex societal influences (IOM, 2000; Smedley and Syme, 2001). Social determinants then encompass various factors: social and economic conditions such as poverty; the conditions of work, and access to health care delivery; the chemical toxicants and pollutants associated with industrial development; and the positive aspects of human settlements that make active living and healthy eating possible (Hiatt and Breen, 2008). The socio-ecologic model also incorporates and augments discoveries in cancer biology and toxicology, in addition to those from the behavioral and social sciences.

Within this framework, the committee's predominant focus was on exposure to physical and chemical toxicants, and on individual behavior related to diet and physical activity. When possible, the committee examined evidence regarding the implications of the timing of those exposures across the life course. Although the committee recognizes that the nature of households, families, workplaces, communities, and societies in which people live play a major role in determining these exposures (Hiatt and Breen, 2008), the focus of this report was on the more proximate environmental exposures that may increase the risk of breast cancer. As understanding of the epidemiology, toxicology, and mechanisms of breast cancer continues to improve, efforts to develop effective interventions to mitigate risk may be aided by approaches that include modification of the social determinants of exposure to various risk factors.

INVESTIGATING WHETHER ENVIRONMENTAL FACTORS ARE RELATED TO BREAST CANCER

Efforts to determine whether exposure to an aspect of the environment is related to the development of breast cancer depend on many types of research, including laboratory analyses of the response of cells or tissues (in vitro testing), experimental studies of effects in laboratory animals (in vivo testing), and epidemiologic studies of human subjects. U.S. regulatory agencies, including the Environmental Protection Agency (EPA) and the Food and Drug Administration (FDA), require a variety of in vitro and animal tests for cancer and other endpoints for licensing or registering pesticides, food additives, and pharmaceuticals (NRC, 2006). In laboratory studies, exposures are determined by the researcher, but in studies of human subjects, exposure assessment becomes a crucial part of the investigation.

Reviewed briefly here are basic features of this range of studies and of exposure assessment. Chapter 4 provides discussion of the challenges in using these various research tools to study breast cancer and draw valid conclusions about environmental risk factors.

In Vitro Testing

In vitro testing makes use of artificial environments to study tissues, cells, and cellular components. In the context of breast cancer, this type of testing allows for detailed examination of behavior of specific parts of larger, more complex organisms. Increasingly, in vitro testing allows for rapid analysis of a large number of variables, such as changes in gene expression. Although in vitro testing does not capture the critical interactions of the multiple systems in an intact organism, it provides a means to explore biological processes that are otherwise difficult to isolate.

In vitro tests for genotoxicity are an integral part of screening chemicals for their potential to cause DNA damage and thereby contribute to tumor formation. Various assays are used to assess gene mutations (e.g., Ames test, mouse lymphoma TK+/– assay) and structural or numerical aberrations in chromosomes (e.g., Chinese hamster ovary cells or mouse lymphoma TK+/– assay). Chemicals that show potential for genotoxicity are often avoided in product development programs for pesticides and pharmaceuticals.

Advances in molecular genetics, proteomics, and immunohistochemistry are fine-tuning investigations of mechanisms of action and treatment for breast cancer through studies of gene amplification, hormone receptor binding, biomolecular analysis of cells derived from tissue microdissection, and genome and transcriptional analysis (Thayer and Foster, 2007; Pasqualini, 2009). For example, such tools have led to the development of selective estrogen receptor modulators (SERMs; e.g., tamoxifen and raloxifene) and down-regulators (SERDs) that have provided both new therapeutic approaches to treating breast cancer and pharmacologic approaches to the prevention of breast cancer in some women (McDonell and Wardell, 2010). Next-generation SERMs and SERDs are now in clinical trials. Such tools will also allow a deeper understanding of the cell signaling events that are disrupted in the process of breast carcinogenesis, providing a rational basis from which to identify potential environmental influences on breast cancer risk. For example, they can aid in studying the potential role of melatonin and circadian disruption as a modulator of breast cancer risk (Blask et al., 2011). High-throughput microarray methods are used to examine various global gene expression changes related to high tumor aggressiveness, potentially leading to a new breast cancer molecular taxonomy and multigene signatures that might predict outcome and response to systemic therapies (Colombo et al., 2011).

Cell cultures from normal breast tissue and from breast tumors are being used to screen for the potential for chemicals to promote the growth of breast cancer cells or to evaluate the effectiveness of various therapeutic agents. Immortalized human breast cell lines (e.g., MCF-10F) have been established to study various aspects of tumorigenicity (e.g., Russo et al., 2002), and immortalized breast cancer cell lines (e.g., MCF-7) to study tumor progress and response to therapeutic agents (Wistuba et al., 1998; Fillmore and Kuperwasser, 2008). In vitro tests of the potential for chemicals to interact with estrogen, androgen, and thyroid hormonal systems may eventually be applied to most pesticides to generate other mechanistic information related to carcinogenicity. At present, while much has been learned about the potential for hormonal activity for some chemicals, data are limited on many others. In 2009, EPA required that about two dozen pesticides be screened for these effects (EPA, 2009).

Whole Animal (In Vivo) Studies of Carcinogenicity

Rodents have long been used to study mammary tumorigenesis. Specific rat and mouse strains have been selected for routine screening of chemicals and pharmaceuticals for carcinogenic effects. This testing is generally intended to detect any indication of carcinogenicity at any site in the body; it is not designed to identify likely sites for specific human cancers, such as breast cancer. EPA's (2005) *Guidelines for Carcinogen Risk Assessment* notes, however, that certain modes of action (e.g., disruption of thyroid function) will have consequences for particular tissues and that this provides a basis for anticipation of site concordance between rodents and humans in certain cases. Rodent models are also widely used by research scientists to investigate mammary carcinogenesis and the effects of timing and combinations of exposure to environmental factors. Challenges in using these models are discussed in Chapter 4.

Scope of Carcinogenicity Testing

Carcinogenicity testing in two species, typically rodents, is part of the standard battery of tests required for most pharmaceuticals, pesticides, and some food additives. Registration or licensing for marketing for products that require such approval involves establishing to the satisfaction of the appropriate government agency that the compound can be safely used under the registered use scenarios or, in the case of a pharmaceutical, that it has an adequate "risk–benefit" ratio.

Premarket testing of chemicals used in consumer products and in industry is rarely undertaken because the federal government has limited authority to require it under the Toxic Substances Control Act, which was enacted in 1976 (GAO, 2009). Only about 15 percent of the notices submitted to EPA for manufacturing or importing new industrial chemicals have any specific health or safety data (GAO, 2009). Instead, considerable reliance is placed on evaluating, qualitatively or through modeling, the similarities in structure to compounds that are carcinogenic or mutagenic (GAO, 2005; NRC, 2006). Each year, the National Toxicology Program (NTP) of the National Institute of Environmental Health Sciences conducts carcinogenicity screening for a few chemicals that would otherwise go untested. These chemicals are selected based on concern about their potential toxicity or the extent of human exposure. In 2007, the European Union began transferring responsibility for safety testing to manufacturers under the REACH program (Registration, Evaluation, Authorisation and Restriction of Chemical Substances) (European Chemicals Agency, 2007).

Carcinogenicity testing is also generally not required before new cosmetics and dietary supplements are marketed (FDA, 2005, 2009). Manufacturers are responsible for identifying ingredients and declaring that they are safe for the intended use. The FDA does have the authority to remove products from the market if they are found to be adulterated or misbranded.

NTP Carcinogenicity Study Protocols

Whole-animal studies are conducted as part of many types of academic and industry research on breast cancer and carcinogenicity. These studies can vary widely in design, depending on their purpose. For formal carcinogenicity reviews by EPA or the International Agency for Research on Cancer (IARC), the NTP study designs for whole-animal bioassays typically represent a recognized standard for carcinogenicity testing.

Under NTP (2006) protocols, carcinogenicity testing is usually based on a 2-year chronic dosing program. Testing uses three or more exposure-level groups and one unexposed control group, with separate test groups for male and female animals. Each group typically has 50 animals. The highest dose used in the assays is usually the maximally tolerated dose, with the aim of maximizing the ability to detect effects in small numbers of animals and minimizing the loss of animals from acutely toxic effects of the test substance. Dosing usually begins when the animals are 5 to 6 weeks of age. Under revised NTP (2010) study designs, rats (but not mice) may receive in utero and lactational exposure to the test substance, which will allow the testing procedures to identify adverse effects associated with exposures at the very earliest times of life.

The NTP currently uses Harlan Sprague Dawley rats, and one strain of mice, the B6C3F1 hybrid. Previously, other rat strains have been used (typically F344/N, although some chemicals were tested in Sprague Dawley and Osborne Mendel strains). Tests of similar design are required for pesticide registration (EPA, 1998) and pharmaceutical testing (FDA, 1997), although the animal strains used typically differ, and in utero testing is rarely performed (EPA, 2002).

At the end of the 2-year test period, the surviving animals are killed and necropsied. Any animals that die during the study period are also necropsied. To date, the NTP (2011) has tested more than 500 chemicals. Overall evaluation of the test results for carcinogenic hazard includes consideration of both malignant and benign tumors found anywhere in the animals.

Assessing the Process of Carcinogenesis and Susceptibility to Environmental Exposures

In addition to the use of experimental animals for standardized carcinogen bioassays, several animal models of chemically induced breast cancer have been used to evaluate (1) the cellular and molecular development and progression of breast cancer, and (2) the ability of environmental and developmental factors to modify breast carcinogenesis. The two most common models use induction of mammary tumors in rodents by the administration of N-methyl-N-nitrosourea (MNU) or 7,12-dimethylbenz[a]anthracene

(DMBA) (Russo and Russo, 1996; Thompson and Singh, 2000; Medina, 2010). In rats, these carcinogen-induced tumors arise from terminal end buds, which are similar in structure to the terminal ductal lobular unit in the human breast. Similar to human breast cancers, these chemically induced mammary carcinomas have altered expression of proteins that regulate cell growth and differentiation (e.g., HER2), and most rat mammary tumors express estrogen and progesterone receptors. For example, rat mammary tumors induced by MNU appear to be similar to low- to intermediate-grade human breast cancers that are ER+ and noninvasive (Chan et al., 2005).

Although these rodent models differ in important ways from human breast cancer (e.g., specific gene mutations, metastatic potential), they have been used extensively to explore mechanisms of mammary carcinogenesis and ways environmental factors influence that process. For example, studies have used DMBA-induction of mammary tumors in rats to demonstrate that obesity enhances tumor incidence and shortens the time to tumor development (e.g., Hakkak et al., 2005). These models make it possible to explore the impact of exposure to environmental agents at different times in life. For example, as discussed in Chapter 3, dioxins do not induce mammary tumors in rats in the 2-year chronic bioassay, but rats with prenatal exposure to 2,3,7,8-tetrachlorodibenzo-p-dioxin (TCDD) have shown altered mammary gland differentiation and an increased susceptibility to DMBA-induced mammary tumors (Jenkins et al., 2007). However, prenatal exposure of mice to TCDD delayed DMBA-induced tumor formation by 4 weeks relative to controls, and resulted in lower tumor incidence throughout the 27-week time course (Wang et al., 2011). The authors suggested that activation of the aryl hydrocarbon receptor (AhR) by TCDD slows the promotion of preneoplastic lesions to overt mammary tumors in mice. Interpreting such differences in response between rats and mice is among the challenges discussed in Chapter 4.

Another example of the use of whole animal models of carcinogen-induced mammary tumors in evaluating environmental risk factors for breast cancer was provided by La Merrill et al. (2009). Because some forms of breast cancer are associated with greater adiposity, these authors used three mouse models of breast cancer to examine the effect of prenatal TCDD exposure and high- or low-fat diet on physical characteristics associated with metabolic syndrome. The models were the DMBA mouse model and two different transgenic models of ER− breast cancer. Each model showed a different response (e.g., increase in body fat with or without changes in fasting glucose), but the TCDD exposure was associated with effects (reduced triglycerides) in only one of the models and only in the animals on the high-fat diet. The variation in response in models such as these may help in exploring the variability in human susceptibility to factors that increase risk of breast cancer.

Epidemiologic Studies[5]

Case–control studies compare exposures to the factor of interest (an "exposure") among individuals who have a disease of interest (cases) and individuals who do not have the disease (controls). The controls should come from a population that is judged comparable to the one from which the cases were identified (e.g., people with similar characteristics from the same community or the same hospital). Because of their more efficient study design, case–control studies are often done when a disease is rare or to explore a suspected association within a shorter period than a cohort approach would require. They are usually retrospective, looking back at exposure histories among the cases and controls. But assessing the timing of the exposures can be challenging. Among cases, it can difficult to be certain that the exposure preceded the disease. Studies with retrospective data collection that involves patient interviews can be subject to recall bias.[6] For example, cases, who have been diagnosed with cancer and who are likely to have thought carefully about why they have it, may be more likely to recall an exposure than controls, who do not have the disease and therefore may not have thought quite as carefully about whether they may have been exposed.

Cohort studies compare the occurrence of health outcomes among groups with different levels of exposure to a factor of interest. These studies may be prospective, beginning before individuals have been diagnosed with a disease and following them for a given period of time, or retrospective, using records or interviews to collect information about past exposures and health outcomes. For example, cohorts of smokers and nonsmokers could be followed to assess the incidence of lung cancer in each group. A prospective study ensures that exposure precedes diagnosis but exposure levels are not controlled by the investigator. Collection of information on exposures that vary over time is difficult and often not carried out with sufficient detail. Cohort studies avoid the problem of recall bias, but they can be subject to other forms of bias. The time frame for prospective cohort studies may be several years or as long as decades, depending on the hypothesized nature of the relation between the exposure(s) and the disease being studied. With breast cancer, for example, the disease may become evident only many years after an exposure of interest, so cohorts must be followed long enough to

---

[5]Additional information about study design and analysis is available from sources such as Rothman (2002) and Szklo and Nieto (2004).

[6]Forms of bias in epidemiologic studies are discussed in Chapter 4.

allow for this interval. If childhood or prenatal exposures play a role, then it could require five or more decades of follow-up. Extended follow-up of a study population can be expensive and administratively challenging. A listing of approximately 50 cohorts in the United States and other countries that have investigated breast cancer risks has been compiled by the Silent Spring Institute (2011). The listing illustrates the variation in characteristics and size of these study populations.

Controlled trials, also referred to as *clinical trials,* are experiments in which the investigator makes the decision as to who is assigned to receive the treatment (exposure) versus being in the comparison group. If the assignment is made at random and the sample size is adequate to ensure that confounding was minimized by the random assignment, then the result of the experiment can have a causal interpretation. For example, to determine if a medication that lowers serum cholesterol prevents heart attacks, it is possible to treat one group of individuals with a cholesterol-lowering medication and compare their cholesterol levels and incidence of heart attacks to those of a control group that did not receive the intervention. If the study is sufficiently large (in this case, takes place over a long enough time period for the number of events in the comparison group to be sufficient) and the assignment to treatment is random, then any reduction in incidence of heart attacks among the treated group, relative to the controls, can be interpreted to be a causal one. The comparison of measurements of cholesterol can also be used in drawing conclusions about the mechanism of action of the medications, although other mechanisms would also need to be taken into account. Studies that are investigating preventive care may be referred to as *intervention trials.* If an exposure is potentially harmful, controlled trials can examine ways to minimize or eliminate the exposure, but studies that deliberately expose participants to something expected to be harmful are not done. An optimal design of a clinical trial includes not only random assignment of study participants to the treatment or comparison group but also blinding of study participants and researchers to those assignments. Such blinding will minimize bias in the assessment of the outcomes.

Exposure Assessment

Studying the potential effects of environmental factors on risk for breast cancer requires some basis for distinguishing the women who have been exposed to the factor from those who have not. Exposure assessment is the process of establishing that an exposure has occurred and determining critical features of the exposure, including who is exposed and the magnitude, route, and timing of exposure. Errors in classifying who is more and who is less exposed (exposure misclassification) can limit the ability of a study to determine whether the environmental factor is associated with an increase or decrease in risk for breast cancer.

The approach to exposure assessment may depend on the type of study, the nature of the environmental factor of interest, the way exposure occurs, and the tools available to measure the exposure. In clinical trials or intervention trials, the population to be exposed and the exposure are determined in advance by the researchers. Even so, study participants may deviate from their prescribed exposures. In cohort and case–control studies, exposure status can sometimes be objectively determined (e.g., by measuring weight), but it often depends on reports by study participants of past or present experience (e.g., exposure to tobacco smoke in childhood or use of specific products in the home). Researchers may also use indirect means to estimate exposures, such as residence in a particular locality or distance from a particular source of concern (e.g., an air pollution source). Exposure to some chemicals can be established with tests of biologic specimens (e.g., blood, urine), but many exposures are not detectable in this manner and collection of specimens may not be possible. Because the first steps in breast cancer may begin decades before the diagnosis, relevant exposures may occur several decades before a cancer is detected.

Historically, studies in occupational settings have been an important means for identifying chemical carcinogens. The types and amounts of chemicals used may be documented, and exposure levels may be higher than in other settings. Studies in an occupational setting may be able to draw on records of job histories, understanding of production processes and chemicals used, or data from personal or area sampling. Exposure of certain workers to some chemicals may be thousands of times greater (or more) than that experienced by the general public, while other workers with different job tasks might experience a wide range of exposures. These pronounced variations in exposure allow for firmer conclusions as to whether exposure is associated with risk of disease. When exposure levels are low, contrasts between the exposed and unexposed are smaller, and associations with differences in disease risk may be more difficult to detect. However, the relatively small number of women in industries with heavy exposures, except during World War II, has limited the opportunity to study risks for breast cancer in those settings.

A potentially hazardous environmental factor can only pose a risk when it can enter the body and interact with tissues where it can do harm. Thus, an understanding of the possible points of entry of a given substance into the body, called "routes of exposure," is fundamental to evaluating its potential effects. These routes of exposure are inhalation, ingestion, or contact with the skin (dermal exposure). In occupational settings, inhalation is frequently the primary route of exposure, with dermal contact as a secondary route. In the general population, ingestion and dermal exposure play a large role, but inhalation is highly relevant for tobacco smoke and other air pollutants. Sometimes potential routes of exposure can be overlooked. For example, when taking showers, people experience dermal exposure to chemicals in the water supply, but showers also present an opportunity to inhale (typically low levels of) any water contaminants that readily volatilize.

The potential effect of an environmental exposure is usually strongly influenced by the magnitude of that exposure—the dose. A higher dose of a hazardous exposure is generally more likely to be associated with adverse health effects than a lower dose is. Factors that influence dose include the duration and frequency of exposure and the biologic processes that govern the absorption, distribution, metabolism, excretion, and storage of a substance in the body. The results of these toxicokinetic processes differ depending on the substance introduced into the body. Some ingested chemicals, for example, are poorly absorbed and rapidly excreted, while others may be readily absorbed, transformed by metabolism into new substances, and possibly stored in body tissues such as fat. The route of exposure may influence how the body responds to a substance. Also, differences among individuals in their genetics or exposure to other risk factors can result in differing responses to equal doses of a substance.

SOME MEASURES OF DISEASE RISK

Estimates of disease risk associated with a factor of interest—such as a personal characteristic (e.g., age), an environmental exposure (e.g., alcohol consumption or radiation exposure), or a medical treatment (e.g., a prescribed medication)—can be measured in multiple ways, including absolute risk, relative risk, hazard ratios, odds ratios, attributable risk, population attributable risk, and number needed to treat (NNT) or number needed to harm (NNH). The measure that is used depends on the study design, the available data, and in some cases the purpose for which the information is presented.[7]

In case–control studies, the prevalence of the factor of interest among cases and controls is compared using an odds ratio: the odds that a case is exposed compared to the odds that a control is exposed. Odds ratios of 1.0 mean that cases and controls were equally likely to have been exposed, and therefore the exposure is not associated with the disease and it is not a risk factor. An odds ratio that is statistically significantly less than 1.0 means that cases were less likely to have been exposed than controls. An odds ratio that is statistically significantly greater than 1.0 indicates that the

---

[7]Additional methodologic information is available from sources such as Rothman (2002) and Jewell (2004).

exposure is more likely to be reported among the case group than among the control group, indicating that the exposure is statistically associated with the disease, and thus is a potential risk factor for the disease.

Cohort studies typically use the measure of relative risk or the hazard ratio. Relative risk is a ratio of the absolute risk (incidence) of disease in an exposed group (or groups with different levels of exposure) to the absolute risk (incidence) of disease in an unexposed group (or some other designated comparison group). A hazard ratio incorporates information on the pace at which events (e.g., cases of breast cancer) occur over the course of a study. Clinical trials also use relative risk and hazard ratios. The relative risk is interpreted in much the same way as the odds ratio. A relative risk of 1.0 means the exposure is not associated with development of disease; a ratio that is statistically significantly less than 1.0 means that those who were exposed were less likely to develop the disease than those who were not (indicating that the exposure is protective); and a ratio that is statistically significantly greater than 1.0 means that the exposure is associated with the disease, indicating that it is potentially a risk factor for the disease.

Relative risk estimates and odds ratios represent an estimate of the strength of the association of a risk factor with breast cancer, but by themselves they do not provide insight into the underlying incidence of the disease and the absolute impact of a given factor. A relative risk of 2.0 means that a factor is associated with a doubling of the incidence of the health outcome in the exposed group compared to the unexposed. But this can mean an increase to 2 cases per 100,000 people or 200 cases per 100,000 people, depending on whether the underlying incidence is 1 case per 100,000 people or 100 cases per 100,000 people. Measures such as NNT and NNH are other ways of relating estimates of risk to absolute numbers. NNT is the number of people who would have to receive a treatment during a given time period for one person to benefit.

Other measures that are used to assess the impact of a risk factor include attributable risk (AR) and population attributable risk (PAR). The AR is defined as the percentage of cases that occur in the exposed group that are in excess of the cases in the comparison group. The PAR is a population-based measure of the percentage of excess cases associated with the exposure of interest that also takes into account the distribution of the risk factor within the population. If a risk factor is rare, it may contribute only a small proportion of a population's disease risk, even if the incidence of the disease is much higher among those who are exposed (which would produce a high relative risk). To adequately estimate the PAR requires high-quality studies in which confounding and overlapping contributions from multiple factors are analyzed appropriately. There are numerous pitfalls in interpreting the PAR (discussed in Chapter 4) (Rockhill et al., 1998). Ideally, the PAR provides information on the percentage of disease that can be eliminated by avoiding the exposure, but the variation in estimates of PAR underscores how difficult it is to separate the effects from multiple risk factors. Because of this problem, and because PARs for individual factors cannot simply be added together, PARs are sometimes calculated for a group of factors rather than single factors. Appendix D shows, for instance, a range of estimated PAR values (see e.g., physical activity or hormone therapy). These ranges may reflect variation in the contribution of a given factor across different populations, or variation in the degree to which the different studies adequately controlled confounding, or a combination of the two.

SUMMARY

Overall, breast cancer becomes increasingly common as women grow older, but the patterns of the disease vary among women in different racial and ethnic groups. These differences are likely to reflect the influence of a mix of genetic and environmental factors. Although the scope of environmental influences can be understood to encompass cultural and societal factors, most of the human, animal, and mechanistic research to date has focused more narrowly on individual exposures and the related biological processes. In the following chapter, the committee examines evidence regarding a set of environmental factors that illustrate varied types of exposures that may occur and the range of evidence available to assess whether exposure is associated with increased risk of breast cancer.

REFERENCES

ACS (American Cancer Society). 2011a. *Breast cancer facts and figures 2011–2012.* Atlanta, GA: ACS. http://www.cancer.org/acs/groups/content/@epidemiologysurveilance/documents/document/acspc-030975.pdf (accessed October 24, 2011).

ACS. 2011b. *Cancer facts and figures 2011.* Atlanta, GA: ACS. http://www.cancer.org/Research/CancerFactsFigures/CancerFactsFigures/cancer-facts-figures-2011 (accessed June 22, 2011).

Allred, D. C. 2010. Ductal carcinoma in situ: Terminology, classification, and natural history. *J Natl Cancer Inst Monogr* 2010(41):134–138.

Althuis, M. D., J. H. Fergenbaum, M. Garcia-Closas, L. A. Brinton, M. P. Madigan, and M. E. Sherman. 2004. Etiology of hormone receptor-defined breast cancer: A systematic review of the literature. *Cancer Epidemiol Biomarkers Prev* 13(10):1558–1568.

Anderson, L. M., and D. S. May. 1995. Has the use of cervical, breast, and colorectal cancer screening increased in the United States? *Am J Public Health* 85(6):840–842.

Anderson, W. F., I. Jatoi, and S. S. Devesa. 2006a. Assessing the impact of screening mammography: Breast cancer incidence and mortality rates in Connecticut (1943–2002). *Breast Cancer Res Treat* 99(3):333–340.

Anderson, W. F., R. M. Pfeiffer, G. M. Dores, and M. E. Sherman. 2006b. Comparison of age distribution patterns for different histopathologic types of breast carcinoma. *Cancer Epidemiol Biomarkers Prev* 15(10):1899–1905.

Anderson, W. F., B. E. Chen, L. A. Brinton, and S. S. Devesa. 2007. Qualitative age interactions (or effect modification) suggest different cancer pathways for early-onset and late-onset breast cancers. *Cancer Causes Control* 18(10):1187–1198.

Anderson, W. F., I. Jatoi, J. Tse, and P. S. Rosenberg. 2010. Male breast cancer: A population-based comparison with female breast cancer. *J Clin Oncol* 28(2):232–239.

Arendt, L. M., J. A. Rudnick, P. J. Keller, and C. Kuperwasser. 2010. Stroma in breast development and disease. *Semin Cell Dev Biol* 21(1):11–18.

Armes, J. E., L. Trute, D. White, M. C. Southey, F. Hammet, A. Tesoriero, A. M. Hutchins, G. S. Dite, et al. 1999. Distinct molecular pathogeneses of early-onset breast cancers in *BRCA1* and *BRCA2* mutation carriers: A population-based study. *Cancer Res* 59(8):2011–2017.

Atchley, D. P., C. T. Albarracin, A. Lopez, V. Valero, C. I. Amos, A. M. Gonzalez-Angulo, G. N. Hortobagyi, and B. K. Arun. 2008. Clinical and pathologic characteristics of patients with *BRCA*-positive and *BRCA*-negative breast cancer. *J Clin Oncol* 26(26):4282–4288.

Bauer, K. R., M. Brown, R. D. Cress, C. A. Parise, and V. Caggiano. 2007. Descriptive analysis of estrogen receptor (ER)-negative, progesterone receptor (PR)-negative, and HER2-negative invasive breast cancer, the so-called triple-negative phenotype: A population-based study from the California Cancer Registry. *Cancer* 109(9):1721–1728.

Bernstein, L., C. R. Teal, S. Joslyn, and J. Wilson. 2003. Ethnicity-related variation in breast cancer risk factors. *Cancer* 97(1 Suppl):222–229.

Blask, D. E., S. M. Hill, R. T. Dauchy, S. Xiang, L. Yuan, T. Duplessis, L. Mao, E. Dauchy, et al. 2011. Circadian regulation of molecular, dietary, and metabolic signaling mechanisms of human breast cancer growth by the nocturnal melatonin signal and the consequences of its disruption by light at night. *J Pineal Res* 51(3):259–269.

Boyd, N. F., G. S. Dite, J. Stone, A. Gunasekara, D. R. English, M. R. McCredie, G. G. Giles, D. Tritchler, et al. 2002. Heritability of mammographic density, a risk factor for breast cancer. *N Engl J Med* 347(12):886–894.

Boyd, N. F., L. J. Martin, M. Bronskill, M. J. Yaffe, N. Duric, and S. Minkin. 2010. Breast tissue composition and susceptibility to breast cancer. *J Natl Cancer Inst* 102(16):1224–1237.

Breen, N., D. K. Wagener, M. L. Brown, W. W. Davis, and R. Ballard-Barbash. 2001. Progress in cancer screening over a decade: Results of cancer screening from the 1987, 1992, and 1998 National Health Interview Surveys. *J Natl Cancer Inst* 93(22):1704–1713.

Breen, N., J. F. Gentleman, and J. S. Schiller. 2011. Update on mammography trends: Comparisons of rates in 2000, 2005, and 2008. *Cancer* 117(10):2209–2218.

Buell, P. 1973. Changing incidence of breast cancer in Japanese-American women. *J Natl Cancer Inst* 51(5):1479–1483.

Burke, W., A. H. Olsen, L. E. Pinsky, S. E. Reynolds, and N. A. Press. 2001. Misleading presentation of breast cancer in popular magazines. *Eff Clin Pract* 4(2):58–64.

Canzian, F., D. G. Cox, V. W. Setiawan, D. O. Stram, R. G. Ziegler, L. Dossus, L. Beckmann, H. Blanche, et al. 2010. Comprehensive analysis of common genetic variation in 61 genes related to steroid hormone and insulin-like growth factor-I metabolism and breast cancer risk in the NCI breast and prostate cancer cohort consortium. *Hum Mol Genet* 19(19):3873–3884.

Carey, L. A., C. M. Perou, C. A. Livasy, L. G. Dressler, D. Cowan, K. Conway, G. Karaca, M. A. Troester, et al. 2006. Race, breast cancer subtypes, and survival in the Carolina Breast Cancer Study. *JAMA* 295(21):2492–2502.

CDC (Centers for Disease Control and Prevention). 2010. *National Program of Cancer Registries.* http://www.cdc.gov/cancer/npcr/about.htm (accessed November 8, 2011).

Chan, M. M., X. Lu, F. M. Merchant, J. D. Iglehart, and P. L. Miron. 2005. Gene expression profiling of NMU-induced rat mammary tumors: Cross species comparison with human breast cancer. *Carcinogenesis* 26(8):1343–1353.

Chumlea, W. C., C. M. Schubert, A. F. Roche, H. E. Kulin, P. A. Lee, J. H. Himes, and S. S. Sun. 2003. Age at menarche and racial comparisons in U.S. girls. *Pediatrics* 111(1):110–113.

Clarke, C. A., S. L. Glaser, C. S. Uratsu, J. V. Selby, L. H. Kushi, and L. J. Herrinton. 2006. Recent declines in hormone therapy utilization and breast cancer incidence: Clinical and population-based evidence. *J Clin Oncol* 24(33):e49–e50.

Colombo, P. E., F. Milanezi, B. Weigelt, and J. S. Reis-Filho. 2011. Microarrays in the 2010s: The contribution of microarray-based gene expression profiling to breast cancer classification, prognostication and prediction. *Breast Cancer Res* 13(3):212.

De Morgan, S., S. Redman, K. J. White, B. Cakir, and J. Boyages. 2002. "Well, have I got cancer or haven't I?" The psycho-social issues for women diagnosed with ductal carcinoma in situ. *Health Expect* 5(4):310–318.

Deapen, D., L. Liu, C. Perkins, L. Bernstein, and R. K. Ross. 2002. Rapidly rising breast cancer incidence rates among Asian-American women. *Int J Cancer* 99(5):747–750.

DeSantis, C., N. Howlader, K. A. Cronin, and A. Jemal. 2011. Breast cancer incidence rates in U.S. women are no longer declining. *Cancer Epidemiol Biomarkers Prev* 20(5):733–739.

Easton, D. F., K. A. Pooley, A. M. Dunning, P. D. Pharoah, D. Thompson, D. G. Ballinger, J. P. Struewing, J. Morrison, et al. 2007. Genome-wide association study identifies novel breast cancer susceptibility loci. *Nature* 447(7148):1087–1093.

EPA (Environmental Protection Agency). 1998. *Health effects test guidelines. OPPTS 870.4200: Carcinogenicity.* EPA 712–C–98–211. Washington, DC: Government Printing Office. http://hero.epa.gov/index.cfm?action=search.view&reference_ID=6378 (accessed November 16, 2011).

EPA. 2002. *A review of the reference dose and reference concentration processes.* EPA/630/P-02/002F. Washington, DC: EPA. http://www.epa.gov/raf/publications/pdfs/rfd-final.pdf (accessed November 17, 2011).

EPA. 2005. *Guidelines for carcinogen risk assessment.* Washington, DC: EPA. http://www.epa.gov/osa/mmoaframework/pdfs/CANCER-GUIDELINES-FINAL-3-25-05%5B1%5D.pdf (accessed October 23, 2011).

EPA. 2009. Final list of initial pesticide active ingredients and pesticide inert ingredients to be screened under the Federal Food, Drug, and Cosmetic Act. *Federal Register* 74(71): 17579–17585. http://www.epa.gov/scipoly/oscpendo/pubs/final_list_frn_041509.pdf (accessed October 25, 2011).

European Chemicals Agency. 2007. *REACH.* http://echa.europa.eu/reach_en.asp (accessed November 8, 2011).

FDA (Food and Drug Administration). 1997. *Guidance for industry: S1B testing for carcinogenicity of pharmaceuticals.* Rockville, MD: FDA. http://www.fda.gov/downloads/Drugs/GuidanceComplianceRegulatoryInformation/Guidances/ucm074916.pdf (accessed November 17, 2011).

FDA. 2005. *Fair Packaging and Labeling Act, Title 15—Commerce and Trade, Chapter 39—Fair Packaging and Labeling Program.* http://www.fda.gov/regulatoryinformation/legislation/ucm148722.htm (accessed November 8, 2011).

FDA. 2009. *Overview of dietary supplements.* http://www.fda.gov/Food/DietarySupplements/ConsumerInformation/ucm110417.htm (accessed November 8, 2011).

Fillmore, C. M., and C. Kuperwasser. 2008. Human breast cancer cell lines contain stem-like cells that self-renew, give rise to phenotypically diverse progeny and survive chemotherapy. *Breast Cancer Res* 10(2):R25.

Foulkes, W. D., I. M. Stefansson, P. O. Chappuis, L. R. Begin, J. R. Goffin, N. Wong, M. Trudel, and L. A. Akslen. 2003. Germline *BRCA1* mutations and a basal epithelial phenotype in breast cancer. *J Natl Cancer Inst* 95(19):1482–1485.

Foulkes, W. D., I. E. Smith, and J. S. Reis-Filho. 2010. Triple-negative breast cancer. *N Engl J Med* 363(20):1938–1948.

GAO (Government Accountability Office). 2005. *Chemical regulation: Options exist to improve EPA's ability to assess health risks and manage its chemical review program.* GAO-05-458. Washington, DC: GAO. http://www.gao.gov/new.items/d05458.pdf (accessed December 12, 2011).

GAO. 2009. *Chemical regulation: Observations on improving the Toxic Substances Control Act.* GAO-10-292T. Washington, DC: GAO. http://www.gao.gov/products/GAO-10-292T (accessed October 24, 2011).

Glass, A. G., J. V. Lacey, Jr., J. D. Carreon, and R. N. Hoover. 2007. Breast cancer incidence, 1980–2006: Combined roles of menopausal hormone therapy, screening mammography, and estrogen receptor status. *J Natl Cancer Inst* 99(15):1152–1161.

Gomez, S. L., C. A. Clarke, S. J. Shema, E. T. Chang, T. H. Keegan, and S. L. Glaser. 2010. Disparities in breast cancer survival among Asian women by ethnicity and immigrant status: A population-based study. *Am J Public Health* 100(5):861–869.

Grieco, E. 2010. *Race and Hispanic origin of the foreign-born population in the United States: 2007.* http://www.census.gov/prod/2010pubs/acs-11.pdf (accessed November 8, 2011).

Gu, F., F. R. Schumacher, F. Canzian, N. E. Allen, D. Albanes, C. D. Berg, S. I. Berndt, H. Boeing, et al. 2010. Eighteen insulin-like growth factor pathway genes, circulating levels of IGF-I and its binding protein, and risk of prostate and breast cancer. *Cancer Epidemiol Biomarkers Prev* 19(11):2877–2887.

Hakkak, R., A. W. Holley, S. L. Macleod, P. M. Simpson, G. J. Fuchs, C. H. Jo, T. Kieber-Emmons, and S. Korourian. 2005. Obesity promotes 7,12-dimethylbenz(a)anthracene-induced mammary tumor development in female Zucker rats. *Breast Cancer Res* 7(5):R627–R633.

Hedeen, A. N., E. White, and V. Taylor. 1999. Ethnicity and birthplace in relation to tumor size and stage in Asian American women with breast cancer. *Am J Public Health* 89(8):1248–1252.

Hersh, A. L., M. L. Stefanick, and R. S. Stafford. 2004. National use of postmenopausal hormone therapy: Annual trends and response to recent evidence. *JAMA* 291(1):47–53.

Hiatt, R. A., and N. Breen. 2008. The social determinants of cancer: A challenge for transdisciplinary science. *Am J Prev Med* 35(2 Suppl):S141–S150.

Hines, L. M., B. Risendal, M. L. Slattery, K. B. Baumgartner, A. R. Giuliano, C. Sweeney, D. E. Rollison, and T. Byers. 2010. Comparative analysis of breast cancer risk factors among Hispanic and non-Hispanic white women. *Cancer* 116(13):3215–3223.

Hulley, S., D. Grady, T. Bush, C. Furberg, D. Herrington, B. Riggs, and E. Vittinghoff. 1998. Randomized trial of estrogen plus progestin for secondary prevention of coronary heart disease in postmenopausal women. *JAMA* 280(7):605–613.

Hunter, D. J., P. Kraft, K. B. Jacobs, D. G. Cox, M. Yeager, S. E. Hankinson, S. Wacholder, Z. Wang, et al. 2007. A genome-wide association study identifies alleles in *FGFR2* associated with risk of sporadic postmenopausal breast cancer. *Nat Genet* 39(7):870–874.

Hunter, D. J., D. Altshuler, and D. J. Rader. 2008. From Darwin's finches to canaries in the coal mine—mining the genome for new biology. *N Engl J Med* 358(26):2760–2763.

Huo, D., F. Ikpatt, A. Khramtsov, J. M. Dangou, R. Nanda, J. Dignam, B. Zhang, T. Grushko, et al. 2009. Population differences in breast cancer: Survey in indigenous African women reveals overrepresentation of triple-negative breast cancer. *J Clin Oncol* 27(27):4515–4521.

IOM (Institute of Medicine). 2000. *Promoting health: Intervention strategies from social and behavioral research.* Washington, DC: National Academy Press.

Jemal, A., T. Murray, E. Ward, A. Samuels, R. C. Tiwari, A. Ghafoor, E. J. Feuer, and M. J. Thun. 2005. Cancer statistics, 2005. *CA Cancer J Clin* 55(1):10–30.

Jemal, A., E. Ward, and M. J. Thun. 2007. Recent trends in breast cancer incidence rates by age and tumor characteristics among U.S. women. *Breast Cancer Res* 9(3):R28.

Jenkins, S., C. Rowell, J. Wang, and C. A. Lamartiniere. 2007. Prenatal TCDD exposure predisposes for mammary cancer in rats. *Reprod Toxicol* 23(3):391–396.

Jewell, N. P. 2004. *Statistics for epidemiology*. Boca Raton, FL: Chapman & Hall/CRC.

Johansen Taber, K. A., L. R. Morisy, A. J. Osbahr, 3rd, and B. D. Dickinson. 2010. Male breast cancer: Risk factors, diagnosis, and management (review). *Oncol Rep* 24(5):1115–1120.

Johnson, M. C. 2010. Anatomy and physiology of the breast. In *Management of breast diseases*, edited by I. Jatoi and M. Kaufmann. Berlin, Germany: Springer-Verlag.

Joslyn, S. A., M. L. Foote, K. Nasseri, S. S. Coughlin, and H. L. Howe. 2005. Racial and ethnic disparities in breast cancer rates by age: NAACCR Breast Cancer Project. *Breast Cancer Res Treat* 92(2):97–105.

Keegan, T. H., S. L. Gomez, C. A. Clarke, J. K. Chan, and S. L. Glaser. 2007. Recent trends in breast cancer incidence among 6 Asian groups in the Greater Bay Area of Northern California. *Int J Cancer* 120(6):1324–1329.

Keegan, T. H., E. M. John, K. M. Fish, T. Alfaro-Velcamp, C. A. Clarke, and S. L. Gomez. 2010. Breast cancer incidence patterns among California Hispanic women: Differences by nativity and residence in an enclave. *Cancer Epidemiol Biomarkers Prev* 19(5):1208–1218.

Kerlikowske, K. 2010. Epidemiology of ductal carcinoma in situ. *J Natl Cancer Inst Monogr* 41:139–141.

Kolonel, L. N., and L. Wilkens. 2006. Migrant studies. In *Cancer epidemiology and prevention*, 3rd ed. Edited by D. Schottenfeld and J. Fraumeni. New York: Oxford University Press.

Korde, L. A., J. A. Zujewski, L. Kamin, S. Giordano, S. Domchek, W. F. Anderson, J. M. Bartlett, K. Gelmon, et al. 2010. Multidisciplinary meeting on male breast cancer: Summary and research recommendations. *J Clin Oncol* 28(12):2114–2122.

Kotsopoulos, J., W. Y. Chen, M. A. Gates, S. S. Tworoger, S. E. Hankinson, and B. A. Rosner. 2010. Risk factors for ductal and lobular breast cancer: Results from the Nurses' Health Study. *Breast Cancer Res* 12(6):R106.

Kravchenko, J., I. Akushevich, V. L. Seewaldt, A. P. Abernethy, and H. K. Lyerly. 2011. Breast cancer as heterogeneous disease: Contributing factors and carcinogenesis mechanisms. *Breast Cancer Res Treat* 128(2):483–493.

Kreike, B., M. van Kouwenhove, H. Horlings, B. Weigelt, H. Peterse, H. Bartelink, and M. J. van de Vijver. 2007. Gene expression profiling and histopathological characterization of triple-negative/basal-like breast carcinomas. *Breast Cancer Res* 9(5):R65.

Kurian, A. W., K. Fish, S. J. Shema, and C. A. Clarke. 2010. Lifetime risks of specific breast cancer subtypes among women in four racial/ethnic groups. *Breast Cancer Res* 12(6):R99.

Kwan, M. L., L. H. Kushi, E. Weltzien, B. Maring, S. E. Kutner, R. S. Fulton, M. M. Lee, C. B. Ambrosone, et al. 2009. Epidemiology of breast cancer subtypes in two prospective cohort studies of breast cancer survivors. *Breast Cancer Res* 11(3):R31.

La Merrill, M., D. S. Baston, M. S. Denison, L. S. Birnbaum, D. Pomp, and D. W. Threadgill. 2009. Mouse breast cancer model-dependent changes in metabolic syndrome-associated phenotypes caused by maternal dioxin exposure and dietary fat. *Am J Physiol Endocrinol Metab* 296(1):E203–E210.

Li, C. I., K. E. Malone, and J. R. Daling. 2003. Differences in breast cancer stage, treatment, and survival by race and ethnicity. *Arch Intern Med* 163(1):49–56.

Liaw, D., D. J. Marsh, J. Li, P. L. Dahia, S. I. Wang, Z. Zheng, S. Bose, K. M. Call, et al. 1997. Germline mutations of the PTEN gene in Cowden disease, an inherited breast and thyroid cancer syndrome. *Nat Genet* 16(1):64–67.

Liu, Y., M. Perez, M. Schootman, R. L. Aft, W. E. Gillanders, M. J. Ellis, and D. B. Jeffe. 2010. A longitudinal study of factors associated with perceived risk of recurrence in women with ductal carcinoma in situ and early-stage invasive breast cancer. *Breast Cancer Res Treat* 124(3):835–844.

Liu, L., J. Zhang, A. H. Wu, M. C. Pike, and D. Deapen. 2011. Invasive breast cancer incidence trends by detailed race/ethnicity and age. *Int J Cancer*. doi: 10.1002/ijc.26004. [Epub ahead of print]

Ma, H., L. Bernstein, M. C. Pike, and G. Ursin. 2006. Reproductive factors and breast cancer risk according to joint estrogen and progesterone receptor status: A meta-analysis of epidemiological studies. *Breast Cancer Res* 8(4):R43.

Malkin, D., F. P. Li, L. C. Strong, J. F. Fraumeni, Jr., C. E. Nelson, D. H. Kim, J. Kassel, M. A. Gryka, et al. 1990. Germline p53 mutations in a familial syndrome of breast cancer, sarcomas, and other neoplasms. *Science* 250(4985):1233–1238.

Manolio, T. A. 2010. Genomewide association studies and assessment of the risk of disease. *N Engl J Med* 363(2):166–176.

McCracken, M., M. Olsen, M. S. Chen, Jr., A. Jemal, M. Thun, V. Cokkinides, D. Deapen, and E. Ward. 2007. Cancer incidence, mortality, and associated risk factors among Asian Americans of Chinese, Filipino, Vietnamese, Korean, and Japanese ethnicities. *CA Cancer J Clin* 57(4):190–205.

McDonnell, D. P., and S. E. Wardell. 2010. The molecular mechanisms underlying the pharmacological actions of ER modulators: Implications for new drug discovery in breast cancer. *Curr Opin Pharmacol* 10(6):620–628.

Medina, D. 2010. Of mice and women: A short history of mouse mammary cancer research with an emphasis on the paradigms inspired by the transplantation method. *Cold Spring Harb Perspect Biol* 2(10):a004523.

Miller, B. A., K. C. Chu, B. F. Hankey, and L. A. Ries. 2008. Cancer incidence and mortality patterns among specific Asian and Pacific Islander populations in the U.S. *Cancer Causes Control* 19(3):227–256.

Morris, C.R., and S. L. Kwong, eds. 2004. *Breast cancer in California, 2003*. Sacramento, CA: California Department of Health Services.

Narod, S. A., and K. Offit. 2005. Prevention and management of hereditary breast cancer. *J Clin Oncol* 23(8):1656–1663.

NCI (National Cancer Institute). 2005. *SEER brochure*. http://seer.cancer.gov/about/SEER_brochure.pdf (accessed November 8, 2011).

NCI. 2006. *Supplemental data for the annual report to the nation (1975–2003)*. http://www.seer.cancer.gov/report_to_nation/1975_2003/supplemental.html (accessed November 8, 2011).

NCI. 2011. *SEER cancer statistics review, 1975–2008*. Edited by N. Howlader, A. M. Noone, M. Krapcho, N. Neyman, R. Aminou, W. Waldron, S. F. Altekruse, C. L. Kosary, J. Ruhl, Z. Tatalovich, H. Cho, A. Mariotto, M. P. Eisner, D. R. Lewis, H. S. Chen, E. J. Feuer, K. A. Cronin, and B. K. Edwards. Bethesda, MD: NCI. (based on November 2010 SEER data submission, posted to the SEER website, 2011). http://seer.cancer.gov/csr/1975_2008/ (accessed June 1, 2011).

Newcomb, P. A., A. Trentham-Dietz, J. M. Hampton, K. M. Egan, L. Titus-Ernstoff, S. Warren Andersen, E. R. Greenberg, and W. C. Willett. 2011. Late age at first full term birth is strongly associated with lobular breast cancer. *Cancer* 117(9):1946–1956.

NRC (National Research Council). 2006. *Toxicity testing for assessment of environmental agents: Interim report*. Washington, DC: The National Academies Press.

NTP (National Toxicology Program). 2006. *Specifications for the conduct of studies to evaluate the toxic and carcinogenic potential of chemical, biological and physical agents in laboratory animals for the National Toxicology Program (NTP)*. http://ntp.niehs.nih.gov/?objectid=72015DAF-BDB7-CEBA-F9A7F9CAA57DD7F5 (accessed November 8, 2011).

NTP. 2010. *Toxicology/carcinogenicity*. http://ntp.niehs.nih.gov/?objectid=72015DAF-BDB7-CEBA-F9A7F9CAA57DD7F5 (accessed November 3, 2011).

NTP. 2011. *Long-term study reports and abstracts.* http://ntp.niehs.nih.gov/index.cfm? objectid=0847DDA0-F261-59BF-FAA04EB1EC032B61 (accessed November 1, 2011).

Oldenburg, R. A., H. Meijers-Heijboer, C. J. Cornelisse, and P. Devilee. 2007. Genetic suscep-tibility for breast cancer: How many more genes to be found? *Crit Rev Oncol Hematol* 63(2):125–149.

Parkin, D. M., S. L. Whelan, J. Ferlay, L. Raymond, and J. Young, eds. 1997. *Cancer inci-dence in five continents.* Vol. VII. Lyon, France: International Agency for Research on Cancer (IARC).

Parkin, D. M., F. Bray, J. Ferlay, and P. Pisani. 2005. Global cancer statistics, 2002. *CA Cancer J Clin* 55(2):74–108.

Partridge, A., K. Adloff, E. Blood, E. C. Dees, C. Kaelin, M. Golshan, J. Ligibel, J. S. de Moor, et al. 2008. Risk perceptions and psychosocial outcomes of women with ductal carcinoma in situ: Longitudinal results from a cohort study. *J Natl Cancer Inst* 100(4):243–251.

Pasqualini, J. R. 2009. Breast cancer and steroid metabolizing enzymes: The role of progesto-gens. *Maturitas* 65(Suppl 1):S17–S21.

Phipps, A. I., C. I. Li, K. Kerlikowske, W. E. Barlow, and D. S. Buist. 2010. Risk factors for ductal, lobular, and mixed ductal-lobular breast cancer in a screening population. *Cancer Epidemiol Biomarkers Prev* 19(6):1643–1654.

Ponce, N. A., S. H. Babey, D. A. Etzioni, B. A. Spencer, E. R. Brown, and N. Chawla. 2003a. *Cancer screening in California: Findings from the 2001 Health Interview Study.* Los Angeles, CA: University of California at Los Angeles, Center for Health Policy Research. http://www.healthpolicy.ucla.edu/pubs/files/Cancer_Screening_Report.pdf (accessed De-cember 12, 2011).

Ponce, N. A., M. Gatchell, and E. R. Brown. 2003b. *Cancer screening rates among Asian eth-nic groups.* Los Angeles, CA: University of California at Los Angeles, Center for Health Policy Research. http://www.healthpolicy.ucla.edu/pubs/files/Asian_Cancer_FactSheet.pdf (accessed December 12, 2011).

Rakha, E. A., M. E. El-Sayed, A. R. Green, A. H. Lee, J. F. Robertson, and I. O. Ellis. 2007. Prognostic markers in triple-negative breast cancer. *Cancer* 109(1):25–32.

Ravdin, P. M., K. A. Cronin, N. Howlader, C. D. Berg, R. T. Chlebowski, E. J. Feuer, B. K. Edwards, and D. A. Berry. 2007. The decrease in breast-cancer incidence in 2003 in the United States. *N Engl J Med* 356(16):1670–1674.

Reynolds, P., S. Hurley, D. Goldberg, T. Quach, R. Rull, and J. Von Behren. 2011. An excess of breast cancer among young California-born Asian women. *Ethn Dis* 21(2):196–201.

Ries, L. A. G., and M. P. Eisner. 2007. Chapter 13: Cancer of the female breast. In *SEER survival monograph: Cancer survival among adults: U.S. SEER Program, 1988–2001, patient and tumor characteristics.* Edited by L. A. G. Ries, J. L. Young, G. E. Keel, M. P. Eisner, Y. D. Lin, and M. J. Horner. Bethesda, MD: National Cancer Institute, SEER Program.

Robbins, A. S., and C. A. Clarke. 2007. Regional changes in hormone therapy use and breast cancer incidence in California from 2001 to 2004. *J Clin Oncol* 25(23):3437–3439.

Rockhill, B., B. Newman, and C. Weinberg. 1998. Use and misuse of population attributable fractions. *Am J Public Health* 88(1):15–19.

Rothman, K. J. 2002. *Epidemiology: An introduction.* New York: Oxford University Press.

Russo, J., and I. H. Russo. 1996. Experimentally induced mammary tumors in rats. *Breast Cancer Res Treat* 39(1):7–20.

Russo, J., Q. Tahin, M. H. Lareef, Y. F. Hu, and I. H. Russo. 2002. Neoplastic transforma-tion of human breast epithelial cells by estrogens and chemical carcinogens. *Environ Mol Mutagen* 39(2–3):254–263.

Silent Spring Institute. 2011. *Silent Spring Institute guide to cohort studies for environmental breast cancer research.* http://www.silentspring.org/pdf/our_tools/CohortStudiesTable.pdf (accessed November 8, 2011).

Smedley, B. D., and S. L. Syme. 2001. Promoting health: Intervention strategies from social and behavioral research. *Am J Health Promot* 15(3):149–166.

Stacey, S. N., A. Manolescu, P. Sulem, T. Rafnar, J. Gudmundsson, S. A. Gudjonsson, G. Masson, M. Jakobsdottir, et al. 2007. Common variants on chromosomes 2q35 and 16q12 confer susceptibility to estrogen receptor-positive breast cancer. *Nat Genet* 39(7): 865–869.

Stanford, J. L., L. J. Herrinton, S. M. Schwartz, and N. S. Weiss. 1995. Breast cancer incidence in Asian migrants to the United States and their descendants. *Epidemiology* 6(2):181–183.

Stead, L. A., T. L. Lash, J. E. Sobieraj, D. D. Chi, J. L. Westrup, M. Charlot, R. A. Blanchard, J. C. Lee, et al. 2009. Triple-negative breast cancers are increased in black women regardless of age or body mass index. *Breast Cancer Res* 11(2):R18.

Szklo, M., and J. Nieto. 2004. *Epidemiology: Beyond the basics,* 2nd ed. Seattle, WA: Jones and Bartlett.

Thayer, K. A., and P. M. Foster. 2007. Workgroup report: National Toxicology Program workshop on hormonally induced reproductive tumors—relevance of rodent bioassays. *Environ Health Perspect* 115(9):1351–1356.

Thomas, D. B., and M. R. Karagas. 1987. Cancer in first and second generation Americans. *Cancer Res* 47(21):5771–5776.

Thompson, H. J., and M. Singh. 2000. Rat models of premalignant breast disease. *J Mammary Gland Biol Neoplasia* 5(4):409–420.

Trivers, K. F., M. J. Lund, P. L. Porter, J. M. Liff, E. W. Flagg, R. J. Coates, and J. W. Eley. 2009. The epidemiology of triple-negative breast cancer, including race. *Cancer Causes Control* 20(7):1071–1082.

Turnbull, C., S. Ahmed, J. Morrison, D. Pernet, A. Renwick, M. Maranian, S. Seal, M. Ghoussaini, et al. 2010. Genome-wide association study identifies five new breast cancer susceptibility loci. *Nat Genet* 42(6):504–507.

Turner, N. C., J. S. Reis-Filho, A. M. Russell, R. J. Springall, K. Ryder, D. Steele, K. Savage, C. E. Gillett, et al. 2007. BRCA1 dysfunction in sporadic basal-like breast cancer. *Oncogene* 26(14):2126–2132.

Ursin, G., E. O. Lillie, E. Lee, M. Cockburn, N. J. Schork, W. Cozen, Y. R. Parisky, A. S. Hamilton, et al. 2009. The relative importance of genetics and environment on mammographic density. *Cancer Epidemiol Biomarkers Prev* 18(1):102–112.

USPSTF (U.S. Preventive Services Task Force). 2002. *Chemoprevention of breast cancer: Recommendations and rationale.* http://www.uspreventiveservicestaskforce.org/uspstf/ uspsbrpv.htm (accessed August 9, 2011).

Wacholder, S., P. Hartge, R. Prentice, M. Garcia-Closas, H. S. Feigelson, W. R. Diver, M. J. Thun, D. G. Cox, et al. 2010. Performance of common genetic variants in breast-cancer risk models. *N Engl J Med* 362(11):986–993.

Walsh, T., S. Casadei, K. H. Coats, E. Swisher, S. M. Stray, J. Higgins, K. C. Roach, J. Mandell, et al. 2006. Spectrum of mutations in BRCA1, BRCA2, CHEK2, and TP53 in families at high risk of breast cancer. *JAMA* 295(12):1379–1388.

Wang, T., H. M. Gavin, V. M. Arlt, B. P. Lawrence, S. E. Fenton, D. Medina, and B. A. Vorderstrasse. 2011. Aryl hydrocarbon receptor activation during pregnancy, and in adult nulliparous mice, delays the subsequent development of DMBA-induced mammary tumors. *Int J Cancer* 128(7):1509–1523.

Wistuba, I. I., C. Behrens, S. Milchgrub, S. Syed, M. Ahmadian, A. K. Virmani, V. Kurvari, T. H. Cunningham, et al. 1998. Comparison of features of human breast cancer cell lines and their corresponding tumors. *Clin Cancer Res* 4(12):2931–2938.

Writing Group for the Women's Health Initiative Investigators. 2002. Risks and benefits of estrogen plus progestin in healthy postmenopausal women: Principal results from the Women's Health Initiative randomized controlled trial. *JAMA* 288(3):321–333.

Yang, X. R., J. Chang-Claude, E. L. Goode, F. J. Couch, H. Nevanlinna, R. L. Milne, M. Gaudet, M. K. Schmidt, et al. 2011. Associations of breast cancer risk factors with tumor subtypes: A pooled analysis from the Breast Cancer Association Consortium studies. *J Natl Cancer Inst* 103(3):250–263.

Zheng, W., J. Long, Y. T. Gao, C. Li, Y. Zheng, Y. B. Xiang, W. Wen, S. Levy, et al. 2009. Genome-wide association study identifies a new breast cancer susceptibility locus at 6q25.1. *Nat Genet* 41(3):324–328.

Ziegler, R. G., R. N. Hoover, M. C. Pike, A. Hildesheim, A. M. Nomura, D. W. West, A. H. Wu-Williams, L. N. Kolonel, et al. 1993. Migration patterns and breast cancer risk in Asian-American women. *J Natl Cancer Inst* 85(22):1819–1827.

3

What We Have Learned from Current Approaches to Studying Environmental Risk Factors

As one of its tasks, the committee was asked to review and assess the strength of the science base regarding the relationship between breast cancer and the environment. This body of evidence has evolved over many years through diverse fields of inquiry, including epidemiologic investigations, experimental studies in laboratory animals, and in vitro laboratory research on questions at the molecular, genetic, cellular, and tissue levels. Indeed, since the rise in breast cancer diagnoses that became particularly steep around 30 years ago, tremendous efforts have been made to identify the causes.

In this chapter the committee reviews approaches to assessing evidence concerning risk for breast cancer, summarizes the existing evidence on a selection of factors, and offers its assessment of the implications of the evidence. For many of the environmental risk factors, the results of the committee's review are far from conclusive. Reasons for the continuing gaps in knowledge are numerous. Chapter 4 discusses some of the challenges to studying causes of breast cancer and why the existing evidence permits few definitive conclusions. In some cases, recent advances using more sensitive tools to examine the pathobiology of breast cancer can be expected to provide new models for research in humans, animals, and in vitro systems.

Although the results of newer approaches to research on risk factors for breast cancer are promising, the extant literature is primarily grounded in older technologies and approaches. In light of this transitional state of the science, the committee nevertheless faced the question, what has been possible to discern from the work done so far? Here the committee outlines the scope of its review, describes evidentiary standards that have been used by leading authoritative bodies, and reviews the evidence on a selected set of risk factors.

SCOPE OF THE REVIEW

As discussed in Chapters 1 and 2, the committee adopted a broad definition of "environment" that includes all factors not directly inherited through DNA. In selecting environmental factors for examination, the committee took into account several considerations, including variety in the types of potential risk factors and routes of exposure, availability of evidence for review, and indications of public concern. From the enormous list of candidates, the committee selected a limited set of factors in order to illustrate a variety of environmental exposures, and to emphasize the need for new approaches to investigate and increase the knowledge base of potential environmental risks for breast cancer. With an evolving understanding of the mechanisms for cancer development and concern about whether the right questions have yet been asked or asked using appropriate study designs, the committee saw limited value in a full review of evidence for an extensive list of environmental factors that is available from a number of other sources (e.g., International Agency for Research in Cancer [IARC], the World Cancer Research Fund/American Institute for Cancer Research [WCRF/AICR], the U.S. Environmental Protection Agency [EPA], and the National Toxicology Program [NTP]), nor was it feasible for the present study. Of the large number of environmental factors with potential but uncertain impact on breast cancer, the committee reviewed only a selected number that illustrated particular types of challenges in assessment. For example, the committee evaluated factors for which extensive epidemiologic evidence and systematic reviews were available (e.g., alcohol consumption), and it also reviewed chemicals for which studies evaluating breast cancer in humans were very limited (e.g., bisphenol A).

Little attention was given to several very familiar topics, such as dietary fat and micronutrients, that are receiving ongoing systematic review by other organizations. The committee also chose not to include established reproductive risk factors, such as age at menarche or first full-term pregnancy, and anthropometric features such as birthweight or attained height in its review of environmental factors. These risk factors have also received considerable attention elsewhere. In Chapter 7 the committee has included recommendations for additional research to confirm the appropriateness of using alterations in such reproductive and anthropometric intermediate endpoints as valid and reliable markers of alterations in risk for breast cancer.

Process for the Evidence Review

Given the scope and time line of the committee's study, it was not feasible to carry out formal, systematic reviews of the scale or depth of those carried out by the WCRF/AICR, IARC, or the Cochrane Collaboration. Such reviews entail examination of the results of exhaustive literature searches and extensive documentation.[1] The committee found that given the changing science and the apparent gaps in the evidence base, it could most fruitfully apply its efforts in reviewing and speaking to a larger picture in the science of breast cancer and the environment.

The committee's process for its review of the evidence was as follows: The committee turned first to the conclusions available from the extensive reviews by authoritative groups (WCRF/AICR, 2007, 2008, 2010; EPA, 2011b; IARC, 2011; NTP, 2011a). Where the results of a systematic review were available for particular risk factors, the committee preferentially drew on these resources. These sources were supplemented by review of additional literature identified by committee members and staff and in targeted searches by an Institute of Medicine (IOM) research librarian. The targeted searches on the committee's selected risk factors discussed in this chapter used the PubMed and Embase databases in searches of the peer-reviewed, English-language literature published between January 2000 and October 2010, expecting that literature available before 2000 had been extensively reviewed by other authoritative reviews or subsequent publications. The searches were designed to identify literature on breast cancer in humans, mammary neoplasms in animals, and related in vitro and mechanistic studies. The process was supplemented by testimony from advocates, expert scientists, and members of the public.

Committee members examined these resources to evaluate the strength of the science base regarding the association of a given risk factor with breast cancer.

Hierarchy of Studies

Widely used standards of evidence for identifying and evaluating hazards or risks from potential carcinogens share several features. They

---

[1]For example, the WCRF/AICR review released in 2007, *Food, Nutrition, Physical Activity, and the Prevention of Cancer: A Global Perspective*, took place over 6 years (WCRF/AICR, 2007, Appendix A, p. 396). It required the work of an expert task force to develop the systematic review methodology, and methods testing at two centers. Next, research teams at nine institutions in Europe and North America carried out systematic literature reviews. Finally, a panel of experts worked to assess the evidence and agree on recommendations. Since then, the Continuous Update Project has been following scientific developments in this field. Its updates capture new evidence since the last systematic literature review to permit review and meta-analysis (WCRF/AICR, 2010).

typically rely only on published and peer-reviewed literature, and they ultimately reach conclusions about factors/agents based on the relevant studies, the strength of the results, and the coherence and plausibility of the evidence base. By virtue of their design, certain study types are given greater weight based on their relevance and freedom from bias.

Randomized controlled trials have an experimental design and, when well conducted, are considered the strongest form of epidemiologic study for directly determining causal associations between interventions or exposures and health outcomes. As discussed further in Chapter 4, randomization for many environmental exposures would be unethical or not feasible. In research on suspected environmental hazards, which is the focus of the committee's work, most epidemiologic studies are observational rather than experimental. Observational studies evaluate the exposures to the factor of interest as they take place in the real world, not based on intervention by any scientist. Thus, the determination of who is and who is not exposed may be related to marketing practices; changes in formulations, regulations, and laws (e.g., for emissions into air, water, or soil, or for chemicals to be used in manufacture of consumer products) at the federal, state, or local level; disposal practices; and personal choices about consumer product use, or behaviors (eating pesticide-free produce or not; leaving windows open to ventilate home). Observational studies can be informative when the comparison populations are appropriately defined and sufficient attention is given to exposure assessment and to confounding.[2] Other characteristics of observational studies that influence their validity are discussed in Chapter 4.

In addition to experimental (when available) and observational epidemiologic studies in humans, the committee drew on information from experimental studies in animals and studies carried out in vitro (in cells or tissues, rather than a whole organism) to inform its assessment of risk factors for breast cancer. As discussed in Chapter 4, these studies are powerful tools for exploring possible health effects, mechanisms of action, and the biologic plausibility of a factor's association with a change in risk for breast cancer. The literature review included reports from experimental studies in animals conducted for regulatory purposes as well as from studies by researchers.

Categories of Evidence

Several organizations have developed methods and criteria to classify the strength of evidence for the carcinogenicity of an exposure or to convey the strength of an association between a risk factor and a particular

---

[2]Confounding can occur when an exposure variable and the disease outcome are both related to one or more other variables not being studied. It is discussed further in Chapter 4.

health effect. The criteria aim to be explicit about the weight, or relative importance, given to studies in humans and in animals or other experimental systems. For example, IARC, EPA, NTP, and the WCRF/AICR each have a set of categories and approaches to applying them that reflect their work to classify potential carcinogens or risk factors. These classification schemes are developed under different mandates and missions with regard to their role in informing decision making. Designations by IARC, NTP, and WCRF/AICR are qualitative and do not attempt to quantify risk in relation to dose, whereas EPA carries out more quantitative evaluations. Various IOM committees have also developed qualitative systems of classification of evidence for their work in evaluating associations between exposures and outcomes (e.g., IOM, 1991, 2001, 2010, 2011).

The IARC, EPA, and NTP classification systems focus on identifying substances that may pose a cancer hazard; that is, whether a given substance is "capable of causing cancer under some circumstances" (IARC, 2006b). These systems work first by separately evaluating and rating the three types of evidence—human, animal, and other relevant data, such as from cell cultures—in categories such as "sufficient evidence in animals" or "limited evidence in humans." Second, the three evidence streams are integrated to reach an overall conclusion about the potential for a substance to be a carcinogen. Terms like "known" or "possible" carcinogen are used for the overall evidence categories. The IOM, WCRF, and Cochrane reviews primarily focus on the human evidence of risk (i.e., that an exposure is associated with an adverse human health outcome) and do not go through the formal exercise of rating the animal or other relevant evidence to reach conclusions about possible human carcinogenicity. The various IOM categories are applied to evidence for any relevant health outcome, not just cancers.

Strong and consistent positive epidemiologic evidence in rigorously conducted studies is prima facie evidence that the substance is a risk factor: People exposed to the agent were affected in sufficient numbers or the associated risk was sufficiently strong that it was possible to detect the breast cancer effect through epidemiologic study. There is a range of views within the scientific community as to whether strong nonhuman evidence of hazard should be a basis for concluding that a human risk exists. Formal translation of a hazard conclusion into a risk conclusion could involve quantitative evaluations of a number of factors, including the extent of the population that is exposed to the factor in question; the magnitude of exposure for specific segments of the population; and the extent to which the exposure to the substance accumulates with other exposures to pose risk to the population. But experimental evidence in nonhuman species or in vitro systems can indicate that the substance is a possible, biologically plausible risk factor, given sufficient dose at a relevant time. At present, in the absence of adequate human data, nonhuman evidence of hazard is used as the basis for regulatory decision making.

A critical difference among the categories and approaches used by IARC, EPA, NTP, IOM, and WCRF/AICR is the role that data from experimental studies in animals and studies employing in vitro systems using human or other cell lines play in determining the category for a substance. Full descriptions of the classifications used by IARC, EPA, NTP, IOM studies of Gulf War exposures, and WCRF are provided in Appendix C. For each organization, strong and convincing evidence from human epidemiologic studies is a basis for concluding that a substance or risk factor is causally associated with human cancer. WCRF includes in its criteria for "convincing causal relationship" that there be strong experimental evidence from human or animal studies that typical human exposure can lead to relevant cancer outcomes.[3] In rare circumstances ("exceptionally") under the EPA, NTP, and IARC schemes, very strong animal and mechanistic evidence (EPA and IARC) or strong human mechanistic evidence (NTP) can lead to a conclusion that a substance causes human cancer when definitive epidemiologic evidence is absent. Also in those schemes, strong experimental evidence alone can lead to a finding that a substance is probably or possibly a human carcinogen. In one case (EPA), suggestive animal evidence is treated as suggestive evidence of carcinogenic potential. In contrast, the approach used in several IOM studies focuses on evaluating the strength of human data, using animal and in vitro studies only as supplemental evidence for considering the biologic plausibility of observed epidemiologic associations in making determinations about causality.

The classifications used by this committee take elements from systems used by IOM, IARC, and EPA. The committee chose to use terms that more explicitly identify the relative strengths of the epidemiologic data for pointing out known and probable risk factors being evaluated, along the lines of approaches used by IOM committees. Factors for which epidemiologic evidence shows a consistently positive association with breast cancer that is not explained by bias or confounding and that falls outside the realm of chance are considered as "risk factors" for breast cancer. Thus, because epidemiologic studies by their very nature include consideration of human exposures, they are able to observe "risk," not just "hazard." In contrast, mechanistic and animal studies address "hazard" (the potential to cause an effect), but are not observations of human "risk factors." As noted above, other steps are needed to make judgments about whether substances identi-

---

[3]Experimental evidence must fall into the WCRF/AICR (2007) Class I category, either in vivo data from studies using human volunteers, genetically modified animal models related to human cancer (e.g., gene knockout or transgenic mouse models), or rodent cancer models designed to investigate modifiers of the cancer process.

fied experimentally in animals or in vitro as cancer hazards should be considered risk factors. However, analogous to IARC and EPA, the committee indicated in certain instances that it is possible or biologically plausible that certain substances are risk factors for breast cancer. The committee's criteria thus reflect the important differences between studies that observe risk factors in human populations and those that evaluate hazard potential.

The committee chose these criteria in part because of its mandate to consider potential evidence-based actions that women could take to reduce their risk of breast cancer. It was conscious of a wish to note "risk factors" and distinguish them from hazards, as described above.

After careful consideration, the committee chose to convey its assessments of the literature using broad groupings that reflect very generally the state of the evidence available. For example, for a factor for which compelling evidence from studies in humans, often distilled by others' systematic reviews, shows it to be an established risk factor for breast cancer, the committee used the designation assigned by the systematic reviews. Similarly, the committee noted as "probable" breast cancer risk factors those with strong but not definitive evidence from epidemiologic studies, sometimes with supporting evidence from animal or in vitro models.

Factors that did not fall into these categories were reviewed and discussed in terms of the need for additional questions to be answered, and some were flagged as possible, biologically plausible risk factors based on the hazard indicated in animal or in vitro studies or other relevant data. "Biologically plausible" meant consistent positive results for mammary tumors in animal bioassays or multiple, consistent in vitro studies demonstrating that a substance can modify a pathway or processes involved in breast carcinogenesis (e.g., modification of hormonal signaling pathways, mutagenesis of oncogenes or tumor suppressor genes, inhibition of apoptosis of precancerous breast cells, etc.). In some instances, concerns about the potential later effects of exposures that may occur at specific (earlier or later) times of life are underscored. For other factors, addressing the remaining uncertainty was considered not to be a high priority, given the limited population exposures to the substance.

In addition to reviewing the extent and strength of evidence indicating an association between a particular risk factor and breast cancer (and its direction: i.e., whether it is associated with an increase or a decrease in risk), the committee also reported on additional dimensions, when information was available. For example, quantitative estimates of the size of the effect in terms of relative risk or absolute risk, and accompanying measures of uncertainty in the form of confidence limits, are presented when available. The committee also noted information relevant to consideration of whether the timing of exposure influenced risk, such as the effects or associations pertaining to in utero or early-life exposures as compared with adult exposures. Similarly, the review notes whether the exposure showed a relationship to a particular tumor type based on hormone receptor status or other molecular markers.

DISCUSSION OF SPECIFIC ENVIRONMENTAL FACTORS

In the remainder of the chapter the committee presents summaries describing the strength of the evidence regarding the association of its selection of environmental factors with breast cancer. These factors are listed in Box 3-1 and grouped by their initial characteristic uses (e.g., industrial chemicals), route of exposure (e.g., ingestion of diet-related substances), or other features. Some of the substances reviewed by the committee are mixtures or classes of chemicals (e.g., tobacco smoke, polychlorinated biphenyls [PCBs]) and others are single chemicals (e.g., ethylene oxide). In either case, the committee typically focused on the literature on that specific mixture, class, or single chemical. It generally did not attempt to evaluate the evidence on interactions among risk factors but recognized that this is an important area to address in advancing knowledge in the field.

These groupings and labels are not definitive; different groupings or group labels may be used when these factors are discussed by others. Also, many additional factors that were not reviewed by the committee could be included in several of these groups; the committee's assessments concern only the specific factors listed.

The committee frequently uses relative risks (RRs) or similar measures in reporting evidence regarding the size of the association or effect for a given risk factor. A relative risk is an estimate of comparative risk derived from a defined population exposed to the factors, compared to an unexposed group. These measures of association do not convey the absolute risk that may be experienced by any one individual or group of individuals exposed to the factors. Chapters 2 and 6 describe these measures of risk further.

Exogenous Hormones

As described in Chapter 2, the breast is a hormonally responsive organ, and the majority of breast cancer that occurs responds to hormonal therapy. Thus it is no surprise that hormonal risk factors have been a major focus of breast cancer research. Prospective cohort studies have clearly shown an association between endogenous estrogen levels and development of breast cancer (Key et al., 2002). Because many of the established risk factors, such as age at menarche and age at first birth, are related to changes in the endogenous hormonal milieu, it was plausible to anticipate that exogenous factors that influence endogenous hormone levels may have an impact on

BOX 3-1
Environmental Factors Included in the
Committee's Evidence Review[a]

Exogenous hormones
- Hormone therapy: androgens, estrogens, combined estrogen–progestin
- Oral contraceptives

Body fatness and abdominal fat

Adult weight gain

Physical activity

Dietary factors
- Alcohol consumption
- Dietary supplements and vitamins
- Zeranol and zearalenone

Tobacco smoke
- Active smoking
- Passive smoking

Radiation
- Ionizing (including X-rays and gamma rays)
- Non-ionizing (extremely low frequency electric and magnetic fields [ELF-EMF])

Shift work

Metals
- Aluminum
- Arsenic
- Cadmium
- Iron
- Lead
- Mercury

Consumer products and constituents
- Alkylphenols
- Bisphenol A (BPA)
- Nail products
- Hair dyes
- Parabens
- Perfluorinated compounds (PFOA, PFOS)
- Phthalates
- Polybrominated diphenyl ethers (PBDEs; flame retardants)

Industrial chemicals
- Benzene
- 1,3-Butadiene
- PCBs
- Ethylene oxide
- Vinyl chloride

Pesticides
- DDT/DDE
- Dieldrin and aldrin
- Atrazine and S-chloro triazine herbicides (atrazine)

Polycyclic aromatic hydrocarbons (PAHs)

Dioxins

[a]The committee reviewed a selected set of factors for illustration; the chemicals were not chosen to be representative of any class. Some epidemiologic, mechanistic, or animal data relevant to mammary tumorigenesis or breast cancer are available for numerous other chemicals.

breast cancer incidence. Although much of the focus has been on influences specifically of estrogen, prospective studies have also shown an association with androgen concentrations and the risk of breast cancer (Helzlsouer et al., 1992; Key et al., 2002; Tworoger et al., 2006). Many factors are thought to affect breast cancer by influencing endogenous hormone levels. Exogenous hormone use is an obvious factor to consider in relation to breast cancer.

Exogenous hormone use by women is fairly common. The oral contraceptive pill was the leading method of contraception in the United States in 2006–2008, used by 10.7 million women (Mosher and Jones, 2010). Use of hormone therapy (HT) for relief of menopausal symptoms has also been widespread, but it has changed as findings have emerged about health risks associated with these products (Haas et al., 2004; Hersh et al., 2004). In a 1995 telephone survey of U.S. households (Keating et al., 1999), current use of menopausal hormone therapy was reported by 37.6 percent of women participating. National Health Interview Survey data from 2008 (DeSantis et al., 2011) report rates of combination HT use for women ages 50 and older of 0.9 to 2.8 percent, depending on race and ethnicity, and of estrogen-only HT from 2.1 to 5.9 percent, depending on race and ethnicity.

Evaluating the hormonal effects of exogenous hormone sources, such as oral contraceptives and hormone therapy, is challenging because of the use of a variety of single or combined hormone preparations and a multitude of dosages and delivery schedules. Additionally, hormones have differential effects on hormonally responsive tissue such as the ovaries, endometrium, and breast. Oral contraceptives are mostly combined hormonal preparations of estrogen and progestins and have been classified by IARC (2007) as Group I carcinogens; however, the effects are not consistent across all cancer types. Oral contraceptives modestly increase the risk of breast cancer among current users, as indicated by the Nurses' Health Study II (multivariate RR = 1.33, 95% CI, 1.03–1.73) (Hunter et al., 2010), but this risk dissipates 4 years following cessation. On the other hand, oral contraceptives are associated with a long-term reduced risk of endometrial and ovarian cancers. The overall evaluation by IARC reflects this mixed risk profile: "Combined oral estrogen–progestogen contraceptives are *carcinogenic to humans (Group 1)*. There is also convincing evidence in humans that these agents confer a protective effect against cancer of the endometrium and ovary" (IARC, 2007, p. 175).

IARC has also classified combined estrogen and progestin postmenopausal HT as "carcinogenic to humans" (Group 1). Data from randomized controlled clinical trials (19 trials involving 41,904 women) have shown that combined long-term menopausal hormone therapy with estrogen and progestins is associated with a significantly increased risk of breast cancer (Farquhar et al., 2009). The largest controlled clinical trial of combined postmenopausal HT with estrogen and progestin was the Women's Health Initiative (WHI), a 5-year randomized trial that was stopped early due to lack of a global health benefit with hormone therapy (Writing Group for the Women's Health Initiative Investigators, 2002). After a mean of 5 years, the RR of invasive breast cancer among the combined HT group compared with the placebo was 1.26 (95% CI, 1.02–1.56). This risk translates into an absolute excess of 8 cases of invasive breast cancer per 10,000 person-years attributed to estrogen and progestin (Writing Group for the Women's Health Initiative Investigators, 2002). After stopping combined hormone therapy, the excess risk declined (Chlebowski et al., 2009) in a manner similar to that observed after stopping combined oral contraceptive therapy. A rapid decline in breast cancer rates has been observed in the United States and several other countries following release of the WHI trial results (DeSantis et al., 2011; NCI, 2011) concomitant with declines in prevalence of combination HT use or prescriptions.

The effects of estrogen-only postmenopausal hormone therapy on breast cancer risk are not as clear as those of combined estrogen–progestin therapy. While estrogen-only therapy has been associated with a modestly increased risk of breast cancer in prospective cohort studies (Beral et al., 2011), this observation was not supported in the large randomized controlled clinical trial of estrogen-only therapy among women who had a hysterectomy (Anderson et al., 2004; LaCroix et al., 2011). The inconsistency in the findings between the observational study and the randomized controlled trial may imply some heterogeneity across subgroups in the population. Or, it may be partially due to misclassification of women in the observational study as taking only estrogen when they may have taken combined estrogen–progestin therapy at some point in their treatment. In addition, the timing of therapy with respect to onset of menopause may influence the magnitude of risk.

In the Million Women Study, women initiating estrogen-only HT more than 5 years after menopause had little or no increase in risk of breast cancer, while those initiating therapy before or within 5 years of onset of menopause had an excess risk of breast cancer compared to never users of hormones (Beral et al., 2011). In the WHI estrogen-only trial, women taking estrogen-only hormone therapy had a decreased risk of breast cancer that was not statistically significant. The magnitude of risk, after a mean follow-up of 7 years, was an RR of 0.77 (95% CI, 0.59–1.01), which would translate to a reduction of 26–33 breast cancers per 10,000 person-years (Anderson et al., 2004). In subsequent follow-up the decreased risk of breast cancer persisted and, when considering the intervention and follow-up periods, was statistically significant (LaCroix et al., 2011). It is important to note that women in the estrogen-only arm of the WHI did not have a uterus and therefore were not at risk for endometrial cancer, which has been clearly established as an increased risk with use of unopposed exogenous estrogen.

Androgenic hormones such as dehydroepiandrosterone (DHEA) are available as supplements that are claimed to enhance muscle performance or provide other health benefits, but they have not been studied in randomized clinical trials in relation to breast cancer. The evidence on the relation between higher endogenous concentrations of DHEA and its sulfated form, DHEAS, and breast cancer risk has been inconsistent in observational studies. Kaaks et al. (2005) observed increased risk of breast cancer with increasing serum measures of testosterone, androstenedione, and DHEAS in premenopausal women, and Tworoger et al. (2006) reported a positive association between endogenous DHEAS and estrogen receptor–positive/ progesterone receptor–positive (ER+/PR+) breast cancer in predominantly premenopausal women. Key et al. (2002) observed increasing breast cancer risk with endogenous levels of all sex hormones examined, including DHEAS in postmenopausal women. Another prospective study showed varying results for DHEA and DHEAS and by menopausal status (Gordon et al., 1990; Helzlsouer et al., 1992). Whether exogenous androgen supplements increase risk of breast cancer is uncertain, but based on studies of endogenous levels, this may depend on timing of supplement use with respect to menopause.

In summary, strong evidence has established that use of certain exogenous hormones affects breast cancer risk, and in particular that use of combined estrogen and progestin menopausal HT increases breast cancer risk. These hormones can have different effects on different tissues, and their effects may also differ depending on the timing of exposure. Additional discussion of the implications of risk associated with HT use appears in Chapter 6.

Body Fatness and Abdominal Fat

A relationship between body weight or body weight adjusted for height (as in the body mass index, or BMI) and breast cancer risk contingent on menopausal status has been observed for decades. Based on the 2007–2008 National Health and Nutrition Examination Survey (NHANES), the combined age-adjusted prevalence of overweight and obesity[4] in U.S. adults was 68 percent (Flegal et al., 2010), and an estimated 32 percent of children

---

[4]Body mass index (BMI) is an approximate measure of body fat based on height (in meters) and weight (in kilograms). BMI is defined as the individual's body weight divided by the square of his or her height. BMI categories are underweight, ≤ 18.5; normal weight, 18.5–24.9; overweight, 25–29.9; and obese, ≥ 30. BMI has shortcomings as a proxy for body fat (Romero-Corral et al., 2008), but is widely used.

and adolescents ages 2–19 were overweight or obese[5] (Ogden et al., 2010). Among subpopulations of adult women (age $\geq$20), data from the 2007–2008 NHANES showed that the prevalence of obesity ranged from 47 to 52 percent among non-Hispanic black women, 31 to 36 percent among non-Hispanic white women, and 38 to 47 percent among Hispanic women (Flegal et al., 2010). For women ages 60 and older, about 50 percent of non-Hispanic black women were obese compared to 31 percent of whites and 47 percent of Hispanic women (Flegal et al., 2010).

Numerous studies have evaluated the risk for breast cancer associated with greater body fatness. A systematic literature review carried out on behalf of WCRF/AICR included 43 cohort studies, 156 case–control studies, and 2 ecological studies examining a relationship between body fatness[6] (as measured by BMI) and breast cancer (WCRF/AICR, 2007). Although data from these studies were inconsistent when grouped for all ages, consistent effects were observed when examined by menopausal status. A meta-analysis found that the premenopausal cohort data indicated a lower risk with greater body fatness, while the postmenopausal cohort data showed greater risk with increasingly greater body fatness. An updated meta-analysis of cohort studies, carried out as part of the continuous update, showed for premenopausal women a 7 percent decrease in risk for breast cancer per 5 kg/m^2 increase in BMI, and for postmenopausal women a 13 percent increase in risk per 5 kg/m^2 increase in BMI (WCRF/AICR, 2010).

In summary, the WCRF/AICR (2007) systematic review found clear and consistent evidence indicating that body fatness protects against premenopausal breast cancer, classifying it as a probable protective factor for cancer, despite limited understanding of the mechanisms involved. With an abundance of consistent epidemiologic evidence as well as an understanding of the mechanisms involved, WCRF/AICR classified the evidence on greater body fatness and increased risk of postmenopausal breast cancer as convincing for a causal association.

These findings require further clarification with regard to body weight or BMI at earlier life stages. Although body fatness is associated with a reduced breast cancer risk in premenopausal women, greater body fatness in prepubertal girls is associated with an earlier age of menarche (Kaplowitz et al., 2001; Lee et al., 2007; Biro et al., 2010), which in turn is a generally

---

[5]In children and adolescents ages 2–19, overweight is defined as being at or above the 85th percentile of BMI for age, and obesity as at or above the 95th percentile of BMI for age, based on the 2000 Centers for Disease Control and Prevention sex-specific, BMI-for-age growth charts derived from nationally representative U.S. samples (Kuczmarski et al., 2002).

[6]This review used the term "body fatness" because of the finding that "the relationship between body fatness and cancer is continuous across the range of BMI" (WCRF/AICR, 2007, p. 214) rather than respecting specific cutpoints.

recognized risk factor for breast cancer, particularly for ER+/PR+ cancers (Ma et al., 2006). But earlier menarche may have less association with risk for breast cancer among Hispanic women than among non-Hispanic white women (Hines et al., 2010). Furthermore, the associations between prepubertal obesity and early menarche may not result in an increased risk of breast cancer in adulthood. Data from the Nurses' Health Study showed that the women with the greatest body fatness during childhood had a reduced risk of breast cancer compared with the women with the least body fatness (odds ratio [OR] = 0.67, 95% CI, 0.52–0.86) (Harris et al., 2011).[7] Similarly, women exposed between ages 2 and 9 to severe caloric deprivation during the 1944–1945 Dutch famine showed indications of increased risk for breast cancer, despite delayed menarche and earlier menopause (van Noord, 2004).

Fat distributed intra-abdominally is more metabolically active than other body fat, and measures of abdominal fat predict "the risk of chronic diseases, such as metabolic disorders and cardiovascular disease, better than overall indicators of body fatness" (WCRF/AICR, 2007, p. 212). Waist circumference or waist-to-hip ratios are sometimes used as indicators of how fat is distributed. The systematic review by WCRF found eight cohort studies and three case–control studies examining waist circumference and postmenopausal breast cancer risk, and eight cohort studies and eight case–control studies looking at waist-to-hip ratio as a measure of abdominal fat. Nearly all of the studies (all of the waist circumference studies and most of the waist-to-hip ratio studies) showed increased risk of postmenopausal breast cancer with more abdominal fatness. The mechanisms of this relationship are thought to be based on increased levels of circulating estrogens and decreased insulin sensitivity in association with greater abdominal fatness independently of overall body fatness. Adipose tissue is the main site of estrogen synthesis in men and postmenopausal women (WCRF/AICR, 2007, p. 39), and increased adipose tissue can thus contribute increased circulating estrogens. Based on its systematic review of the literature, WCRF classified abdominal fatness as a probable cause of postmenopausal breast cancer.

Body fatness and abdominal fatness could influence cancer risk through several mechanisms (see additional discussion in Chapter 5). These include changes in circulating hormones such as estrogens, insulin, and insulin-like growth factors; decreases in insulin sensitivity; and increases in inflammatory responses. The mechanism through which body fatness might decrease breast cancer risk in premenopausal women is not well established, but potential clues might lie in the different tumor markers observed in pre- and

---

[7]Body fatness in childhood was assessed using line drawings of nine figures illustrating a scale of increasing fatness (Harris et al., 2011).

postmenopausal breast cancer. A meta-analysis of 9 cohort and 22 case–control studies assessed the association between body weight and ER and PR status (Suzuki et al., 2009). No associations were observed for estrogen receptor–negative/progesterone receptor–negative (ER–/PR–) or ER+/PR– tumors.[8] The risk for ER+/PR+ tumors was 20 percent lower among premenopausal women and 82 percent higher among postmenopausal women in comparisons between the highest category of body weight and the reference group. The authors concluded that "the relation between body weight and breast cancer risk is critically dependent on the tumor's ER/PR status and the woman's menopausal status" (Suzuki et al., 2009, p. 698). A case series reported by Stark et al. (2009) found that excess body weight significantly decreased the diagnostic risk of triple-negative (ER–/PR–, HER2–) and ER–/PR–, HER+ disease relative to ER+ and/or PR+/HER2– subtypes. This association was not observed in African American participants.

An analysis of pooled tumor marker and epidemiologic risk factor data from 34 studies of the Breast Cancer Consortium (Yang et al., 2011) found increased BMI not to be associated with the risk of core basal phenotype (ER–/PR–/HER2–/[CK5 or CK5/6]+ or EGFR+). The analysis found obesity in women younger than age 50 to be a more frequent finding in ER–/PR– than in ER+/PR+ tumors, and obesity in women over age 50 was less frequent in PR– than in PR+ findings. These results support the hypothesis that different subtypes of breast cancer may have different etiologies.

Data are inconsistent on whether these associations, derived mostly from white populations, are also seen in African American populations. Palmer et al. (2007) found a reduced risk of breast cancer in African American women with BMIs of 25 or more at age 18 relative to those with BMIs of less than 20 for both pre- and postmenopausal breast cancer, and a lack of association of obesity with receptor-negative tumors. A recent case–control study using data from the Women's Contraceptive and Reproductive Experiences Study found a high recent BMI to be associated with an increased risk of ER+/PR+ tumors among postmenopausal African American women (Berstad et al., 2010). BMI did not have a statistically significant association with breast cancer risk among postmenopausal African American women with ER–/PR– tumors in this study. However, Trivers et al. (2009) found a positive association between obesity and triple-negative disease (ER–/PR–/HER2–). ER–/PR– tumors were associated with black race, young age at first birth, having a recent birth, and being overweight.

In conclusion, data are still needed to shed light on the differences in the apparent effects of body fatness with regard to pre- and postmenopausal breast cancer, but it is likely that these differences can be explained by the differences in the likelihood of different tumor types at different life stages,

---

[8]Tumor markers such as ER, PR, and HER2 are described in Chapter 2.

and that obesity is primarily a risk factor for ER+/PR+ breast cancers. An additional focus on tumor types and ethnicities in ongoing research on body fatness as a risk factor for breast cancer may better refine understanding of these associations and help target preventive action. Many other aspects also remain to be understood. The conundrum remains how to reconcile the decreased risk associated with greater body fatness in premenopausal women and the increased risk for breast cancer associated with earlier menarche, which itself appears to be associated with greater body fatness in young girls.

Adult Weight Gain

WCRF included 7 cohort studies and 17 case–control studies of adult weight gain and postmenopausal breast cancer in their review. They classified adult weight gain as a probable cause of postmenopausal breast cancer (WCRF/AICR, 2007). Evidence added via the continuous update (WCRF/AICR, 2010) also provided plentiful, consistent epidemiologic evidence for this relationship, with a dose–response relationship apparent. Again, this relationship may be different for nonwhite populations. Palmer et al. (2007) did not find an association between adult weight gain and postmenopausal breast cancer risk in data from the Black Women's Health study.

Preventing weight gain may be particularly important because it is not yet clear whether overweight and obese women can reduce their risk of postmenopausal breast cancer by losing weight. The Iowa Women's Health Study (Harvie et al., 2005) and the Nurses' Health Study (NHS) (Eliassen et al., 2006) observed reduced risk for women who lost weight compared with those who maintained a stable weight. However, other studies (Ahn et al., 2007; Teras et al., 2011) did not find reduced risk among women who lost weight. Additional research is needed to help focus prevention strategies.

Physical Activity

Physical activity has been defined as "bodily movement that is produced by the contraction of skeletal muscle that substantially increases energy expenditure" (HHS, 1996; IARC, 2002b, p. 6). It can be performed in various ways—as a part of one's occupational duties; as a component of housework; through gardening, sports, or other recreational activities; or transport, such as the commute to and from a destination (IARC, 2002b; WCRF/AICR, 2007). Because of the wide range of types of physical activities, it is difficult to measure exposure consistently. Approaches include calorimetry, physiological markers, monitors (e.g., pedometers or heart rate monitors), behavioral observation, or surveys involving subject recall

(IARC, 2002b). Other challenges in studies of physical activity include differences in study design, confounding due to other variables that may influence engagement in physical activity, and a tendency for exaggerated recall of vigorous recreational activity compared to other total daily activity.

Despite these difficulties, the relationship between physical activity and breast cancer has been extensively studied. Systematic reviews have been carried out by IARC (2002b) and WCRF/AICR (2007, 2010). Of the 33 separate studies reviewed by IARC, 22 (8 of 14 cohort studies, 14 of 19 case–control studies) found reduced risk for the most physically active participants compared with the least active. The average observed relative decrease in risk was about 20 to 40 percent between the most active and the most sedentary, with some studies observing up to 70 percent risk reductions. Most of the studies that examined a dose–response relationship found evidence of a linear trend whereby risk of breast cancer decreased with increasing duration of activity, regardless of type of activity (recreational or occupational), menopausal status, time period in life, or level of intensity of activity.

In its more recent review, WCRF/AICR (2007) considered pre- and postmenopausal breast cancer separately. From its review of studies of physical activity (studies of total physical activity as well as occupational and recreational activity) in premenopausal women, the panel found ample evidence to review, but inconsistent results. For premenopausal breast cancer, WCRF/AICR found limited evidence supporting protection from physical activity. For postmenopausal breast cancer, the review found stronger evidence of a protective effect, noting

> ample evidence from prospective studies showing lower risk of postmenopausal breast cancer with higher levels of physical activity, with a dose response relationship, although there was some heterogeneity. There was little evidence on frequency, duration, or intensity of activity. There is robust evidence for mechanisms operating in humans. (WCRF/AICR, 2007, p. 205)

They concluded that physical activity is a probable preventative factor against postmenopausal breast cancer.

Because of the abundance of human studies addressing physical activity and breast cancer incidence, systematic reviews have not relied heavily on experimental animal models to address a reduction of carcinogenicity after physical activity. However, many mechanisms have been proposed for physical activity's protective effect against breast cancer and other cancers as well. Physical activity is closely tied to body fatness and weight gain, and it has a beneficial effect on an individual's fat distribution. Physical activity is also thought to affect endogenous steroid hormone metabolism, reduce circulating estrogen and androgen levels, and strengthen the immune system (WCRF/AICR, 2007).

Although the level of risk reduction for breast cancer that is achieved by performing physical activity varies widely among studies, the body of research on cancer as well as the broader literature on health, particularly on cardiovascular outcomes, suggests that being active can be of great benefit to pre- and postmenopausal women (IARC, 2002b; Thompson and Lim, 2003; Warburton et al., 2006; WCRF/AICR, 2007).

Dietary Factors

Alcohol Consumption

Consumption of alcoholic beverages is widespread in the United States. In the 2008 National Health Interview Survey, 58 percent of women over 18 identified themselves as current drinkers,[9] and 15 percent as former drinkers (NIAAA, 2009). As stated by IARC (2010a), household income, education, and employment status are associated with current drinking status and more frequent drinking, but these factors have an inverse relationship with heavier drinking measures such as weekly heavy drinking (Midanik and Clark, 1994; Greenfield et al., 2000).

The association of alcohol consumption with breast cancer risk has been well studied. More than 100 epidemiologic studies have been conducted in all regions of the world, using both cohort and case–control epidemiologic designs. Recent systematic reviews of the scientific evidence have found a consistent association between greater self-reported consumption of alcohol and an increased risk for breast cancer (WCRF/AICR, 2007, 2008, 2010; IARC, 2010a). IARC (2010a) classified alcohol consumption as "carcinogenic to humans" (Group 1), based on evidence regarding cancer at several sites including the female breast, and WCRF/AICR classified the evidence that consumption of alcoholic drinks increases breast cancer risk for both pre- and postmenopausal women as "convincing" (WCRF/AICR, 2007, p. 157, 2008, 2010). Alcoholic beverages of all types (e.g., beer, liquor, wine) confer similar levels of risk after accounting for their differences in ethanol content.

The Collaborative Group on Hormonal Factors in Breast Cancer (2002) carried out a pooled analysis of 53 studies that included a total of 58,515 women with breast cancer. It found a linear increase in risk with increasing

---

[9]41.7 percent of women reported as abstaining from drinking, 45.1 percent reported as light drinkers (on average, three or fewer drinks per week in the past year), 8.3 percent as moderate drinkers (on average, more than three but no more than seven drinks per week), and 5 percent as heavier drinkers (on average, more than one drink per day in the past year) (NIAAA, 2009).

consumption of alcoholic beverages. Results suggested an RR of about 1.5 (95% CI, 1.3–1.6) associated with consuming 45 g or more alcohol per day (one U.S. drink includes approximately 14 g of ethanol [CDC, 2011a], so 45 g is more than three typical drinks). Even self-reported alcohol intake of about 18 g per day is associated with some increase in risk (RR = 1.13, 95% CI, 1.07–1.20), with increasing risk of 7 percent corresponding to each increase of 10 g per day (Collaborative Group on Hormonal Factors in Breast Cancer, 2002; cited by IARC, 2010a). These results were consistent with an earlier meta-analysis of data from 38 epidemiologic studies that reported an 11 percent increase in risk of breast cancer for daily consumption of 13 g compared to nondrinkers (Longnecker, 1994). Most recently, the WCRF review (2007, p. 168) also found "ample, generally consistent evidence from case–control and cohort studies" and noted that a dose-response relationship is apparent, with no threshold identified. According to IARC (2010a, p. 1277), "the effects of duration or cessation of consumption of alcoholic beverages on the risk for breast cancer are uncertain."

Studies measuring levels of alcohol consumption, like all observational epidemiologic studies, rely on subject recall and reporting. Because self-reports of current or past consumption of alcohol are generally believed to underestimate consumption, the relationships observed in multiple studies are noteworthy for the consistency of the positive association. Self-reported alcohol consumption has been evaluated against reports from the remote past and been found to be "reasonably reliable" for ranking subjects consistently by repeated measures (Longnecker et al., 1992). Such reliability, however, does not preclude differential reporting by cases versus controls. The main evidence against recall bias is the positive relationship of self-reported alcohol consumption with breast cancer in many large cohort studies where recall bias would not be a factor. These cohort studies go as far back as 1984 (Hiatt and Bawol, 1984) and have been confirmed repeatedly since then. For instance, a pooled analysis of six cohort studies with 322,647 women and 4,335 incident invasive breast cancers found that consumption of each additional 10 g of alcohol was associated with a 9 percent relative increase in risk (95% CI, 1.04–1.13) (Smith-Warner et al., 1998). Thus, the findings are probably not due to differential misclassification. If instead there is a tendency among all participants to underreport high levels of alcohol consumption, estimates of risk at lower levels of alcohol consumption may be overstated, and a threshold would be difficult to detect or identify. Estimates of risk at higher levels, which represent a relatively small proportion of women, may also be overestimated by underreporting of dose, but are more likely to represent increased breast cancer risks for this group.

The effects of alcohol consumption at various times in life have been examined by multiple case–control and cohort studies. Several earlier case–control studies suggested that risk might be elevated for women who were first exposed to alcohol as young adults ages 18–35 (Harvey et al., 1987; van't Veer et al., 1989; Young, 1989), but these were generally small studies and did not distinguish between early first exposure and exposure only at earlier ages. Other earlier studies (Hiatt et al., 1988; La Vecchia et al., 1989; Nasca et al., 1990) did not support higher risk associated with earlier exposures.

More recent studies (four cohort and four case–control studies), all of substantial size and conducted in a variety of populations worldwide (Freudenheim et al., 1995; Holmberg et al., 1995; Garland et al., 1999; Lenz et al., 2002; Horn-Ross et al., 2004; Tjonneland et al., 2004; Lin et al., 2005; Terry et al., 2006b), have examined exposure to alcohol at various times along the life course. All except one, in a population in Western New York with low overall alcohol consumption (Freudenheim et al., 1995), confirmed the modest relationship between alcohol consumption and increased risk of postmenopausal breast cancer (an RR of about 1.3, or a 30 percent increase with 1–2 drinks per day). Likewise, all except one found no evidence that alcohol consumption early in life was associated with an increased risk (Holmberg et al., 1995; Lenz et al., 2002; Horn-Ross et al., 2004; Tjonneland et al., 2004; Lin et al., 2005; Terry et al., 2006b). The exception was the NHS, which found that women who reported higher levels of alcohol consumption when they were ages 23–30 had a nonsignificant positive association with premenopausal breast cancer risk (Garland et al., 1999). In the other recent studies, it appears, if anything, that current alcohol consumption at older ages is more highly associated with breast cancer than consumption at younger ages (Holmberg et al., 1995; Horn-Ross et al., 2004; Tjonneland et al., 2004).

However, all of these were studies of adult women, and they relied on self-reported recall of alcohol consumption in adolescence and young adulthood. In contrast, a prospective study of the daughters of nurses who were asked to report their alcohol consumption confidentially at ages 16–23 years found an increased risk of benign breast disease (BBD) in surveys conducted 2 and 4 years later (OR = 1.5 per drink/day, 95% CI, 1.19–1.90) (Berkey et al., 2010). These results suggest that alcohol consumption early in life may increase breast cancer incidence in adulthood, given that BBD is an established risk factor for breast cancer.

As reported by IARC (2010a), risk related to alcohol consumption does not vary substantially by menopausal status, childbearing patterns, use of hormones, or family history of breast cancer. While this appears to be true for the evidence from most case–control studies, suggestive evidence from at least three large cohort studies indicates there may be a significant interaction between alcohol consumption and use of HT (Gapstur et al., 1992; Chen et al., 2002; Horn-Ross et al., 2004). Among 41,873 postmenopausal women in the Iowa Women's Health Study, there was an 80 to 90 percent higher risk of breast cancer for moderate (5–14.9 g/day) alcohol consumption (RR = 1.88, 95% CI, 1.3–2.72) and heavy (15 g/day or more) alcohol consumption (RR = 1.83, 95% CI, 1.18–2.85), but no association for alcohol consumption and breast cancer among women who never used estrogen (Gapstur et al., 1992). Similarly, in a follow-up of 44,187 postmenopausal women in the NHS, alcohol consumption was significantly associated with breast cancer risk in women taking postmenopausal hormones, but not in women who previously or never used HT (Chen et al., 2002). In the California Teachers Study (CTS), women whose alcohol consumption was an average of 20 g/day or more and who used estrogen plus progestin HT had more than twice the risk of developing breast cancer (RR = 2.24, 95% CI, 1.59–3.14), while never users of HT had no elevated breast cancer risk associated with alcohol consumption (RR = 0.94, 95% CI, 0.54–1.65) (Horn-Ross et al., 2004).

WCRF/AICR (2008, p. 83) reported findings from the Swedish Mammography Cohort that alcohol intake was associated with increased risk for ER+/PR+ tumors, but not for ER−/PR− or ER+/PR− tumors (Suzuki et al., 2005). The Iowa Women's Health Study found alcohol intake to be most strongly associated with ER−/PR− tumors (Gapstur et al., 1995). A dose–response meta-analysis by Suzuki et al. (2008) indicated a statistically significant increased risk for all ER+, all ER−, ER+/PR+, and ER+/PR− tumors, but not for ER−/PR− tumors. This analysis indicated a 27 percent higher risk (95% CI, 1.17–1.38) of developing ER+ tumors and 14 percent higher risk (95% CI, 1.03–1.26) of developing ER− tumors in the highest versus lowest alcohol consumption group (Suzuki et al., 2008, as summarized by AHRQ, 2010). Barnes et al. (2010) noted an inverse relationship between alcohol consumption and ER−/PR− tumors.

Studies in laboratory animals provide additional evidence of the effect of alcohol exposure on mammary tumor formation. The Agency for Healthcare Research and Quality, or AHRQ (2010), reviewed nine experimental animal studies evaluating mammary tumorigenesis caused or enhanced by alcohol. Of these studies, six (four of which administered a cocarcinogen) reported increased tumorigenesis, and three studies (one of which administered a cocarcinogen) did not support a link between ethanol and increased mammary cancer risk.

Alcohol may increase breast cancer incidence through numerous possible mechanisms. Studies in humans indicate that alcohol may affect breast cancer risk through formation of genotoxic metabolites (particularly acetaldehyde), as well as by inducing changes in levels of hormones such as estrogens, prolactin, or dehydroepiandrosterone (Seitz and Maurer, 2007; AHRQ, 2010). Mechanistic studies in animals have investigated the effects of alcohol on alteration in levels of hormones or hormone receptors, biotransformation and accumulation of genotoxic metabolites such as acetaldehyde, DNA adduct formation, suppression of cellular immunity, increase in terminal-end bud density and decrease in alveolar bud structures, enhanced tumor progression, and effect on DNA synthesis (AHRQ, 2010). In vitro studies reviewed by AHRQ (2010) further suggested increased cyclic adenosine monophosphate, change in potassium channels, and modulation of gene expression. In summary, alcohol may contribute to breast cancer risk through multiple mechanisms, although the relative importance of these mechanisms is unclear (AHRQ, 2010).

Regarding timing of exposure, Hilakivi-Clarke et al. (2004, reviewed by AHRQ, 2010) reported that in utero exposure to alcohol resulted in increases in mammary tumor incidence and multiplicity when animals were later exposed to the laboratory carcinogen 7,12-dimethylbenz[a]anthracene (DMBA). In a study by Polanco et al. (2010), 6.7 percent alcohol in diet was administered to pregnant rats on days 11–21 of gestation, and offspring received an intraperitoneal injection of N-nitroso-N-methylurea (MNU) at day 50. Compared with controls that did not receive alcohol exposure in utero, the alcohol-exposed offspring had greater numbers of tumors, decreased latency, more malignant tumors, more ER-alpha negative tumors (50 percent compared to approximately 15 percent in controls), and increased estradiol levels.

Some evidence shows gene–environment interactions in the risk for breast cancer from alcohol consumption. Polymorphisms in genes that control key enzymes involved in metabolism of alcohol (alcohol dehydrogenase [ADH], aldehyde dehydrogenase [ALDH], cytochrome P-450 [CYP2E1], xanthine oxidoreductase [XOR]) may result in increased levels of reactive intermediates and thereby result in altered risk for breast cancer in certain populations (AHRQ, 2010). ADH and CYP2E1 catalyze the conversion of alcohol to aldehyde, whereas ALDH and XOR catalyze the conversion of acetaldehyde to acetate, which is further metabolized (AHRQ, 2010). Polymorphisms that result in the forms of these enzymes that increase the rate of conversion of alcohol to acetaldehyde or decrease the metabolism of aldehyde result in higher levels of acetaldehyde, a cytotoxic and genotoxic metabolite that has been implicated in oral, colon, breast, and other cancers from alcohol exposure (AHRQ, 2010).

One variant of the *ALDH2* gene that results in a nearly inactive form of ALDH is found only in Asian populations (Seitz and Stickel, 2010). Approximately 10 percent of the Japanese population are reported to be homozygous for the inactive form of ALDH, and about 40 percent of Asians are reported to be heterozygous, resulting in greatly reduced (10 percent of normal) ALDH activity (Seitz and Stickel, 2010). However, both Caucasian and Asian populations have a polymorphism in *ADH* that results in variants with faster conversion of alcohol to acetaldehyde (Seitz and Stickel, 2010). The enzyme encoded by the ADH1C*1 allele not only catalyzes a higher rate of alcohol conversion to acetaldehyde, but also affects the metabolism of estrogen and other steroid hormones (Seitz and Maurer, 2007).

Several studies have reported increased risk for women with more active variants of ADH, primarily in women who consumed high amounts of alcohol. Seitz and Stickel (2010) report that high alcohol intake (60 g/day) and the ADH1C*1 allele in Caucasians (n = 400) were associated with increased risk of breast cancer as well as cancer of the digestive tract, liver, and colon, although the studies cited do not mention breast cancer. In other studies, increased cancer risks were reported primarily for moderate to heavy drinkers, for those who are homozygous for this allele, and for premenopausal women. Terry et al. (2006a) reported increased breast cancer risk in premenopausal women with a lifetime alcohol intake rate of 15–30 g/day who were homozygous for the more active ADH1C*1 allele (ADH1C*1,1) compared to nondrinkers with intermediate or slow ADH1C genotypes (OR = 2.9, 95% CI, 1.2–7.1). At the same alcohol intake rate, risks were not significantly elevated for postmenopausal women with the ADH1C*1,1 genotype or for women who were intermediate (ADH1C*1,2) or slow metabolizers (ADH1C*2,2). Similarly, Freudenheim et al. (1999) reported the highest increase in breast cancer risk (OR = 3.6, 95% CI, 1.5–8.8) for premenopausal women with the ADH1C*1,1 genotype who consumed more than the median number of drinks per month (>6.5/month averaged over the past 20 years) compared to those who consumed less alcohol and did not have this genotype. Coutelle et al. (2004) did not specifically examine pre- versus postmenopausal women, but reported that frequency of the ADH1C*1 allele was greater in breast cancer cases than in controls who were heavy drinkers but who did not have cancer (62% compared to 41.9%). This study also reported that women with the ADH1C*1,1 genotype had a greater risk of breast cancer than those with ADH1C*1,2 or ADH1C*2,2 genotypes (OR = 1.8, 95% CI, 1.4 –2.3). In addition, women with the ADH1C*1,1 genotype who consumed more than 20 g/day of alcohol had a greater risk of breast cancer than those with this genotype who consumed <20 g/day (OR = 1.4, 95% CI, 1.0–3.35).

Studies that have not found increased risks for more active ADH variants appear to be those involving lower alcohol intake, small sample size, or postmenopausal women. Benzon Larsen et al. (2010) reported that among 809 postmenopausal breast cancer cases and 809 controls within the prospective Diet, Cancer, and Health Study, women with ADH polymorphisms with faster conversion of alcohol to aldehyde did not have higher breast cancer risks; in fact, variants for slow metabolizers were associated with slightly higher risks (14% per 10 g of alcohol intake/day). A study as a part of the NHS (Hines et al., 2000) did not find an effect of alcohol on breast cancer risk or interaction with ADH1C polymorphism for pre- or postmenopausal women. However, this study had relatively small sample sizes, particularly for premenopausal women (88 cases versus 94 controls), for an analysis that considered alcohol intake (none, ≤10 g/day, >10 g/day) and ADH polymorphism groups (slow, fast, intermediate). Visvanathan et al. (2007) likewise stated that the lack of increase in breast cancer risk for more active variants of ADH may be attributed to low alcohol consumption in their study population (median of 13 g/wk).

Polymorphisms in ADH are also thought to affect breast cancer risk through the involvement of ADH in metabolism of estrogens as well as by acetaldehyde formation (Seitz and Maurer, 2007). The effect of ADH on estrogen and acetaldehyde production may be combined as indicated by evidence of particularly high blood acetaldehyde levels for women consuming alcohol during the period of the menstrual cycle when estradiol levels peaked (Seitz and Maurer, 2007).

By contrast, Kawase et al. (2009) did not find an increased risk of breast cancer in Japanese women (456 breast cancer cases versus 912 age- and menopausal status-matched controls) for alcohol drinking and polymorphisms in *ADHI1B* or *ALDH2*. Kawase et al. hypothesized that Japanese women may not drink enough alcohol or other factors may cause different outcomes among populations. Studies in Japanese populations have, however, found that the inactive form of *ALDH2* is associated with increased risks of other cancers such as oropharyngolaryngeal and esophageal cancers. Evidence reviewed by AHRQ (2010) indicates that breast tissue contains ADH, CYP2E1, and XOR rather than ALDH2 for metabolism of acetaldehyde, indicating that acetaldehyde is metabolized in breast tissue by XOR rather than enzymes associated with *ALDH2*.

Other gene–environmental interactions for alcohol and breast cancer have been reported. High alcohol intake and a homozygous variant of enzymes related to the one-carbon metabolism enzyme methylenetetrahydrofolate reductase have been associated with increased risk in postmenopausal but not premenopausal women (Platek et al., 2009). A study has also found an association with increased risk of breast cancer for women with a specific polymorphism in the mitochondria genome and who consumed alcohol compared with those who did not drink (Pezzotti et al., 2009).

A further issue pertains to confounding from other ingredients and contaminants in alcoholic beverages, which may have associations with cancer risk (Seitz and Simanowski, 1988; HHS, 2000; Baan et al., 2007; Monteiro et al., 2008).

In conclusion, evidence from human, animal, and in vitro studies supports a modest but causal relationship between alcohol consumption and breast cancer for both premenopausal and postmenopausal women. The reduction of alcohol consumption is an action women can take to reduce their breast cancer risk, even though the overall risk is rather small for lighter drinkers. Also entering into the choices women must make is the well-documented protective effect of low-level alcohol consumption (<3 drinks/day) on coronary artery disease, a more common cause of death in postmenopausal women (Klatsky, 2010). There is no clear threshold for the onset of increased risk of breast cancer. The choice of whether to consume alcohol, or how much, must remain an individual one.

Vitamins and Dietary Supplements

In the *Dietary Supplement Health and Education Act of 1994*, Congress defined dietary supplements as products, other than tobacco, that (1) supplement the diet; (2) contain one or more dietary ingredients "(including vitamins; minerals; herbs or other botanicals; amino acids; and other substances) or their constituents"; (3) are intended to be taken orally in pill, capsule, or liquid form; and (4) are clearly labeled as dietary supplements on the front panel of their packaging (NIH, 2011). Dietary supplements can come in many forms—as combinations of ingredients such as botanicals or herbs, as multivitamin supplements, or as supplements containing individual vitamins or ingredients. The committee focused primarily on studies of multivitamin and single-substance supplements. However, terms such as "multivitamin" have "no standard scientific, regulatory or marketplace definitions" (Yetley, 2007, p. 269S). Formulations of combination vitamin and mineral supplements therefore vary in content, which presents challenges in conducting and comparing studies.

Unlike many factors, the evidence on dietary supplements includes results from experimental studies in humans. In large-scale, randomized, double-blind, placebo-controlled trials, neither antioxidant supplementation (Hercberg et al., 2004) nor supplements of folic acid plus vitamins B6 and B12 (Zhang et al., 2008) showed an association with differences in risk for breast cancer.[10] However, mandatory folate fortification in the United States since 1998 (NIH, 2011) may have made it difficult to detect an effect associated with the additional supplementation in the study. Large-scale observational studies of multivitamin use have been inconsistent. The Swedish Mammography Cohort with 974 incident cases of breast cancer among 35,329 women, ages 49–83 over a mean 9.5-year follow-up, found an RR of 1.19 (95% CI, 1.04–1.37) among those reporting use of multivitamins (Larsson et al., 2010). Women reporting "ever" use of multivitamins in the Prostate, Lung, Colorectal, and Ovarian Cancer Trial (Stolzenberg-Solomon

---

[10]Hercberg et al. (2004) followed 7,876 women ages 35–60 from the general population for a median of 7.5 years for cancer incidence and mortality, and Zhang et al. (2008) followed 5,442 women ages 42 and older with preexisting cardiovascular disease or three or more coronary risk factors for 7.3 years for cancer incidence.

et al., 2006) had an RR for postmenopausal breast cancer of 1.18 (95% CI, 0.95–1.48); this study reported a statistically significant increase in breast cancer risk with folate supplementation (RR = 1.19, 95% CI, 1.01–1.41). Other studies, mostly in premenopausal women, found no statistically significant association of multivitamin use with breast cancer risk (Feigelson et al., 2003; Ishitani et al., 2008; Maruti et al., 2009; Neuhouser et al., 2009). Studies of use of individual supplements, such as vitamins C, D, E, and A, have also shown conflicting results or no differences in risk with supplement intake (Verhoeven et al., 1997; Nissen et al., 2003; Stolzenberg-Solomon et al., 2006; Robien et al., 2007). In its systematic review, the WCRF/AICR included vitamins A, B6, B12, C, D, and E and riboflavin, folate, calcium, iron, selenium, carotenoids, and isoflavones in its evidence category of "limited–no conclusion" for both premenopausal and postmenopausal breast cancer (WCRF/AICR, 2007).

Nondietary phytoestrogen-containing supplements are widely used by women for the treatment of menopausal symptoms. A meta-analysis of 92 randomized controlled trials that studied women undergoing treatment of menopausal symptoms with phytoestrogen-containing supplements showed no statistically significant increase in breast cancer risk in any of the individual studies or in the meta-analysis of all 92 studies (Tempfer et al., 2009). However, the median duration of the studies included in the meta-analysis was only 6.2 months and breast cancer was not a primary endpoint, so even substantial effects of long-term use of phytoestrogen supplements cannot be ruled out. Although there was no statistically significant risk for breast cancer with increasing duration of supplement use, the duration of use in most studies was far too short to make a confident statement about risk. In addition, the exact composition of the phytoestrogens studied varied among supplements and was poorly characterized.

Multiple factors make dietary supplementation a challenging focus of study. In general, multivitamin and mineral supplements are used by women who practice healthier lifestyles and are therefore more likely to have regular breast cancer screening. This clustering of characteristics makes observational studies of the relationship between use of these supplements and health outcomes difficult to interpret. Healthier lifestyles might result in downward bias (fewer cancers, decreased likelihood of observing an association), while regular screening might result in upward bias (more cancers diagnosed). Furthermore, dietary supplement use was often assessed through self-administered questionnaires, which can introduce errors resulting from poor recall. In addition, not all studies collected specific information about brand names and product names of the supplements. The specific ingredients and the amount of each ingredient in a supplement vary widely, and if researchers combine a wide range of types of supplements, the analysis may not be meaningful. These and other limitations pose serious challenges for the conduct of studies on the effects of multivitamins, or their initiation and duration of use at various life stages.

In vitro studies aimed at evaluating the relationship between dietary supplement products or ingredients and breast cancer risk are nearly all carried out using established breast cancer cell lines or in cellular assay systems with immortalized cells treated with chemical carcinogens or ionizing radiation, in addition to the vitamin or supplement of interest. The relevance of these studies to human carcinogenesis is difficult to interpret.

Because of the widespread use of dietary supplements, and the variety of substances involved, it is important that continued attention be paid to the potential risks or benefits they may pose for breast cancer. However, as noted above, refined research approaches will be needed because of the multiplicity of challenges to this type of research.

Zeranol and Zearalenone

Zearalenone is a mycotoxin product from fungi of the genus *Fusarium*. It is a common contaminant of grains and thus is present in the diet, albeit at low levels. A synthetic form of a reduction product of zearalenone, called zeranol (Ralgro), is one of six growth promoters approved by the Food and Drug Administration (FDA) and is widely used in feedlot beef production in the United States and many other countries. Both zeranol and zearalenone are relatively potent nonsteroidal estrogens (Peters, 1972), with the estrogenic activity of zeranol substantially greater than that of zearalenone (Mirocha et al., 1979; Shier et al., 2001).

The primary route of exposure is oral, via diet. Zearalenone does not degrade during the cooking and processing of foods (European Commission, 2000). Low-level exposure from contamination of cereal grains occurs, and mean daily U.S. exposures have been estimated as 0.03 µg/kg/day (Zinedine et al., 2007). Outbreaks of *Fusarium* contamination of corn and other commodities can occasionally lead to very high levels in foods (Zinedine et al., 2007). In the United States, the largest potential source of exposure is likely to be through residues of zeranol in meat from sheep and cattle implanted with Ralgro pellets, a process that is monitored by the FDA. Although use of zeranol is permitted in the United States, the European Union prohibits the use of hormones or the import of hormone-treated beef products from the United States or Canada where zeranol is used as a growth promoter (European Commission, 2007).

No epidemiologic studies are known to have addressed whether exposures to zearalenone or zeranol could contribute to breast cancer (or any cancer) risk.

Studies in animals include a 2-year bioassay of zearalenone by the NTP (1982b). The final evaluation concluded that zearalenone was carcinogenic in a specific strain of mice, but not in the rat strain tested. There was no report of increased mammary tumors in either rats or mice, although in mice "estrogen-related, dose-dependent effects were seen in several tissues (fibrosis in the uterus, cystic ducts in mammary glands)" (European Commission, 2000, p. 5). The European Commission (2000) concluded the tumors observed in the NTP bioassay (liver and pituitary) were related to the estrogenic effects of the compound. The IARC (1993) evaluation of zearalenone drew on the NTP bioassay data and concluded that "there is *limited* evidence in experimental animals for the carcinogenicity of zearalenone," with an overall characterization as "not classifiable as to its carcinogenicity to humans" (Group 3).

Sheep and pigs appear to be more sensitive than rodents to the estrogenic effects of zearalenone (European Commission, 2000). Large differences among species in sensitivity to the estrogenicity of zearalenone are thought to result from differences in metabolic capacity and presence of various estrogenic metabolites of zearalenone (Ueno et al., 1983; Pompa et al., 1988; Malekinejad et al., 2006). How human sensitivity compares with the pig, sheep, or rodent requires further study.

Mechanistic research shows that zeranol binds to the ligand-binding domain of human estrogen receptor alpha and beta in a manner similar to estradiol-17β (E2) (Takemura et al., 2007). Zeranol has also been demonstrated to stimulate the growth of human MCF-7 breast cancer cell lines in vitro (Makela et al., 1994; Zava et al., 1997), and to enlarge existing mammary tumors in mice (Schoental, 1974), reflecting its estrogen receptor-agonist properties.

Because zearalenone and zeranol are rather potent (nonsteroidal) xenoestrogens, timing of exposure may be important. A few recent studies have explored early-life exposures to low levels of zearalenone and biological effects. For example, fetal and neonatal exposure of rats to levels near those of human exposure (0.2 µg/kg in utero and first 5 days of life) was observed to affect terminal end bud length (Belli et al., 2010), and uterine hyperplasia was induced in young pigs fed relatively low levels (20 µg/kg) of zearalenone for 48 days (Gajecka et al., 2011). While these and other recent studies are intriguing, they are too few to reach any firm conclusions regarding the potential impact of low-level exposure to these compounds early in life on breast cancer risk in humans, or at other specific life stages. Interestingly, prepubertal exposure of rats to a low dose (20 µg, or about 1 mg/kg) of zearalenone was shown to significantly reduce the incidence of mammary adenocarcinomas induced by treatment with MNU or DMBA, possibly by increasing differentiation of the mammary epithelial tree (Hilakivi-Clarke et al., 1999; Nikaido et al., 2003).

Due to a paucity of epidemiologic studies and of animal bioassays and mechanistic studies that address mammary tumor endpoints and explore the impact of timing of exposure, at this point, although it is biologically plausible, no conclusion can be reached on the role of zearalenone or zeranol in the etiology of breast cancer. It remains an area for further study.

Tobacco Smoke

That tobacco smoke may be implicated as a possible risk factor in breast cancer etiology is not surprising; smoking has wide-ranging impacts on general health and is established as a carcinogen and causal agent in many forms of cancer (IARC, 2004). Tobacco smoke is a complex mixture that includes many toxic substances, more than 50 of which are known, probable, or possible human carcinogens (e.g., polonium-210, benzene, several metals, and vinyl chloride) (IARC, 2004; NTP, 2011a). Exposure to tobacco smoke occurs through active smoking, with smoke directly inhaled by the smoker, and through what is termed passive smoking or secondhand smoke exposures.[11] Many of the same compounds are present in both directly inhaled and secondhand smoke, but their amounts and proportions differ (IARC, 2004), which results in differing toxicities.

Before 1993, more than 50 epidemiologic studies examined the relationship between breast cancer and exposure to tobacco smoke. Although the quality of studies was highly variable, the better conducted studies did not suggest a causal relationship (Palmer and Rosenberg, 1993). An IARC review published in 2004 included studies conducted before 2002, and it relied heavily on a pooled analysis of 53 case–control and cohort studies by the Collaborative Group on Hormonal Factors in Breast Cancer Study (2002) that contended that apparent associations with smoking were confounded by alcohol consumption. The IARC (2004) conclusions were that neither active nor passive smoking was associated with increased risk of breast cancer.

Since 2004, two scientific consensus reviews concluded, based on high-quality studies, that the available evidence supports causal associations between breast cancer and active smoking or premenopausal breast cancer and exposure to secondhand smoke, or both (CalEPA, 2005; Collishaw et al., 2009). A 2006 U.S. Surgeon General's report concluded that the evidence on passive smoking was suggestive but not conclusive for a causal relationship with increased risk of breast cancer (HHS, 2006). The most recent IARC review characterized the evidence on active smoking as limited and the evidence on passive smoking as inconclusive (Secretan et al., 2009).

---

[11]Throughout the report the phrases "passive smoking" and "secondhand smoke" are used interchangeably to refer to exposure to smoke emitted by the burning end of a cigarette, cigar, or pipe or smoke exhaled by a smoker.

Active Smoking

The epidemiologic literature on active smoking is often characterized as mixed, with some studies finding statistically significant associations between smoking and breast cancer while others do not. Many earlier studies were limited by the use of crude measures of exposure, small sample sizes, and lack of control for key covariates. Moreover, some of these studies of risks to smokers included women with passive smoke exposure in their "unexposed" referent groups, potentially reducing statistical power to distinguish the impact of active smoking. Over time, assessments of exposure to tobacco smoke have been refined in many studies.

Age at smoking initiation may play an important role in the tobacco smoke–breast cancer association, and tobacco smoke may be one of the carcinogens that is more potent at certain stages of life. As noted in Chapter 2, the breast does not fully mature until after a first full-term pregnancy. A meta-analysis examined the effect of smoking before a first pregnancy in 23 studies published from 1988 through 2009 (DeRoo et al., 2011). The summary risk ratio was 1.10 (95% CI, 1.07–1.14), indicating a weak association with increased risk for early initiation of smoking. For women who smoked only after a first pregnancy, the summary risk ratio was 1.07, but it was not a statistically significant increase in risk (95% CI, 0.99–1.15) (DeRoo et al., 2011).

A subsequent report from the NHS found a statistically significant increase in risk associated with greater smoking intensity (i.e., pack-years of smoking) from menarche to a first birth (p for trend <.001) (Xue et al., 2011). At 1–5 pack-years of smoking before a first birth, the hazard ratio (HR) is 1.11 (95% CI, 1.04–1.20); for 16 or more pack-years, the HR is 1.25 (95% CI, 1.11–1.40). No increase in risk was evident for pack-years smoked from after a first pregnancy to menopause. For 31 or more pack-years, the HR was 1.05 (95% CI, 0.92–1.19). However, pack-years of smoking after menopause may be associated with a slight reduction in risk (p for trend = .02) (Xue et al., 2011). For 16 or more pack-years of postmenopausal smoking, the HR was 0.88 (95% CI, 0.79–0.99).

Recent reports from the Women's Health Initiative that were not included in the meta-analysis by DeRoo et al. (2011) have also examined the effects of smoking on postmenopausal breast cancer risks. Using data from the observational arm of the Women's Health Initiative, Luo et al. (2011b) found a higher risk with younger age at initiation of smoking. For women who started smoking between ages 15 and 19, the HR was 1.21 (95% CI, 1.01–1.44); whereas for those who initiated smoking after age 30, the HR was 1.00 (95% CI, 0.76–1.32). Similarly, initiation of smoking before first full-term pregnancy was associated with a statistically significant increase in risk (HR = 1.28, 95% CI, 1.06–1.55); the risk with initiation after a first pregnancy was elevated but not statistically significant (HR = 1.17, 95% CI, 0.90–1.52). These results suggest that failure to stratify by age at initiation of smoking, or during critical windows of time, may obscure evidence of an association between smoking and breast cancer.

An additional analysis of data from the observational portion of the Women's Health Initiative found that postmenopausal obesity may modify the association between smoking and breast cancer risk (Luo et al., 2011a). For women who were obese based on BMI at entry into the study (BMI ≥ 30), smoking did not increase breast cancer risk on the basis of age at initiation of smoking (< age 20: HR = 1.00, 95% CI, 0.85–1.18; p for trend = .73), pack-years of smoking (≥ 50 pack-years: HR = 1.15, 95% CI, 0.89–1.48; p for trend = .84), or other measures. By comparison, women who were not obese (BMI <30) had an increased risk with both earlier initiation of smoking (e.g., < age 20: HR = 1.19, 95% CI, 1.08–1.31) and pack-years (e.g., ≥ 50 pack-years: HR = 1.20, 95% CI, 1.00–1.43), supported by statistically significant trends.

Other recent reports have considered smoking in relation to ethnicity or particular types of breast cancer. Brown et al. (2010) concluded that their data did not show a consistent association between smoking and significant increases in breast cancer risk among U.S.- or foreign-born Asian women. For example, the results for current smokers showed an OR of 0.9 (95% CI, 0.6–1.3) while ex-smokers had an OR of 1.6 (95% CI, 1.1–2.2). The small number of women who started smoking before age 16 (11 cases, 9 controls) had an OR of 2.92 (95% CI, 1.1–7.9) whereas women who began smoking at ages 16–18 had an elevated but not statistically significant risk (OR = 1.18, 95% CI, 0.7–1.9) compared with women who had never smoked.

A study that examined risk for triple negative breast cancer found no statistically significant increase in risk over nonsmokers based on smoking status, age at initiation, or duration of smoking (Kabat et al., 2011). By comparison, women with estrogen receptor–positive (ER+) cancers were at significantly increased risk with earlier initiation (< age 20: HR = 1.16, 95% CI, 1.05–1.28) and longer duration of smoking (≥ 30 years: HR = 1.14, 95% CI, 1.01–1.28). In a study focused on DCIS, smoking was not associated with an increased risk based on smoking status, age at initiation, or duration of smoking (Kabat et al., 2010).

A growing body of epidemiologic research is investigating genetic susceptibilities to effects from active smoking. One area of study is risk differences according to women's N-acetyltransferase 2 (*NAT2*) gene alleles. *NAT2* codes for enzymes responsible for metabolism of chemicals not normally present in the body, including the detoxification of aromatic amines, which are present in tobacco smoke (Ambrosone at al., 2008). Genetic variations in *NAT2* result in what are broadly described as slow or fast acetylator types. Although the specific alleles used to determine acetylator status may vary among studies, meta-analyses found a fairly consistent positive association (overall relative risk of 1.4–1.5) between active smoking and breast cancer risk for women, perhaps especially postmenopausal women, who have been long-term heavy smokers and have a slow acetylator form of *NAT2* (Terry and Goodman, 2006; Ambrosone et al., 2008; Zhang et al., 2010). However, a recent Canadian study not included in these meta-analyses found that heavy smoking (>20 pack-years) was associated with a statistically significant increase in risk among fast acetylators (OR = 1.93, 95% CI, 1.01–3.69) but not slow acetylators (OR = 1.27, 95% CI, 0.75–2.15) (Conlon et al., 2010).

An analysis that compared data on Hispanic and non-Hispanic white women found that Hispanic women were less likely to have slow-acetylator forms of *NAT2* and had no change in breast cancer risk based on smoking and *NAT2* status (Baumgartner et al., 2009). Among the non-Hispanic white women who were categorized as very slow acetylators (i.e., carrying two from among the *NAT2*5A, *5B,* and **5C* alleles), ever, former, or current smokers were at statistically significant increased risk over never smokers, with odds ratios of more than 2.0 (Baumgartner et al., 2009). Risks for those characterized as slow acetylators (but not "very slow") were generally elevated but not statistically significantly so.

Thus the evidence generally appears to indicate a gene–environment interaction involving women genetically predisposed to inefficient detoxification of carcinogenic exposures in tobacco smoke, although this is an evolving area of research.

Passive Smoking

Ideally, studies of the effects of secondhand smoke compare the breast cancer experience of exposed women to that of women who have never been exposed. Early studies of the relationship between breast cancer and secondhand smoke exposure are likely to have underestimated exposure by relying only on measures such as spousal smoking status. This approach neglects exposures in the workplace or public settings, which may equal or exceed exposure in the home (Reynolds et al., 2009), and exposure in childhood and adolescence, which may be a particularly vulnerable period, based on evidence for active smoking. To the extent that exposure to secondhand smoke alters breast cancer risk, underestimation of exposure by neglecting exposure sources such as these contributes to a bias toward no association.

A 2005 review by the California Environmental Protection Agency of various health hazards associated with exposure to secondhand smoke included a meta-analysis of 19 epidemiologic studies of breast cancer

(CalEPA, 2005). The conclusion of the review group was that the epidemiologic and toxicologic evidence was consistent with a causal association between exposure to secondhand smoke and breast cancer in "younger, primarily premenopausal women," but that the evidence for older or postmenopausal women was inconclusive (CalEPA, 2005, p. ES-8). The meta-analysis produced an overall estimate for exposed women of RR = 1.25 (95% CI, 1.08–1.44) (CalEPA, 2005; also reported in Miller et al., 2007). When the analysis was restricted to five studies with more comprehensive exposure assessment, the overall estimate was RR = 1.91 (95% CI, 1.53–2.39). An analysis of the 14 studies that had data on younger, primarily premenopausal women produced an overall estimate of RR = 1.68 (95% CI, 1.31–2.15); the estimate for the five studies with more comprehensive exposure assessment was RR = 2.20 (95%CI, 1.69–2.87) (CalEPA, 2005; Miller et al., 2007).

In 2006, the U.S. Surgeon General's report *The Health Consequences of Involuntary Exposure to Tobacco Smoke*, which included consideration of many of the same studies as the California review, concluded, "The evidence is suggestive but not sufficient to infer a causal relationship between secondhand smoke and breast cancer" (HHS, 2006, p. 13). The conclusion was based on a review of the findings from seven prospective cohort studies, 14 case–control studies, and a meta-analysis of all of these studies. The meta-analysis found that women who had ever been exposed to secondhand smoke (10 studies) were at increased risk of breast cancer (RR = 1.40, 95% CI, 1.12–1.76). With stratification by menopausal status, the increase in risk was statistically significant for premenopausal women (6 studies; RR = 1.85, 95% CI, 1.19–2.87) but not for postmenopausal women (5 studies; RR = 1.04, 95% CI, 0.84–1.30). This report noted that its conclusion reflected, in part, an assessment that the biological plausibility of the association was weak (HHS, 2006).

A 2009 Canadian review considered the assessments in both the California report and the Surgeon General's report, as well as three later studies that had not been included in the analyses for those previous reports (Collishaw et al., 2009; also summarized in Johnson et al., 2011). The Canadian review group noted the similarity of the results for the meta-analyses in the California and Surgeon General's reports and found an association between increased risk for breast cancer and exposure to secondhand smoke biologically plausible. The conclusion was that "the association between [second hand smoke] and breast cancer in younger, primarily premenopausal women who have never smoked is consistent with causality" but that the evidence was insufficient to reach conclusions regarding postmenopausal breast cancer and secondhand smoke (Collishaw et al., 2009, p. 3).

The results from two large cohort studies published after the expert reviews from California, the Surgeon General, and Canada have suggested a small but statistically significant increased risk for breast cancer among postmenopausal women with higher levels of secondhand smoke exposure (Reynolds et al., 2009; Luo et al., 2011b). A prospective study of 57,523 women enrolled in the California Teachers Study who were lifetime non-smokers found indications that postmenopausal women reporting high levels of secondhand smoke exposure may be at higher risk of developing breast cancer (Reynolds et al., 2009). Similar results were recently reported for secondhand smoke exposures in a cohort of 41,022 postmenopausal women enrolled in the WHI Observational Study who never smoked (Luo et al., 2011b). For those with the highest secondhand smoke exposures ($\geq$10 years in childhood, $\geq$20 years at home in adulthood, and $\geq$10 years at work in adulthood), the OR for postmenopausal invasive breast cancer was 1.32 (95% CI, 1.04–1.67) as compared with those who never smoked and never experienced secondhand smoke exposures (Luo et al., 2011b). Since 1982, the NHS has followed 36,017 women who never smoked. Follow-up to 2006 has not shown a significant association between breast cancer risk and passive smoking in childhood or adulthood (Xue et al., 2011).

Although epidemiologic studies have suggested that early age of initiation of active smoking and smoking before a first full-term pregnancy are associated with higher breast cancer risk, there is little evidence for risk from exposure to secondhand smoke only in childhood (e.g., HHS, 2006; Chuang et al., 2011; Luo et al., 2011b).

Animal and In Vitro Studies

At least 20 components of tobacco smoke have been classified by IARC as known or suspected human carcinogens and have induced mammary tumors in rodents (Collishaw et al., 2009). Tobacco smoke is also known to contain carcinogens that distribute to breast tissue. Several metabolites of these compounds have been shown to cause DNA damage, reflected by the presence of DNA adducts, in the breast tissue of current smokers, former smokers, and those passively exposed to tobacco smoke (Morabia, 2002).

One of the few animal studies that have tested the effect of exposure to tobacco smoke rather than its components explored the effects of exposure in virgin and pregnant Sprague Dawley rats subsequently treated with the carcinogen MNU (Steinetz et al., 2006). Groups of 50-day-old animals (25 animals each) were exposed to either filtered air or cigarette smoke (described as equivalent to smoking 2.7 packs per day). At 100 days, the animals were given doses of the carcinogen MNU. Smaller groups of control animals (10 animals each) had the same exposures to air or smoke but received no MNU. Among those exposed to MNU, tumor development was earliest and greatest in the virgin rats exposed to cigarette smoke and latest and least in the pregnant rats exposed to air. Pregnancy was protective against the effects of MNU, but exposure to cigarette smoke resulted in increased tumor development. However, rats exposed to cigarette smoke without subsequent MNU exposure did not develop mammary tumors.

Conclusions

Recent scientific consensus reviews have been able to draw on newer studies with better assessments of tobacco smoke exposure than in the past. A 2009 IARC review declared that limited evidence exists to support a causal association between active smoking and breast cancer (Secretan et al., 2009), which constitutes a change from the organization's 2004 conclusion that the evidence on tobacco smoke suggested a lack of breast carcinogenicity (IARC, 2004). Others have concluded that the current evidence is consistent with a causal association between active smoking and breast cancer (Collishaw et al., 2009). Some studies implicate active smoking as a risk factor for breast cancer in two subgroups: women who initiated smoking at an early age or before their first full-term pregnancy, and women with genetic characteristics that result in slow metabolism and detoxification of components of tobacco smoke (*NAT2* slow acetylators).

For exposure to secondhand smoke, IARC found the evidence inconclusive (Secretan et al., 2009), while others have found the evidence to be suggestive of an association (HHS, 2006) or even consistent with a causal association with breast cancer in younger, premenopausal women (CalEPA, 2005; Collishaw et al., 2009). Within the committee there were differing interpretations of the existing data. Some were persuaded that the available evidence supports a causal association between exposure to secondhand smoke and risk for breast cancer, while others view the data as indicating a possible but not conclusive relationship. For most other smoking-related diseases, the relative risks are much stronger for active smoking than passive smoking. Thus findings of equivalent or stronger relative risks for breast cancer with passive smoking than with active smoking are difficult to explain mechanistically.

Because smoking is known to increase the risk of many types of cancer and has numerous negative health effects, substantial efforts to minimize exposure through public health interventions already exist. Although the overall magnitude of the reported effect of exposure to active or passive smoking on risk for breast cancer is not large, some susceptible subgroups appear to have a relative risk that is elevated over that of never smokers. Evidence of an increased risk for breast cancer reinforces the importance of smoking prevention and cessation programs and policies supporting smoke-free environments.

Radiation

The term "radiation" encompasses a broad spectrum of energies and can be divided into ionizing and non-ionizing radiation. Ionizing radiation has the ability to remove electrons from an atom, creating ions. Non-ionizing radiation, on the other hand, is lower in frequency and has insufficient energy to eject electrons from the atom.

Ionizing Radiation

There are two main types of ionizing radiation: photons, including X-rays and γ (gamma)-rays, and particulate radiation, including α (alpha) and β (beta) particles. Alpha and beta particles deliver their energy over shorter distances than photons and tend to pose most carcinogenic risk at very short distances once they enter the body. X-ray and γ-ray exposure are well documented as carcinogens, with sufficient evidence substantiating their role as risk factors for breast cancer. This evidence includes the experience of increased risk for breast cancer among younger members of the population of atomic bomb survivors (IARC, 2000).

In the general population, the most prominent source of exposure to ionizing radiation is from medical diagnostic procedures. X-rays are an important component of diagnostic imaging and are used in procedures ranging from radiographs to fluoroscopy to computed tomography (CT) scans. γ-rays are often delivered in nuclear medical examinations that may use radioactive tracers. X-rays and γ-rays are breast carcinogens in premenopausal women (IARC, 2000). Furthermore, risk of breast cancer is significantly increased following treatment to the chest in pediatric or young adult cancer patients (Henderson et al., 2010). Although it is widely accepted that carcinogenic sensitivity is highest when ionizing radiation exposure occurs in childhood (Carmichael et al., 2003), risk persists even for women of postmenopausal age (Berrington de Gonzalez et al., 2009). Chapter 5 further discusses some of the findings regarding effects from medical treatments at different life stages. Sufficient literature exists on early-life exposure to ionizing radiation (specifically including breast cancer), but exposure in later years may be an area for further research.

Animal data also support the evidence for ionizing radiation–induced breast cancer, with evidence of mammary adenocarcinomas observed in Sprague Dawley rats (IARC, 2000). In vitro data have helped to elucidate the carcinogenic mechanisms behind ionizing radiation. Radiation-induced breast cancer is a complex phenomenon, most likely influenced by the accumulation of genetic and epigenetic alterations (Carmichael et al., 2003). Rather than acting as a single carcinogenetic event, exposure to ionizing radiation is thought to give rise to cancer through the combined effects of induced genetic instability, cellular transformation, and chromosomal damage (IARC, 2000).

Important contributions to breast cancer risk from exposure to ionizing radiation are examined in detail in Appendix F of this document. As an established risk factor for breast cancer, exposures need to be minimized. The committee discusses opportunities for action to reduce risk from ionizing radiation in Chapter 6 and research needs in Chapter 7.

Non-Ionizing Radiation (ELF-EMF)

Non-ionizing radiation can be found as microwave (microwave appliances and telecommunications), infrared (heat lamps), or radiofrequency (radio) radiation. Lower still on the energy or frequency scale is radiation from extremely low frequency electromagnetic fields, ELF-EMF, which arises from electrical current and is of very low energy (energy is proportional to frequency). Non-ionizing radiation may interact with biological systems, and it is therefore of interest to environmental scientists and biologists. Most of the epidemiologic studies on the possible relationship of non-ionizing radiation to breast cancer have examined ELF-EMF.

Non-ionizing radiation is particularly challenging to study. ELF-EMF exposure is "ubiquitous and unmemorable," with all individuals who live near or use electricity exposed to it in some form. Because one cannot see or feel its presence, it is virtually impossible for an individual to record or quantify the frequency of exposure (IARC, 2002a). It is often difficult to distinguish high exposures from low exposures when they differ by only an order of magnitude (Kheifets et al., 1995; IARC, 2002a). Various researchers have postulated different metrics of exposure as being most relevant, such as average, peak, or rate of oscillation. The relatively small range of ELF-EMF exposures and the choice of different exposure metrics can affect the statistical power of epidemiologic studies.

Meta-analyses that have synthesized the findings from studies of breast cancer are consistent in showing no association, and they exclude the possibility of all but very small associations between ELF-EMF and breast cancer. A 2010 meta-analysis of 15 case–control studies from 2000 to 2009, involving 24,338 cases and 60,628 controls, found no significant association between breast cancer risk in relation to ELF-EMF exposure, even when stratifying by menopausal status or the source of exposure (Chen et al., 2010). This conclusion is consistent with a previous meta-analysis that looked at studies from 1996 to 2000 (Erren, 2001).

Studies have also assessed risk associated with specific modes of exposure. For example, case–control studies (Davis et al., 2002; London et al., 2003; Schoenfeld et al., 2003) found no association between ELF-EMF exposure from household exposures and appliances and breast cancer.

Although electric blankets were once raised as a source of concern as a potential risk factor for breast cancer, studies found no apparent associations between electric blanket use and breast cancer (Vena et al., 1991, 1994; Laden et al., 2000; Zheng et al., 2000; McElroy et al., 2001; Kabat et al., 2003). Furthermore, none of these studies found associations based on menopausal status, parity, estrogen receptor status, or hours of use. Early studies looking at occupational exposures to magnetic fields have shown little or no overall effect of ELF-EMF exposure on breast cancer, although some studies have linked exposure with a slight increase in risk for ER+ breast cancer (Van Wijngaarden et al., 2001; Kliukiene et al., 2003; Labreche et al., 2003). These findings, some researchers argue, are primarily the result of faulty study design; many of these studies were small and had little information on potential confounding factors (Forssen et al., 2005).

Occupational ELF-EMF exposure has been raised as a potential risk factor among men with breast cancer. Some studies in the early 1990s found an association between ELF-EMF fields and breast cancer in men (Demers et al., 1991; Matanoski et al., 1991; Loomis, 1992; Guenel et al., 1993; Floderus et al., 1994), while others (Rosenbaum et al., 1994; Theriault et al., 1994; Cantor et al., 1995; Stenlund and Floderus, 1997; Forssen et al., 2000) found no correlation. Studies of non-ionizing radiation and male breast cancer have generally been restricted to small cohorts and are largely inconclusive. Occupational exposure to ELF-EMF as a risk factor for male breast cancer is a potential area for future research; men are often exposed to occupational ELF-EMF in higher doses than women, and do not have confounding factors such as hormonal cycles or pregnancies.

Animal studies have examined the effects of the various forms of non-ionizing radiation. Results have been largely inconclusive; difficulty in interpreting the data is compounded by the fact that results may vary from strain to strain, based on diet, housing conditions, lighting, or laboratory (Anderson et al., 2000b; Fedrowitz et al., 2004). A proposed mechanism for non-ionizing radiation-induced carcinogenicity involves the hormone melatonin. Melatonin, produced by the pineal gland, is thought to inhibit estrogen-mediated cell proliferation. ELF-EMF exposure is hypothesized to suppress melatonin and thereby inhibit its protective effects.

Although IARC (2002a) has classified ELF-EMF as possibly carcinogenic to humans (Group 2B), few studies have assessed whether ELF-EMF has differential effects at various life stages, and the committee is not aware of studies that have examined the effect of timing of exposure through the life course on breast cancer risks. This is a potential area for future research.

Shift Work

According to IARC (2010b), the average prevalence of shift work involving night work in the United States is 14.8 percent (16.7% in men and 12.4% in women). It is most common among those in health care, transportation, communication, leisure and hospitality, and the service, mining, and industrial manufacturing sectors. It is more common in younger workers, decreasing to a prevalence of about 10 percent after age 55 (IARC, 2010b).

It has been proposed that shift work is a risk factor in breast cancer etiology. This phenomenon has been studied through epidemiologic, animal and in vitro studies, and was reviewed extensively by IARC in 2010. In the past decade, eight major epidemiologic studies have examined the relation between shift work and risk for breast cancer among female workers, although these studies had vastly differing definitions of shift work (IARC, 2010b). Among the two prospective cohort studies (Schernhammer et al., 2001; Schernhammer and Hankinson, 2005), one nationwide census-based cohort study (Schwartzbaum et al., 2007), three nested case–control studies (Tynes et al., 1996; Hansen, 2001; Lie et al., 2006), and two case–control studies (Davis et al., 2001; O'Leary et al., 2006), the majority studied postmenopausal women (IARC, 2010b). A notable limitation of the data from these studies is the lack of racial diversity, with only one study including a small subset of Latina and African American women (O'Leary et al., 2006). Despite differences in study methodologies, meta-analyses and systematic reviews of the literature consistently note an increase in relative risk of breast cancer associated with shift work (Megdal et al., 2005; Hansen, 2006; Kolstad, 2008; IARC, 2010b). Megdal et al. (2005) reported an aggregate RR estimate based on 13 combined studies of 1.48 (95% CI, 1.36–1.61).

Animal and in vitro studies on shift work–induced breast cancer are more difficult to design and conduct. Because "shift work" itself cannot be imposed on animals, experimental studies have used models of alteration of light and dark environments, which affect circadian pacemaker function. The exposure to light during the night, and the altered sleep cycle that ensues, has been proposed as the mechanism for shift work–induced breast cancer (Straif et al., 2007).

Numerous animal studies have evaluated the effect of varying light cycles on mammary tumorigenesis in animal models. In CBA mice, continuous light exposure increased the incidence of different spontaneous tumors from a variety of tissues in females, and also reduced overall life span (Anisimov et al., 2004). However, the numbers for mammary adenocarcinomas were very small—one spontaneous adenocarcinoma in light/dark exposed mice, and two adenocarcinomas in the light/light exposed group, with 50 animals in each group (Anisimov et al., 2004). Anderson et al. (2000a) demonstrated in rats that constant light exposure followed by exposure to a chemical carcinogen such as DMBA results in an increased incidence of mammary tumors when compared to an alternating light/dark cycle. Cos et al. (2006) examined the effects of constant light or different patterns of "light at night" on established DMBA-induced mammary carcinomas in female Sprague Dawley rats. They found that female rats exposed to light at night, especially those under a constant dim light during the darkness phase, showed (1) significantly higher rates of tumor growth as well as lower survival than controls (typical 12-hour light–dark cycle), (2) elevated serum estradiol concentration, and (3) decreased nocturnal excretion of 6-sulfatoxymelatonin, but no differences between nocturnal and diurnal levels. They concluded from this that circadian and endocrine disruption induced by light pollution could induce the growth of mammary tumors. The role of stress induced from the constant light exposure cannot be ruled out. It could be a fundamental part of the mechanism of action, and stress would also be relevant to humans with constant disruption of light at night/circadian rhythm. Other studies have shown that light exposure at night increases the growth of different kinds of transplantable tumors in rats (Dauchy et al., 1997, 1999; Blask et al., 2002).

Melatonin is hypothesized to play an important role in shift work–induced breast cancer; this hormone transmits informational cues of environmental light and darkness from the eye to the hypothalamus, to all tissues of the body, helping to set an organism's biological clock. Importantly, "melatonin has anti-proliferative effects on human cancer cells cultured in vitro" (IARC, 2010b, p. 663). According to the melatonin hypothesis, light exposure at night results in a reduction in the circulating levels of melatonin, which removes its check on estrogen, allowing for rising levels of estrogen to promote cell proliferation and increase the risk for malignant transformation (Graham et al., 2001). As an antiestrogen, melatonin down-regulates ERα transcription and alters its functional activity (Molis et al., 1994; Rato et al., 1999; del Río et al., 2004; Cini et al., 2005). Despite numerous in vitro studies on the oncostatic effects of melatonin, there is insufficient evidence regarding the use of melatonin supplements to determine their impact on risk of breast cancer, making this a potential subject area for future studies.

IARC (2010b, p. 764) concluded that "shift work that involves circadian disruption is probably carcinogenic to humans." To understand the role of "light at night" in breast cancer etiology, further studies are needed on its influence on women who do not perform shift work, but who are exposed to light at night in their homes.

Metals

Metals are ubiquitous in the environment and human exposures derive from natural background sources in food, water, and air as well as from extraction, manufacture, and uses in multiple tools, products, medical devices, and building materials. Exposures to metals in the workplace and to the general population were reduced in the latter part of the 20th century with occupational health and safety standards and reductions in environmental emissions and levels. The most notable results of these restrictions include the dramatic declines in blood lead levels since the early 1970s with the ban on lead in gasoline and a reduction in lead use in other consumer products such as paint. The revised drinking water standard for arsenic in 2001 reduced a main source of arsenic exposure from natural occurrence in some parts of the United States. Primary sources of cadmium exposure include cigarette smoke and shellfish consumption. Recent reductions have been made in allowable levels of cadmium and lead in consumer products.

A systematic review of evidence by IARC (Straif et al., 2009) has classified several different metals (arsenic and inorganic arsenic compounds, beryllium, cadmium, chromium, nickel, and their related compounds) as "carcinogenic to humans" (Group 1). These classifications are based on sufficient evidence from human studies that these metals cause tumors in the lung and some other sites, and not on findings regarding breast cancer.

Despite the considerable evidence linking certain metals (e.g., nickel, hexavalent chromium, cadmium, and arsenic) to lung cancer from inhalation and arsenic to internal cancers (primarily bladder, lung, and liver) and skin cancer by ingestion, no clear epidemiologic data have indicated metal exposures to be a risk factor for breast cancer (ATSDR, 2005, 2007, 2008a,b; NTP, 2011a). Much of the evidence for metals and lung cancer in humans arises from studies of worker populations, which have historically included few women. Other than for lung cancer, however, worker studies have shown little evidence for other cancers from metals exposure, indicating insufficient systemic exposure to produce tumors at distant sites. Exposures through routes other than inhalation likewise provide little evidence of breast cancer. With the exception of arsenic, general population exposures to metals through consumer products, medicinal applications, implanted medical devices (McGregor et al., 2000), and elevated levels in drinking water or food have been associated with health effects other than cancer risks, although relatively few studies examining cancer risks have been published for nonoccupational populations (ATSDR, 2005, 2007, 2008a,b; NTP, 2011a).

Arsenic has the most epidemiologic data for evaluating breast cancer risk. A number of large population studies on cancer rates from exposure to elevated arsenic levels in drinking water are available, although most focus on target sites that have consistently shown increased cancer rates (e.g., lung, bladder, skin) and only a few report risks for breast cancer (summarized by ATSDR, 2007). Tsai et al. (1999) reported no increase in breast cancer mortality rates for women exposed to high levels of arsenic in well water in Southwest Taiwan, based on 8,874 breast cancer deaths and over 1.4 million person-years of exposure (SMR compared to local reference = 1.01, 95% CI, 0.74–1.34; SMR compared to national reference = 0.67, 95% CI, 0.48–0.89). Likewise, for a region in northern Chile with over 400,000 people exposed to high arsenic levels in drinking water and in air, breast cancer was used as a control cancer because it has not shown increased risks from arsenic in other studies (Rivara et al., 1997). Compared to a control region with 1.7 million people, breast cancer mortality risks were lower, although not significantly, for the arsenic-exposed region (RR = 0.7, 95% CI, 0.39–1.08) (Rivara et al., 1997).

Studies have reported correlations between levels of various metals (either increased or decreased) in tissues or specimens (e.g., hair, blood, urine) from breast cancer patients or in tumor cells, but most of these studies have small sample sizes and little control for confounding factors. A somewhat larger case–control study of urinary cadmium levels in 246 breast cancer cases in a population in Wisconsin found a two-fold higher risk of breast cancer for women in the highest urinary cadmium quartile compared with those in the lowest fourth (OR = 2.29, 95% CI, 1.3–4.2) after adjustment for several other risk factors (e.g., age, family history of breast cancer, postmenopausal hormone use) (McElroy et al., 2006). Adjustment for smoking (never, former, current) had no effect on the results, although quantifying smoking duration and intensity and hence cadmium exposure from smoking (e.g., pack-years) may have been better able to distinguish an effect. In this study and others that measure levels of metals at the time of the study, it is not possible to distinguish whether metals have a role in the causal pathway for breast cancer or if breast cancer patients or tumor cells have abnormal metal absorption, distribution, metabolism, or excretion. Little evidence associates cadmium exposure with excess cancers of the breast for populations that have much higher cadmium exposures from environmental contamination such as those in Japan, England, or Belgium, although ATSDR (2008a) notes that the statistical power of these studies to detect cancers was not high. These populations have been well studied for kidney and other noncancer effects. A recent IARC review concluded that there was limited evidence from epidemiologic sources for kidney and prostate cancer (Straif et al., 2009).

A few studies in humans have examined associations between levels of urinary cadmium, blood lead, and hormone levels or hormonal effects at different life stages. Urinary cadmium was associated with elevated levels of testosterone, but not estrone, in postmenopausal women (Nagata et al., 2005). Higher blood lead levels, alone or in combination with urinary cadmium levels, were reported to be related to markers of delay of menarche in prepubertal girls. Interestingly, the association was considerably stronger in girls with elevated urinary cadmium in addition to high blood lead (Gollenberg et al., 2010). A delay in age of menarche runs counter to an elevation in risk because early (not late) age of menarche is a risk factor for breast cancer.

Overall, animal studies examining the carcinogenicity of metals such as arsenic, hexavalent chromium, nickel, cadmium, or cobalt have not reported increases in mammary tumors (ATSDR, 2005, 2007, 2008a,b; NTP, 2011a). One animal study found that cadmium administered by injection to pregnant rats mimicked the effects of estrogen in the uterus and mammary glands of the offspring, supporting the hypothesis that cadmium exposure is a potential risk factor for breast cancer (Johnson et al., 2003). The type and magnitude of dosing, however, was not comparable to what would likely occur in humans through environmental exposure. Johnson et al. (2003) compared intraperitoneal injection of up to 5 μg/kg of cadmium on gestation days 12 and 17 to the World Health Organization provisional tolerable intake from the diet of 7 μg/kg/week. However, the systemic dose from dietary intake is reduced by gastrointestinal bioavailability, and the amount absorbed from the diet over 7 days is spread out over much smaller incremental doses than the high acute dose that would result from injection.[12] Lower dosing may also result in less fetal exposure because of more efficient maternal sequestering.

High levels of arsenic, cadmium, and the transition metals such as iron, nickel, chromium, copper, and lead have been associated with free radical generation and oxidative stress, particularly in studies carried out in vitro (Davidson et al., 2007). Prolonged or repeated oxidative stress is a well-known mechanism of carcinogenicity in general. In vitro studies indicate that many metals such as cadmium, arsenic, aluminum, and a number of divalent metals can interact with the estrogen receptor, thereby potentially affecting breast cancer risk; however, with the possible exception of cadmium, little research has investigated this issue. For cadmium, the findings are not entirely consistent on whether estrogen receptor binding results in the expected downstream effects such as expression of genes involved in cell signaling and proliferation critical to breast cancer cell growth. Studies observing such effects have indicated the relative potency of cadmium to be 100 to 1,000 times less than that of estradiol; other studies reported little

---

[12]Adjusting for body weight scaling by a factor of 3/4 (Rhomberg and Lewandowski, 2004) results in the 5 μg/kg dose to a 0.250 kg rat over 5 days being 4 times lower than the 7 μg/kg dose to a 70 kg human over 7 days. However, the dose to the rat is injected at one time and the dose to the human would be spread out over 7 days.

estrogenic effect in vitro over a wide range of concentrations (Silva et al., 2006, and studies reviewed therein). Cadmium has also been reported to transform normal cultured breast epithelial cells in vitro through an estrogen-independent mechanism into cells with characteristics of malignant breast tumor cells (Benbrahim-Tallaa et al., 2009). Concentrations used, however, exceeded those reported to be cytotoxic in other studies (Choe et al., 2003; Silva et al., 2006). Other possible mechanisms suggested include indirect effects such as interactions with other essential enzyme pathways or by depletion of essential metals (e.g., those protective of oxidative stress) or nutrients (e.g., antioxidants).

All told, the evidence available for metals as risk factors for breast cancer indicates biologic plausibility for increased risk of breast cancer in association with exposure to certain metals, particularly cadmium and possibly arsenic, but metals are unlikely to be a major risk factor at environmentally relevant doses. Much of the evidence is from in vitro studies using concentrations of metals that are considerably higher than would occur in humans from environmental exposures.

Consumer Products and Constituents

Alkylphenols

Alkylphenols are a group of chemical intermediates as well as degradation products of alkylphenol ethoxylates. Alkylphenol ethoxylates, and particularly nonylphenol ethoxylate, are widely used non-ionic surfactants added for foam control, wetting, and antifog/antistatic, and as stabilizers in a variety of household, industrial/commercial, and agricultural products such as adhesives, sealants, detergents/cleaners, and pesticides (Lani, 2010). Nonoxynol-9 is an alkylphenol used as a spermicide in contraceptives. Alkylphenols are also plastic or resin additives. The U.S. Department of Health and Human Services Household Products Database lists nonylphenyl polyethoxylate in paints, certain cleaners, and hair color products (HHS, 2010b), and nonylphenol in hardeners and epoxy for household maintenance products (HHS, 2010a).

As a result of widespread use and degradation of alkylphenol ethoxylates, alkylphenols have been detected in municipal and industrial discharges and in receiving water bodies and sediment (Fenet et al., 2003; Gross et al., 2004). NHANES reported urinary levels of 4-tert-octylphenol in survey years 2003–2004 at approximately 0.3 to 0.4 µg/L at the 50th percentile and 1.3 to 2.5 µg/L at the 95th percentile, depending on age or ethnic grouping (CDC, 2009a). Urinary levels of orthophenylphenol measured in the 1999–2000 survey were similar to those of octylphenol in 2003–2004, but had decreased to undetectable levels at the 50th and 75th percentiles in 2001–2002 (no data for 2003–2004). No data were reported for nonylphenol.

Alkylphenols, particularly larger compounds such as octylphenol and nonylphenol, are more lipophilic and persistent in the environment than their parent compounds (Lani, 2010). In vitro studies in breast cancer cell lines also indicate a trend for increasing estrogenic effects with larger alkyl groups such as for octyl- and nonylphenol (Terasaka et al., 2006; Sun et al., 2008). Among the alkylphenol compounds, 4-nonylphenol accounts for 80 percent of the alkylphenol in the environment (Oh et al., 2008) and is the most studied alkylphenol for its potential endocrine disrupting effects. The amount of research on this compound, however, is considerably less than for the related compound, bisphenol A. Alkylphenols have not been evaluated for carcinogenicity by regulatory agencies in the United States (e.g., NTP, EPA) or international groups (e.g., IARC). NTP has immunotoxicology, behavioral toxicology, and multigenerational reproductive studies under way (HHS, 2010b). In addition to estrogenic effects demonstrated in vitro, a few studies in laboratory animals indicate the potential of nonylphenol to alter mammary gland development and increase mammary tumor formation.

Nonylphenol administered by oral gavage at 100 mg/kg (but not at 10 mg/kg) on gestational days 15–19 in rats resulted in advanced lobular development of the mammary glands of the offspring on postnatal day 22 (Moon et al., 2007). Transgenic mice consuming nonylphenol in honey for 32 weeks, beginning at 5–6 weeks of age, showed increased mammary tumor rates at a dose of 45 mg/kg/day, but not at a dose of 30 mg/kg/day (Acevedo et al., 2005). By comparison, an equivalently estrogenic dose of estradiol-17β (E_2) of 0.01 mg/kg/day (based on higher estrogen receptor binding affinity of E_2 relative to nonylphenol) did not increase mammary cancer risk. Acevedo et al. (2005) thus concluded that nonylphenol may be a more potent mammary gland carcinogen than predicted by its relative binding affinity to E_2. Given widespread exposure and hazards identified from in vitro and animal studies at relatively high doses, alkylphenols are candidates for further investigation. Similar to many other relatively unstudied chemicals with endocrine activity, more research is needed to define what risks are posed to the population exposed at low levels in the environment.

Bisphenol A

Bisphenol A, or BPA, is a plasticizer and one of the highest volume chemicals produced worldwide (Vandenberg et al., 2007). It is used in the production of products such as polycarbonate plastics, epoxy resins that line metal cans, dental appliances and composite fillings, and also as a component in thermal paper used for certain receipts. BPA is characterized by widespread use and frequent exposure in developed countries. Human exposure is most likely through the oral route, although transdermal exposure (bathing in contaminated water, handling cash register receipts) (Biedermann et al., 2010), and inhalation are also possible (Stahlhut et al., 2009). Concern has arisen about BPA's leaching from medical products or consumer products such as cans, plastic food wrap, paper towels, paper receipts, and especially from polycarbonate baby bottles. Although studies on BPA are numerous, they are difficult to interpret, and they illustrate the complexities of breast cancer risk research.

BPA has not been evaluated for carcinogenicity by IARC, WCRF/ AICR, or EPA (2011a), although EPA (2010b) has summarized the existing literature as part of an Action Plan to be implemented. Several panels have reviewed toxicological findings about BPA (EFSA, 2006, 2008; vom Saal et al., 2007; FDA, 2008; NTP, 2008; JECFA, 2010). Some of NTP's findings are noted below.

Human studies on BPA have focused primarily on exposure, and exposure is ubiquitous. NHANES data showed that 90 to 95 percent of the U.S. population has detectable levels in urine (Calafat et al., 2008). BPA has been found in virtually all human tissues and in follicular fluid, maternal serum, fetal serum, umbilical cord blood, amniotic fluid, and the placenta (Vandenberg et al., 2007, 2010). Furthermore, the short half-life of BPA means that any detectable exposure was recent, implying that BPA exposure is also continuous. Indeed, the cessation of consumption of packaged food for 3 days resulted in a 66 percent reduction of urinary BPA, which returned to pre-intervention levels once consumption resumed (Rudel et al., 2011). A study of the pharmacokinetics of BPA in adult volunteers with a controlled high dietary exposure[13] suggests that serum concentrations are roughly 42 times lower than urinary levels and below the limit of detection of 1.3 nM (Teeguarden et al., 2011).

Epidemiologic studies on the potential health effects of BPA exposure are limited in both quantity and quality for various reasons. Because of its short half-life, current measurements may not be a sound basis for estimating past exposures. In addition, exposure studies may be unable to distinguish the potential effects of BPA from those of the myriad of other estrogenic compounds that are present in most people examined (Vandenberg et al., 2007). Concern for early-life exposure and developmental effects also further complicates studies; the exposure, pharmacokinetics, and metabolism of BPA in adults cannot always be extrapolated to make predictions for the fetus, infant, or child.

---

[13]The estimated average consumption of BPA was 0.27 µg/kg body weight (range, 0.03–0.86), 21 percent greater than the 95th percentile of aggregate exposure in the adult U.S. population (Teeguarden et al., 2011).

Various studies have been conducted to assess the potential for BPA to induce cancer in rodents, including one set of NTP studies as well as studies that are not cancer bioassays, but that evaluate the structure and function of the mammary gland in pubertal or young adult animals following early-life exposure. The in vivo data have been difficult to interpret. Two-year dietary cancer bioassays were conducted by NTP in 1982 using the standard protocol at that time. Rats and mice of both sexes beginning at 5 weeks of age were exposed for 104 weeks to high levels of BPA in feed. BPA was not shown to induce neoplastic or non-neoplastic lesions in the mammary glands of female rats or mice (NTP, 1982a), although suggestive carcinogenicity observations were reported for other sites (hematopoietic and testicular cancers). The NTP (1982a, p. vii) concluded, "Under the conditions of this bioassay, there was no convincing evidence that bisphenol A was carcinogenic for F344 rats or B6C3F1 mice of either sex."

The study has received criticism by Keri et al. (2007) for many reasons common to studying estrogenic compounds using standard cancer bioassay protocols (see Chapter 4). Prenatal exposure was not included, as also noted recently by NTP (2008). Also, the high dosing can be problematic when assessing endocrine disrupting compounds such as BPA, where dose response often defies conventional toxicological relationships; in some cases, low doses may have important physiological effects, while in others, high doses may be inhibitory (Watson et al., 2007; Kochukov et al., 2009). It also can be difficult to completely eliminate exposure to endocrine disrupting or estrogenic compounds in the control group; cages are often made of BPA polymers, and phytoestrogen-free diets must be followed (Keri et al., 2007).

Animal studies have suggested that perinatal subcutaneous exposure (via osmotic minipumps) to low doses of BPA can cause a variety of tissue changes in the peripubertal mammary gland that may signal an increased susceptibility to tumors in later life (Muñoz-de-Toro et al., 2005; Durando et al., 2007; Vandenberg et al., 2007). Furthermore, low-level exposure subcutaneously administered to pregnant rats has led to preneoplastic lesions—ductal hyperplasia and carcinoma in situ—in their offspring in adulthood (Durando et al., 2007; Murray et al., 2007). However, no data currently exist to determine whether lesions of the severity and extent seen in these studies contribute to the occurrence of invasive carcinoma (NTP, 2008). Because most of the existing data are based on subcutaneous exposure rather than oral dosing, it is difficult to determine whether the pharmacokinetics in animals are informative for human oral or dermal exposure. However, oral exposure of pregnant rats to BPA at a dose of 250 µg/kg body weight has also been studied and observed to similarly affect mammary gland development of offspring in the peripubertal period; exposure at a lower dose (25 µg/kg) showed effects on relevant gene expression (Moral et al., 2008).

NTP (2008) found "minimal concern" for BPA's effects on the mammary gland for females at the fetal, infant, and child stages at current levels of human exposure. In doing so, it noted that "[t]hese studies in laboratory animals provide only limited evidence for adverse effects on development and more research is needed to better understand their implications for human health. However, because these effects in animals occur at bisphenol A exposure levels similar to those experienced by humans, the possibility that bisphenol A may alter human development cannot be dismissed" (NTP, 2008, p. 7).

Currently, the in vivo data are insufficient to determine BPA's effects in adult organisms.

Because of the lack of epidemiologic evidence on BPA and breast cancer risk and limitations of in vivo study designs, current BPA data primarily come from in vitro models. Although such data often do not speak specifically to breast cancer endpoints, they have shed some light on BPA's mechanisms of action. BPA is a well-established xenoestrogen and endocrine disruptor, and it has been shown to mimic, enhance, or inhibit endogenous estrogen activity (Wetherill et al., 2007). BPA selectively binds to both estrogen receptors (ERα and -β), with a higher affinity for ERβ (Kuiper et al., 1997; Routledge et al., 2000; Matthews et al., 2001). Although endocrine disruption is an indirect mechanism for cancer, it has been hypothesized that it is important because of the morphogenic nature of hormones; exposure to even low doses of hormonally active chemicals, especially during development, can alter cellular or tissue organization over time, creating an environment susceptible to diseases such as cancer (Soto and Sonnenschein, 2010). Evidence implicating BPA as genotoxic is conflicting and difficult to interpret. A number of in vitro assays have shown no mutagenic activity (Tennant et al., 1987; Schweikl et al., 1998; Schrader et al., 2002; Keri et al., 2007), but others have shown genotoxic activity correlated with morphological transformation or aneuploidy (Galloway et al., 1998; Hilliard et al., 1998), DNA adduct formation (Tsutsui et al., 1998), or double-stranded breaks (Iso et al., 2006).

Another emerging facet of BPA mechanistic research involves susceptibility at various life stages. It has been proposed that BPA can epigenetically alter or suppress gene expression through endocrine receptor mediated pathways, with effects accumulating over time to increase risk of neoplasia (Doherty et al., 2010; Weng et al., 2010).

In sum, the role of BPA in human breast cancer is not known. Current researchers have not reached a consensus on the effects of BPA in breast cancer etiology, but the effects of BPA extend to other systems, with potentially harmful effects to the fetal, infant, or child brain and behavior (NTP, 2008). The results of a large body of research have shown that BPA has estrogenic effects and effects on the androgen receptor, the thyroid gland, male and female reproductive systems, and immunity. It has also been associated with abnormal liver enzyme concentrations and self-reported cardiovascular disease and diabetes (Lang et al., 2008). Active research efforts are continuing to further clarify its health effects (NIEHS, 2009; FDA, 2010). Because of the complex nature of BPA's action and mechanisms of activity that overlap with those of other xenobiotics, further research should take a mechanistic and systems biology approach to address additive or other cumulative actions of estrogenic compounds and their roles in overall health.

Nail Products

Potential health risks from exposures to chemicals of concern in consumer nail products have attracted public attention. Nail products contain a number of chemicals that are known or suspected carcinogens, as well as agents implicated for risk of breast cancer by virtue of their endocrine disrupting properties. Nail product constituents may include toluene, benzoyl peroxide, formaldehyde, and phthalates (California Department of Health Services, 1999; EPA, 2004).

Relatively little human health research has been done in this area. An early occupational mortality study in California indicated that cosmetologists, including manicurists, had significantly elevated risks for breast cancer mortality (Singleton and Beaumont, 1989), although a U.S. mortality study covering a decade later failed to find such an association (Robinson and Walker, 1999). In terms of incidence, a 1984 study linking licensed cosmetologists to the Connecticut cancer registry noted that women licensed between 1925 and 1934, before the dramatic increase in the nail salon sector, experienced a significant excess of breast cancer compared to the general population in Connecticut (Teta et al., 1984).

The nail salon industry in the United States, now dominated by female Asian immigrant workers, has expanded rapidly over the past two decades (Quach et al., 2008). Although few studies have explicitly addressed cancer risks from use of nail products, a recent California study of nail salon workers suggested that, despite lack of evidence of an excess of breast cancer in the nail salon workforce, the industry is young and further follow-up of workers is needed (Quach et al., 2010). Notably, evidence shows that nail salon workers are exposed to several chemicals of concern, including toluene, methyl methacrylate, and total volatile organic compounds at levels higher than recommended guidelines (Quach et al., 2011).

Animal studies have been carried out on many of the individual chemical components of nail products, some of which are established as known or reasonably anticipated to be human carcinogens (e.g., formaldehyde [IARC, 2006a; NTP, 2011a], styrene [NTP, 2011a]). Some of these chemicals (e.g., toluene, dibutyl phthalate) are also being tested for endocrine disrupting properties based on widespread exposure to them (EPA, 2009b). The committee is not aware of animal data evaluating the effects of mixtures similar to those in nail care products.

Nail care products represent a range of easily obtainable and widely used over-the-counter commodities for which there is sparse information on formulations, chemical exposures, and health risks. Women in the nail salon workforce may be the most highly exposed, but widespread lower level exposure of consumers suggests that this is an area for further inquiry.

Hair Dyes for Personal Use

Hair dyes can be classified as oxidative or non-oxidative. Oxidative hair dyes are permanent dyes and make up the majority (about 80 percent or higher) of the hair dyes that are sold (Baan et al., 2008; IARC, 2010c). They are complex chemical mixtures: several ingredients (particularly para- and ortho-aminophenols, phenylenediamines, meta-aminophenols, and metadiaminobenzenes) are mixed in the presence of hydrogen peroxide to produce the color through a chemical reaction within the hair shaft. The darker the hair dye color, the higher the concentration of chemical ingredients. Non-oxidative hair dyes are semipermanent or temporary dyes, and they may also be called direct dyes. With non-oxidative coloring products, there is no chemical reaction to produce the hair color, and the color will wash out with repeated shampooing. They use high–molecular weight compounds that may contain multiple different dyes to obtain the specific color. Because of the chemical process involved with oxidative dye products and the potential to produce reactive species during the process, it has been previously hypothesized that permanent hair dyes would be more likely than non-oxidative dye products to be associated with cancer (Bolt and Golka, 2007).

Oxidative hair dyes were introduced at the end of the 19th century, and their formulations have changed over time (IARC, 2010c). The use of some chemical ingredients in permanent hair dyes was discontinued in the 1970s. Thus, the association of cancer outcomes with product use before and after 1980 has been examined in some studies. Occupational exposures to hair dyes by hairdressers and barbers have also been examined (IARC, 2010c).

A meta-analysis by Takkouche et al. (2005) included 12 case–control studies (involving 5,019 cases and 8,486 controls) and 2 cohort studies on personal use of hair dyes. All but two case–control studies examined the association of breast cancer with permanent hair dyes, and all of the case–control studies explored an association of breast cancer risk with any type of hair dye use. Intensive exposure, defined as more than 200 lifetime exposures to hair dye, was examined in the 2 cohort studies and 7 of the 12 case–control studies. Among all studies, no statistically significant association was seen between risk of breast cancer and any hair dye use (RR = 1.06, 95% CI, 0.95–1.18) or, from 9 studies, for intensive use (RR = 0.99, 95% CI, 0.89–1.11) (Takkouche et al., 2005). Additionally, a study reporting detailed information on type of hair dye use and color reported no statistically significant association for use of either dark color products or light color products, or age at first use, duration of use, number of applications, or years since first use (Zheng et al., 2002).

An IARC (2010c) review examined hair dyes as occupational and personal exposures. For cancer in general, there was inadequate evidence in humans for the carcinogenicity of personal use of hair dyes; the overall evaluation was that personal use of hair dyes was not classifiable as to its carcinogenicity. For breast cancer, no association was seen for occupational exposures, and the epidemiologic evidence on breast cancer and personal use of hair dyes was considered "inadequate" to reach a conclusion on carcinogenicity (IARC, 2010c, p. 644).

IARC (2010c) categorized the animal evidence regarding carcinogenicity in general as "limited," but noted some studies in rats showed benign lesions of the mammary glands after exposure to oxidative hair dye formulations or components. The majority of rodent studies have exposed adult animals by skin painting: shaving a patch of fur, followed by a direct application of the hair dye. The studies are difficult to interpret because of the variety of product formulations and strengths that may be in use. Most of the animal studies reviewed in the most recent IARC review were conducted in the 1970s and 1980s, and product formulations change over time.

Epidemiologic evidence from case–control and cohort studies does not suggest an association between hair dye use and breast cancer. Limitations of some of the studies include lack of specificity for type of hair dyes used (oxidative versus non-oxidative) and details on color, type, or duration of use. In addition, formulations have changed over time, and they differ based on the region of the world in which they are produced and sold. Strengths of the epidemiologic evidence include studies conducted in a variety of populations, including those with exposure to dark hair colors, examinations by intensity of exposure, and consistent findings of no association among those studies with detailed exposure information. Based on the available human evidence, personal use of hair dyes is unlikely to be an important risk factor for breast cancer.

Parabens

Parabens are a class of synthetic chemicals called para-hydroxybenzoates. They are the most widely used preservatives in cosmetic products, and they are also used in a wide variety of foods and drugs. They can be found in some underarm deodorants and antiperspirants, but most major brands do not currently contain them (NCI, 2008, citing FDA). They meet several criteria of an "ideal preservative": a broad spectrum of antimicrobial activity, especially against yeasts and molds; virtual lack of color and taste; stability over a wide pH range; and extremely low acute and chronic toxicity (Soni et al., 2005). They have, however, been found to be weakly estrogenic (Golden et al., 2005) and concerns have been raised about their effects in combination with other potentially estrogenic compounds (Darbre and Harvey, 2008).

Few epidemiologic studies are of relevance to paraben exposure and breast cancer. A population-based case–control study with response rates of 75 to 78 percent showed no evidence of an association between breast cancer and the use of underarm deodorant or antiperspirant, with or without underarm shaving (Mirick et al., 2002). However, because parabens also have other uses—in other personal care products, as antimicrobials to food products up to concentrations of 0.1 percent, and as preservatives in drugs—the extent to which women using antiperspirants or deodorants were more exposed than the study controls is unclear. The only other study specifically addressing cancer endpoints is a case-only survey with a very low response rate (32.5%) that reported that frequency and earlier onset of antiperspirant or deodorant use were associated with an earlier age of breast cancer diagnosis (McGrath, 2003). With no control subjects and lack of age adjustment, the study design does not permit reliable assessment of breast cancer risk associated with underarm deodorant use. For example, a possible interpretation of the survey of cases is that younger women use more antiperspirant than older women. As shown by this same study, underarm antiperspirant use in women increased dramatically from the 1960s up to 2000. As a result, younger women are more likely to use deodorant at an earlier age and more frequently than older women. Breast cancer rates also increased during this period, but are not necessarily related. A cohort study of girls ages 6–8 at entry showed no association between urinary concentrations of benzophenone-3 (a sunscreen) or parabens and signs of early puberty (Wolff et al., 2010). The 1 year of follow-up of this young population is too short to have breast cancer endpoints. The National Cancer Institute (NCI, 2008, p. 1) states, "there is no conclusive research linking the use of underarm antiperspirants or deodorants and the subsequent development of breast cancer."

In an extensive review of the clinical, experimental animal, and in vitro mechanistic studies of parabens, Golden and colleagues (2005) concluded that in the aggregate, the evidence is extremely weak that parabens, acting through endocrine or estrogenic or endocrine disruption mechanisms, have adverse effects on human health, including breast cancer. The review notes that parabens are 1 thousand to 1 million times less potent than 17β-estradiol and the likelihood of exposure to concentrations that could exert hormonal effects is remote. They conclude that it is "biologically implausible" that exposure to parabens (in utero, or by transdermal, oral, or any other route) increases the risk of any estrogen-mediated endpoint in humans. However, the authors did not make comparisons taking into account pharmacokinetics, persistence, and other aspects of exposure related to the amount of active compound available for interaction with the receptor.

A researcher from the Procter and Gamble Company proposed a new method to refine estimates of exposure to parabens through topically applied cosmetics and food (Cowan-Ellsberry and Robison, 2009). Use of conservative estimates of parabens concentrations in products, application or ingestion frequency, dwell time of topical substances, absorption, and clearance/metabolism led to an aggregate exposure estimate of 1.3 mg/kg/day, and cruder estimates of up to 4.1 mg/kg/day; these levels are below the acceptable daily intake (ADI) for parabens of 10 mg/kg/day (Soni et al., 2005). Regarding estrogenicity, the more branched and longer chained the paraben, the greater the estrogen binding activity (FAO/WHO, 2005; Integrated Laboratory Systems, 2005).

The Cosmetic Ingredient Review, a group established by the cosmetic industry in collaboration with the FDA, has concluded, based on an expert panel review of the epidemiologic evidence in combination with animal toxicology and in vitro mechanistic studies, that use of parabens in cosmetics is safe and is not carcinogenic (Cosmetic Ingredient Review, 2008). The FDA (2007) has concluded that "at the present time there is no reason for consumers to be concerned about the use of cosmetics containing parabens."

On the other hand, in 2005 the European Food Safety Authority withdrew propyl paraben from an ADI, for parabens as a group, because of concerns about the estrogenic and reproductive effects (FAO/WHO, 2007). Male reproductive toxicity was discovered for propyl paraben in animal studies at the same dose as the ADI. Similar toxicity was seen with butyl paraben (not used in Europe as a food additive). The Food and Agriculture Organization of the United Nations (FAO) and the World Health Organization (WHO) in 2007 also withdrew the compound from the group ADI. After review of the toxicological literature it noted, "There are insufficient data to conclude whether the effects observed with parabens of higher alkyl chain length [butyl and propyl] in males are mediated via an estrogenic, anti-androgenic or some other mechanism" (FAO/WHO, 2007, p. 29).

A comprehensive toxicological profile sponsored by the NTP reported butyl paraben to be noncarcinogenic to rats and mice (Integrated Laboratory Systems, 2005). However, because of data gaps, the NTP selected the compound for carcinogenicity evaluation and other toxicological studies.

Perfluorinated Compounds

Perfluorinated compounds such as perfluorooctanoic acid (PFOA) and perfluorooctane sulfonate (PFOS) have been produced since the 1950s and used extensively in the production of industrial chemicals and in surfactants and surface protectors for products such as nonstick cookware and fabric stain and water repellants. The majority of human exposure is probably through diet and drinking water, possibly related to wastewater treatment plants that may concentrate perfluorinated compounds (Steenland et al., 2010). They may also be ingested in dust from treated products (Trudel et al., 2008; Steenland et al., 2010). Testing through NHANES has shown recently declining but nearly universal exposure to PFOA and PFOS in the United States (Calafat et al., 2007). With this widespread exposure, these chemicals have garnered attention for potential long-term adverse health outcomes (White et al., 2011a,b). EPA has not yet completed an assessment of their health risks, and they have not been reviewed by IARC.

The epidemiologic studies to date are limited in number and scope. Grice et al. (2007) surveyed 1,895 past and present workers in perfluorooocatanesulfonyl flouride production and used a job exposure matrix to estimate PFOS exposure in women reporting breast cancer and other conditions validated from medical records. Only 263 women were among 1,400 workers returning questionnaires, with 4 breast cancers reported among them (the expected number of breast cancers for this age distribution of women was not reported). According to the authors, the PFOS exposures of study participants were "substantially higher than exposures in the general population" (Grice et al., 2007, p. 728). This study was limited in its ability to detect health effects, but no association was found with breast cancer or the other conditions of interest. Other studies (reviewed in Olsen et al., 2009) have found no consistent relationship between PFOA and PFOS exposure and human fetal development (e.g., birthweight, ponderal index); no cancer endpoints were evaluated. Studies to assess the impact of PFOA on the onset of puberty as a risk factor for breast cancer are under way as part of the NIH-supported Breast Cancer and the Environment Research Centers (Hiatt et al., 2009).

Few studies have been conducted to assess PFOA or PFOS tumorigenesis in animals. Various tumors have been observed in animals, including equivocal findings of mammary tumors in an early unpublished study by a producer of the compound (Sibinski, 1987; EPA, 2005a). Recent animal studies indicate that PFOA exposure at critical developmental stages can alter mammary gland growth in mice, among other developmental effects (Macon et al., 2011; White et al., 2011a,b). For example, effects were seen in mice exposed to PFOA in utero or chronically to low levels in drinking water before adulthood (White et al., 2011b). The second half of gestation is an especially sensitive period (White et al., 2007). Effects on the in utero development of mammary glands in CD-1 mice have been observed at fairly low doses (0.01 mg/kg/d to dams during gestation) (Macon et al., 2011). Effects on mammary gland development have been also observed in peroxisome proliferator-activated receptor alpha (PPARα) knockout mice, indicating that it is unlikely that PPARα plays any role in adverse impacts on mammary development (Zhao et al., 2010). Stimulatory effects on mammary development from peripubertal exposure to PFOA were associated with increased ovarian steroid hormone production, and with increased growth factors in mammary glands, independent of PPARα (Zhao et al., 2010), indicating that PFOA may act through an endocrine-disruption mechanism.

The potential carcinogenicity of PFOS/PFOA in the mammary gland and effects of exposure during various stages of life provide biologic plausibility to the hypothesis that PFOA may impact breast cancer and remain important topics for future research.

Phthalates

Phthalates, known as "plasticizers," are added to plastics to increase flexibility, and are widely found in consumer products, including plastics used in food packaging, rain gear, footwear, and toys (NTP, 2006a; Rudel et al., 2011). They are also used in cosmetics and personal care products because of their viscosity and lipophilicity, and they are used in perfumes, lotions, suspension agents for aerosols, deodorants, and nail polish (Witorsch and Thomas, 2010). They are also present in some medical devices, blood storage bags, and intravenous tubing (CDC, 2011b). Human exposure to phthalates occurs through ingestion, inhalation, and dermal contact. They have been found to be metabolized and excreted quickly (Anderson et al., 2001). Human studies have identified phthalates in amniotic fluid (Silva et al., 2004), in breast milk (Parmar et al., 1985; Dostal et al., 1987), and in urine of people of all ages (CDC, 2003, 2005; Sathyanarayana et al., 2008).

Concerns have been raised about phthalates because of evidence from laboratory animals that they can act as anti-androgens to affect the development of the male reproductive system at low levels (NRC, 2008). The age of the animal is important for the development of health effects, with the fetus being the most sensitive life stage (NRC, 2008). In 2011, diethylhexyl phthalate was reevaluated by IARC and assigned to category 2B—possibly carcinogenic to humans—because of evidence that it induces Leydig cell tumors of the testes, liver tumors, and pancreatic tumors (Grosse et al., 2011). EPA (2011b) has made a similar classification. The European Union (EU) has banned several phthalates from cosmetics, and both the EU and the United States have restricted the concentration of several phthalates in children's toys.

Data relevant to the possible role of phthalates as a risk factor for breast cancer are limited. A case–control study of 233 women with breast cancer and 221 age-matched controls in Mexico measured urinary levels of phthalates prior to treatment (Lopez-Carillo et al., 2010). After adjustment for other breast cancer risk factors, a significantly elevated risk was found with higher urinary concentrations of monoethyl phthalate (MEP), the main metabolite of diethyl phthalate (DEP) (OR = 2.20, 95% CI, 1.33–3.63). The association was stronger for younger women with premenopausal breast cancer (OR = 4.13, 95% 1.60–10.70). Statistically significant negative or inverse associations were noted for exposure to monobenzyl phthalate (MBzP) (OR = 0.46, 95% CI, 0.27–0.79) and mono (3-carboxylpropyl) phthalate (MCPP) (OR = 0.46, 95% CI, 0.27–0.79). The findings in this study may have been influenced by the fact that the measurements were from urine collected from controls at home and from cases in the hospital, where exposures to phthalates could have been greater.

Some studies have observed effects on timing of puberty, attributed to phthalates' hypothesized action as hormonally active environmental agents. Chou et al. (2009) studied pubertal timing in 30 Taiwanese girls with early thelarche (breast development), 26 with central precocious puberty,[14] and 33 normal controls. Girls with premature pubertal timing had higher (p = .005) levels of monomethyl phthalate (MMP) than controls. Monobutyl phthalate and mono-(2-ethylhexl) phthalate were not associated with premature thelarche. Wolff et al. (2010) measured a panel of nine phthalates and other endocrine disruptors prior to pubertal onset in a cohort of 1,149 ethnically diverse American girls. There was a weak and statistically nonsignificant association between early puberty and a group of low–molecular weight phthalates and a weak association with later pubic hair development and a group of high–molecular weight phthalates.

Few animal and in vitro studies have assessed the effects of phthalates in females, and few directly assess mammary tumors as endpoints, particularly for in utero and early-life exposure. Standard carcinogenesis assays that expose adult rodents to di(2-ethylhexyl)phthalate (DEHP) or diisononyl phthalate (DINP) find tumors at multiple sites, including the testes, but not the mammary gland (EPA, 1997; CPSC, 2001). A study looking at in vivo and in vitro effects of phthalates found conflicting results regarding their estrogenicity; phthalates were able to induce an estrogenic effect in breast cancer cells in vitro, but were unable to do so in an immature rat

---

[14]Girls with central precocious puberty had maturation of the breasts and external genitalia, advanced bone age, and obvious pituitary gonadotropin activity stimulating the gonads (Chou et al., 2009).

model (Hong et al., 2005). More recent studies have found that DEHP is a potent and effective ligand for activation of the constitutive androstate receptor (CAR), a ligand-activated nuclear hormone receptor. The implications of these findings are not yet clear, but they do raise a new mechanism of action for this class of compounds that might be viewed as "endocrine disrupting" in a genetic subset of the population (those with certain CAR splice variants) (DeKeyser et al., 2009, 2011). Butyl benzyl phthalate has been shown to induce genomic changes in the rat mammary gland after neonatal and prepubertal exposure (Moral et al., 2007). In utero exposure in rats affected gene expression and proliferation in the mammary gland, mainly at the beginning of puberty, and also induced more proliferating terminal end buds by age 35 days (Moral et al., 2011). Effects on male and female mammary development were also observed in rats exposed to dibutylphthalate in utero and via lactation (Lee et al., 2004). The generalizability of these findings to other phthalates is not known. Further studies regarding early-life exposures and mammary lesions related to carcinogenesis and the potential mechanisms of the effects of phthalates are necessary to understand their role as a potential risk factor for breast cancer.

Polybrominated Diphenyl Ethers

Polybrominated diphenyl ethers (PBDEs) and other brominated and chlorinated flame retardants (BFRs/CFRs) represent a large class of organohalogenated compounds that were introduced in the 1970s (ATSDR, 2004) and are widely used as flame retardants in plastics, foams, textiles, electronic devices, and building materials (Darnerud et al., 2001; Costa et al., 2008; Lorber, 2008). In the 1970s, some flame retardants were voluntarily removed from the market. This action included polybrominated biphenyls (PBBs) after humans and livestock were accidentally poisoned and brominated tris (tris(2,3-dibromopropyl) phosphate) because of concerns about children's exposures from its use in children's pajamas. Two commercial mixtures of PBDEs have recently been phased out in the United States and banned in California: penta-BDE, which was used in commercial foam, and octa-BDE, which was used in textile coatings and in certain plastics. However, a variety of mostly untested halogenated flame retardants remain on the market, some in frequent use. IARC has not evaluated PBDEs. An EPA toxicological review on one of the penta-BDEs noted, "No studies currently exist on the potential carcinogenicity of BDE-99 [2,2' 4,4'5-pentabromodiphenyl ether] in humans or experimental animals. Under the *Guidelines for Carcinogen Risk Assessment* (EPA, 2005b), there is 'inadequate information to assess the carcinogenic potential' of BDE-99 at this time" (EPA, 2008, p. 66).

Although routes of human exposure have not been well characterized

(Lorber, 2008), reports began to emerge in the late 1990s of high and rapidly rising body burden levels of PDBEs in humans (Hites, 2004; Sjodin et al., 2004; Suvorov and Takser, 2008), particularly in California (Petreas et al., 2003, 2011; Sjodin et al., 2008; Zota et al., 2008; Windham et al., 2010). The few studies of contemporary body burden levels appear to show considerable variation. Early studies concluded that all age groups had fairly similar levels of serum PBDE, except for infants and children from 0 to 4 years (Thomsen et al., 2002). Other studies have demonstrated an inverse association between age and PBDE body burden, with higher levels at younger ages, thought to be associated with breastfeeding and hand-to-mouth behavior in young children (Schecter et al., 2005; Betts, 2008; Costa et al., 2008; Rose et al., 2010). However, the NHANES data also provide some evidence for high exposure among Americans over age 60 (Sjodin et al., 2008), a disproportionate relationship that may be a result of consumers retaining PBDE-treated furniture over long periods of time (Betts, 2008).

Data on the carcinogenic potential of PBDEs in humans are extremely sparse, and to date, there have been few studies related to breast cancer. Elevated rates of total cancer, although not specifically breast cancer, have been reported among populations living in the Zhejiang province of China, an area with documented high levels of PBDE environmental contamination (Yuan et al., 2008; Zhao et al., 2008, 2009; Wen et al., 2009). Otherwise, only three small case–control studies have been published. Two Swedish studies found a modest, but statistically nonsignificant, increase in risk for non-Hodgkin's lymphoma (Hardell et al., 1998), and a statistically significant three-fold increase in risk of testicular cancer in men whose mothers had serum levels of total PBDEs above the 75th percentile (Hardell et al., 2006). A California hospital-based case–control study of breast cancer failed to find an association between measured adipose levels of total PBDEs and breast cancer, although the study was small and the use of benign breast disease controls may have resulted in overmatching, hence making it more difficult to detect an association if one existed (Hurley et al., 2011).

Deca-BDE is believed to have a lower range of toxicities than the phased-out PBDEs, but it degrades to lower brominated forms that have much longer half-lives and greater toxicity. Deca-BDE has been classified by EPA (2008) as having suggestive evidence of carcinogenic potential, based on bioassays conducted 25 years ago showing statistically significant increases in male mice of hepatocellular carcinomas and adenomas (combined incidence) and marginal increases in thyroid gland follicular cell adenomas, as well as liver nodules in male and female rats (NTP, 1986). Standard 2-year carcinogenicity bioassays for the octa- and penta-BDEs have not been conducted, but NTP plans to test hexa-BDE 153 in long-term carcinogenesis studies (NTP, 2011b).

PBDEs and their hydroxylated metabolites and breakdown products have well-established endocrine-disrupting effects (Darnerud, 2008; Legler, 2008; Mercado-Feliciano and Bigsby, 2008a,b; Talsness et al., 2008). They also may modulate sex hormone activity. For example, several PBDE congeners and hydroxylated PBDEs have been found to be estrogen agonists in cell line assays based on ER-dependent luciferase reporter gene expression (Meerts et al., 2001), and other findings have also been indicative of estrogenic activity (Mercado-Feliciano and Bigsby, 2008a,b). Antiestrogenic activity for PBDEs and metabolites has been suggested and is currently an ongoing topic of research. For example, 22 hydroxylated PBDEs were found to significantly inhibit human placental aromatase activity (Cantón et al., 2008).

At present, the epidemiologic, animal, and in vitro evidence is insufficient to assess whether PBDEs are a risk factor for breast cancer. Despite phase-out or banning of certain formulations, the ubiquitousness and persistence of many PBDEs and continuing exposures to the deca-BDEs and their degradation products indicate the need for future research on their potential relationship to breast cancer.

Industrial Chemicals

Benzene

Benzene is a colorless, highly flammable liquid of both naturally occurring and man-made origins, and it is widely used in the United States for industrial purposes. It is present in gasoline and used as a gasoline additive (ATSDR, 2011a). It is also present in tobacco smoke. Commercial production dates back to the mid-1800s (NTP, 2011a). Benzene can evaporate rapidly into the air, where it can react with other chemicals, and it is also found in water and in soil, where it can persist for longer periods (ATSDR, 2011a). Early case reports and case studies indicated an increased risk of cancer in humans, particularly acute myeloid leukemia, and repeated epidemiologic findings of associations between benzene exposure and increased risk of acute myeloid leukemia have established benzene as a known human leukemogen (IARC, 1982, 1987; Baan et al., 2009; NTP, 2011a; Zhang et al., 2011). Associations of cigarette smoking with leukemias may be due to the benzene in tobacco smoke (Korte et al., 2000). More recent epidemiologic studies have also found an association between benzene exposure and increased risk of lymphatic and hematopoietic cancers (ATSDR, 2011a).

Benzene is classified as a human carcinogen by IARC (1982, 1987). In general, however, epidemiologic studies of benzene have focused on exposure in male workers and on the risk for hematopoietic cancers; few studies have examined risks for breast cancer. A study of a cohort of 797 benzene-exposed women working in an Italian shoe factory found elevated standardized incidence and mortality ratios for breast cancer based on small numbers of cases (standardized mortality ratio 151.1, 95% CI, 78.6–290.3 for latency period ≥30 years), lending "moderate support to the hypothesis that benzene constitutes a risk factor for breast cancer" (Costantini et al., 2009, p. 8). A case–referent study of premenopausal women (ages 40 and older) in western New York state found an increased risk for women considered likely to have had moderate to high exposure to benzene (OR = 1.95, 95% CI, 1.14–3.33) (Petralia et al., 1999). Petralia et al. also found risk increased with duration of exposure. Exposure as calculated was estimated based on employment histories and job-exposure matrixes. Two studies addressed breast cancer in exposed men. A study of Danish men occupationally exposed to gasoline and combustion products found an association with the development of breast cancer, especially if time of first employment occurred before age 40 (OR = 5.4, 95% CI, 2.4–11.9) (Hansen, 2000). An increased risk was also seen among male motor vehicle mechanics in a multination European case–control study (OR = 2.1, 95% CI, 1.0–4.4) (Villeneuve et al., 2010).

In animal studies, an increase in malignant mammary tumors was observed in rats and mice exposed to benzene by inhalation (Cronkite et al., 1984; Maltoni et al., 1989) and oral routes (Maltoni et al., 1989). Benzene is metabolized to an epoxide and other active metabolites. It has been proposed to operate through a genotoxic mechanism, eliciting clastogenic effects (causing disruption or breakage of chromosomes) (Dean, 1978, 1985; IARC, 1982; ATSDR, 1997). Evidence of this phenomenon has also been demonstrated in benzene-exposed workers, with more than 20 cytogenetic studies reporting changes in structural or numerical chromosomal aberrations (ATSDR, 1997; CalEPA, 2001).

In summary, evidence in animals suggests a basis for concern regarding increased risk for breast cancer from exposure to benzene, and there is also suggestive evidence from human studies. Because benzene is a known carcinogen for other endpoints, some efforts to minimize exposure of the public and workers are in place through various regulations, including the Clean Air Act, the Clean Water Act, and occupational safety standards (NTP, 2011a). Nonetheless, benzene from ambient and indoor air can be a significant contributor to low-level environmental risk estimates for leukemia. Further research will be needed to clarify the relationship between benzene exposure and risk of human breast cancer and relevant mechanisms that may be operating. If it can be developed, stronger human evidence of increased risk for breast cancer and the mechanisms involved would have important implications for its regulation, and also would provide insights relevant for other environmental contaminants.

1,3-Butadiene

1,3-Butadiene is a gaseous hydrocarbon used primarily to make synthetic rubber and plastics such as acrylics. It is also present in gasoline, automobile exhaust, and cigarette smoke (NTP, 2011a). Exposure occurs primarily through inhalation of contaminated air and can result in effects on the nervous system or serious irritation of the eyes, nose, and throat (ATSDR, 2009). Levels are generally low in urban and suburban environments, unless near a factory producing the substance (ATSDR, 2009). 1,3-Butadiene is classified as a known human carcinogen, inducing hematopoietic cancers in occupational settings (IARC, 2008a; Baan et al., 2009; NTP, 2011a).

No human studies have evaluated the risk of breast cancer from exposure to 1,3-butadiene. Existing studies of butadiene are primarily of male workers in butadiene production and styrene butadiene rubber production. Other population-based studies have not evaluated breast cancer as an endpoint.

1,3-Butadiene causes malignant and benign mammary tumors in both mice and rats, at high and low doses (IARC, 2008). IARC (2008) found strong evidence that genotoxicity is the main mechanism for carcinogenesis. Butadiene is metabolized to DNA-reactive epoxides, and the urinary metabolites of these epoxides are observed in exposed humans. DNA adducts are observed in the lymphocytes of workers (IARC, 2008). Mutations in ras proto-oncogenes and p53 tumor suppressor genes were also identified in various butadiene tumors in mice.

Evidence in animals suggests biologic plausibility of increased risk for breast cancer from exposure to 1,3-butadiene. Because it is a known human hematopoietic carcinogen, efforts to control exposure are already in place (NTP, 2011a). While a finding of breast cancer in occupationally exposed women would have a significant impact for understanding the potential for chemicals to cause cancer, cohorts of heavily exposed women would be difficult to find and study.

Polychlorinated Biphenyls

PCBs are considered persistent organochlorines, and they include 209 possible forms or congeners (Calle et al., 2002). PCBs have been extensively used in the United States as industrial chemicals for purposes ranging from dielectric fluids to plasticizers to pesticide extenders to lubricants, and in consumer goods, but their U.S. production was ended in 1977 (Calle et al., 2002). Although PCBs are no longer produced, environmental contamination remains from old sealants, paints, transformers, and waste material (EFSA, 2010). PCBs bind strongly to soil and can also be taken up by small organisms and fish (ATSDR, 2001). The lipophilicity of PCBs allows them to concentrate in the food chain, accumulate in the body, and resist metabolism (Hunter et al., 1997). IARC (1987) has classified PCBs as probably carcinogenic to humans and characterized the epidemiologic evidence as "limited."

The interest in PCBs as a potential risk factor for breast cancer is because of their (1) persistence in the body, (2) estrogenic and endocrine disrupting properties, and (3) tumorigenic effects in animals (Moysich et al., 2002). Although PCBs have been extensively studied, the epidemiologic evidence for a link to breast cancer is inconsistent (Helzlsouer et al., 1999; Snedeker, 2001; Laden et al., 2002; Negri et al., 2003; Starek, 2003; Lopez-Cervantes et al., 2004; Brody et al., 2007; Gatto et al., 2007; Iwasaki et al., 2008; Salehi et al., 2008; Golden and Kimbrough, 2009; Itoh et al., 2009; Silver et al., 2009; Xu et al., 2010). A number of meta-analyses (Laden et al., 2002; Lopez-Cervantes et al., 2004; Salehi et al., 2008) have concluded that overall, there is no association. It is not clear whether the exposure periods studied, usually from adult life and a relatively short time before the diagnosis of breast cancer, are the most plausible from a life course perspective on breast development.

A more consistent pattern is emerging from studies addressing the degree to which polymorphisms in the cytochrome P-450 1A1 (CYP1A1) gene may influence the relation between PCB exposure and breast cancer risk. Several studies have reported elevated risks associated with high PCB levels among women with the CYPA1-m2 genotype (Moysich et al., 1999; Laden et al., 2002; Zhang et al., 2004; Li et al., 2005). Such polymorphisms were associated with a statistically significant increased breast cancer risk among women with elevated body burdens of PCBs; no correlation was found in women with low serum levels (Moysich et al., 1999). Findings regarding genetic polymorphisms and susceptibility to breast cancer risk are still preliminary and require further study; they are discussed further in Chapter 4.

Some evidence shows that PCB exposures in utero or in early life may influence pubertal development, but these relationships are not clear. Some studies have suggested delayed menarche and breast development in girls with higher blood levels of some PCB congeners (Den Hond et al., 2002; Wolff et al., 2008), but others have suggested no association with maternal levels (Gladen et al., 2000; Vasiliu et al., 2004).

Study of PCBs is complicated by the abundance of congeners, some with estrogenic and some with antiestrogenic properties. Epidemiologic studies have not been able to adequately consider ways in which different forms of PCBs might interact synergistically or antagonistically to influence breast cancer risk (Calle et al., 2002; Brody et al., 2007; Salehi et al., 2008). Furthermore, measurement of PCB levels at the time a breast cancer is diagnosed or at any single point will not adequately represent past exposure history because factors such as weight change and lactation history will influence metabolism and excretion (Verner et al., 2011), while changes in behaviors could alter exposures, particularly through the food chain.

Long-term animal carcinogenesis studies on mixtures of PCBs or specific congeners have found associations with increased liver tumors, but they have not found increases in mammary tumors (NTP, 2006a,b). However, the studies have been conducted with adult rats, and most studies have not assessed the effect of PCB exposure at earlier ages.

The large number of epidemiologic studies on this topic demonstrates consistency in showing no overall effect of PCB exposures on breast cancer risk. However, exposure was assessed in most cases in adult life, often during the period after PCB production ceased, when body burdens were declining. Some recent work suggests that women inheriting a variant of the cytochrome P-450 1A1 gene may be at higher risk for breast cancer from elevated PCB levels. A few investigations into early-life exposures have examined intermediate outcomes. Further research on early-life exposures and/or genetically defined subsets may be warranted.

Ethylene Oxide

Ethylene oxide, a colorless gas with a distinct odor, is used primarily for industrial and medical sterilization (IARC, 2008). Exposure to ethylene oxide occurs mainly in the workplace or in hospital settings. It is classified as a human carcinogen by both IARC (2008; Baan et al., 2009) and NTP (2011a) on the basis of a mix of evidence from epidemiologic, animal, and mechanistic studies. Mechanistic evidence of genotoxicity was a critical component of the IARC assessment.

IARC's review characterized the overall body of epidemiologic evidence on the carcinogenicity of ethylene oxide as "limited" (IARC, 2008; Baan et al., 2009). The studies specifically concerning breast cancer incidence had varied results, with some finding no association and others finding a borderline significant excess risk (Norman et al., 1995). The study considered the most informative (Steenland et al., 2003) examined the breast cancer experience of a large occupational cohort. Risk among women with the highest level of exposure was significantly higher (OR = 1.74, 95% CI, 1.16–2.65) compared with women who had no exposure. This risk remained high (OR = 1.87, 95% CI, 1.12–3.10) among a subset of women for whom information on parity and history of breast cancer in a first degree relative was available for calculation of an adjusted odds ratio.

In peer-reviewed inhalation studies by NTP (1987), incidence of adenocarcinoma or adenosquamous carcinoma of the mammary gland were found elevated in female mice in the low-dose group. The finding in the high-dose group was marginally increased. In vitro and mechanistic findings have been extensive. Ethylene oxide is an epoxide, and various epoxides, or chemicals metabolized to epoxides, have been found to cause malignant mammary tumors in laboratory animal studies (Melnick and Sills, 2001). Ethylene oxide has been shown to cause point mutations in ras proto-oncogenes and the p53 tumor suppressor gene (Houle et al., 2006). IARC (2008, p. 286) concluded that "the genotoxicity data in experimental systems consistently demonstrate that ethylene oxide is a mutagen and clastogen across all phylogenetic levels tested."

There are insufficient data to determine whether ethylene oxide exposure during different life stages has a role in altering breast cancer risk. Nevertheless, the limited epidemiologic research on this compound does provide some support for an effect from adult exposures, and the animal bioassay data and the compound's mechanism of action provide biological plausibility for the compound being a risk factor for breast cancer.

Vinyl Chloride

Vinyl chloride, also known as chloroethene, chloroethylene, and ethylene monochloride, is a colorless gas with a mild odor that is used in the production of plastics. Exposure occurs primarily in occupational settings via inhalation (ATSDR, 2006), and low-level environmental exposures occur through contaminated drinking water and in ambient air near manufacturing facilities. Vinyl chloride was once used as a propellant in hair sprays, deodorants, and other consumer products, but this use was phased out in the 1970s. IARC (2008) classifies vinyl chloride as carcinogenic to humans, with the human evidence showing cancers in the liver.

Data from human studies have not been adequate to evaluate a relationship between vinyl chloride and breast cancer. IARC (2008, p. 372) stated that "although concern has been raised about a potential association between exposure to vinyl chloride and the risk for breast cancer, human studies to date are not informative on this issue because of the very small numbers of women included." An earlier review by the WHO International Program on Chemical Safety similarly concluded that a substantial body of epidemiologic studies with which to assess vinyl chloride is not available and would be difficult to conduct because women in most Western countries have little or no exposure to vinyl chloride, occupational or otherwise (IPCS, 1999). Vinyl chloride has been extensively tested for carcinogenicity in laboratory animals. The animal evidence on vinyl chloride was recently summarized by IARC (2008). Many of the papers from 1976 to 1983 found mammary adenocarcinomas in mice and mammary tumors in rats upon inhalation of vinyl chloride.

Mechanistic studies show that vinyl chloride is oxidized to chloroethylene oxide, which can rearrange to chloroacetaldehyde, and that these metabolites can react with nucleic acid bases to form DNA adducts in animals, which can initiate the genotoxic damage leading to carcinogenesis (IARC, 2008). There is, however, a paucity of data on the occurrence of such adducts in vinyl chloride-exposed humans. The mechanism that leads to base misincorporation following adduct formation is still unclear. Similarly, data are insufficient to draw a conclusion about the effects of timing of exposure to vinyl chloride on breast cancer.

Although considerable animal evidence indicates that the potential for induction of breast cancer from vinyl chloride is biologically plausible, the lack of substantial exposure opportunities for women makes this compound a low priority for future research.

Pesticides

DDT/DDE

Dichlorodiphenyltrichloroethane (DDT), an insecticide used extensively over the past century, was banned in the United States and other developed countries in the early 1970s because of its adverse ecological impacts. DDT and its major metabolite dichlorodiphenyldichloroethylene (DDE) have persistent and lipophilic properties that led to bioconcentration through the food chain. Because of continued use of DDT in developing countries for malarial control and the very long environmental half-life of DDE, these compounds remain present in the environment and in the population today (Petreas et al., 2004; CDC, 2008; Woodruff and Morello-Frosch, 2011).

Neither DDT nor DDE are mutagenic, but both possess estrogenic properties. Although structurally similar, there are substantial differences in the endocrine activity of DDT and DDE. Li et al. (2008) demonstrated that both p,p'-DDE and p,p'-DDT exhibited agonist activity toward ER-alpha, but DDE acted as an antagonist to both androgen and progesterone receptors, and p,p'-DDT had no effect on the progesterone receptor. There is good consensus among expert agencies regarding DDT's potential carcinogenicity. IARC (1991) has classified DDT as "possibly carcinogenic to humans" (Group 2B); NTP has classified it as "reasonably anticipated" to be a human carcinogen (NTP, 2011a); and EPA has classified it as a "probable" human carcinogen (EPA, 2011c). Such classifications, however, are not specific to breast cancer.

Of the organochlorine pesticides, DDT/DDE has been one of the most studied for risk of breast cancer in humans, with numerous epidemiologic studies over the past decade. Several reviews (Snedeker, 2001; Calle et al.,

2002; Brody and Rudel, 2003; Brody et al., 2007) and a careful meta-analysis of 22 studies (Lopez-Cervantes et al., 2004) concluded that evidence was insufficient to infer a risk of breast cancer from DDT exposure.

As in most studies of cancer in relation to environmental chemicals, DDT/DDE exposure levels for most of the studies exploring risk for breast cancer were based on measurements in biologic samples taken near the time of diagnosis for cases, or at a similar time for noncases. A much-cited exception is a California study of 129 women with and 129 women without a diagnosis of breast cancer for whom archived blood samples drawn in the 1960s were assayed for levels of DDT/DDE (Cohn et al., 2007). In this study, although there was no evidence of an association between DDT/DDE exposure and breast cancer in general, the small subset of women who would have been under age 14 in 1945 (a time of peak DDT use) had a statistically significant five-fold higher risk. Although provocative in the context of potential windows of exposure, some skepticism has been expressed about the interpretation of these results because the high exposures that the "baby boomer" generation would have experienced might be expected to predict increasing rates of breast cancer in that birth cohort, but on the contrary their rates have been declining (Tarone, 2008). A previous nested case–control study that also used serum specimens drawn in the 1960s found no association with breast cancer risk, but the analysis did not stratify by birth cohort (Krieger et al., 1994). Similarly, no association was seen in a study that used blood samples obtained in 1974, 20 years before case status (Helzlsouer et al., 1999). A prospective study from Japan found no evidence for higher levels of DDT/DDE at a baseline measurement among the 144 women who had developed breast cancer during follow-up than among the controls (Iwasaki et al., 2008). Neither the Helzlsouer nor the Iwasaki study reported data on exposure concentrations assessed before adulthood.

In vivo animal data provide little support for the hypothesis that DDT or its metabolites could increase breast cancer risk in humans (NTP, 1978; IARC, 1991). However, such studies typically do not include early-life exposures. DDT and DDE are not mutagenic, but both have estrogenic activity (Andersen et al., 1999; Snedeker, 2001). Evidence also shows that administering DDT together with the known carcinogen DMBA can induce cellular and chromosomal alterations in the rat mammary gland (Uppala et al., 2005).

While the role of DDT in breast cancer risk remains unclear, it is possible that early-life exposures to this legacy chemical may play a role in the development of disease.

Dieldrin and Aldrin

Dieldrin and aldrin are persistent organochlorines. Aldrin breaks down to dieldrin in the body and in the environment, and they are closely related in structure. Until the 1970s, they were widely used as insecticides to control damage to crops, but concerns about damaging effects to the environment and health led EPA to ban dieldrin and aldrin for agricultural uses in 1970 and for all uses in 1987 (CDC, 2009b; ATSDR, 2011b). Because dieldrin persists in soil and is a water contaminant, exposure may occur by eating contaminated food (Snedeker, 2001; ATSDR, 2011b). Body burdens of dieldrin have declined, but are still measurable in U.S. adults (CDC, 2009b) due to its high lipophilicity and long biological half-life.

Epidemiologic evidence regarding exposure to dieldrin and subsequent risk of breast cancer is limited and often conflicting. Much of the early interest in dieldrin as a potential risk factor for breast cancer followed publication of the Copenhagen City Heart Study, a prospective study of 7,712 women with 268 cases of breast cancer in 17 years of follow-up (Hoyer et al., 1998). On the basis of serum samples from women who were exposed in the late 1970s, the women in the highest quartile of exposure had twice the risk of breast cancer when compared to the women in the lowest quartile. However, a prospective cohort study of 7,224 Missouri women serum donors was unable to find a similar association with breast cancer risk among 105 breast cancer cases identified during up to 9.5 years of follow-up (Dorgan et al., 1999). A subsequent population-based, case–control study found no substantial elevation in breast cancer risk in relation to the highest quintile of lipid-adjusted serum levels of dieldrin (Gammon et al., 2002).

Animal evidence on dieldrin exposure and mammary gland cancer is also insufficient to reach conclusions regarding hazard. Studies of carcinogenicity in mice via oral administration tend to demonstrate hepatic carcinogenicity as the primary effect (IARC, 1987, p. 185). Although xenoestrogenic potential has been a hypothesized mechanism for dieldrin, it is at best a weak estrogen whose estrogenic potential has not been adequately demonstrated (Snedeker, 2001). With the E-SCREEN assay, which assesses cellular proliferation in estrogen-dependent breast tumor cells, dieldrin is able to induce cellular proliferation only in the highest concentration that can be tested (Snedeker, 2001).

The potential influence of timing of exposure to dieldrin is difficult to assess. Most of the epidemiologic studies have relied on the levels of dieldrin in serum collected after the subject had developed breast cancer, so there is little information that can address whether timing of exposure is important. However, because dieldrin is no longer used and tissue and environmental levels are declining, the committee does not see this area as a priority for additional research.

Atrazine and S-Chloro Triazine Herbicides

Atrazine is an S-chloro triazine herbicide used extensively in U.S. agriculture. Low-level contamination of groundwater with atrazine and other triazine herbicides is fairly common; as a result, its potential health effects have been the subject of substantial scrutiny. Exposure to atrazine via diet is very low. The primary nonoccupational route of exposure is through contamination of drinking water supplies. Such contamination is common, but based on monitoring carried out by EPA (2010a), it is usually at levels that are very low from a population risk perspective. The NHANES III study failed to identify atrazine metabolites in the urine in any of more than 4,000 samples collected between 1999 and 2004 (CDC, 2009b). Although many contaminants of groundwater persist for long periods once present, repeated analysis of atrazine-contaminated aquifers demonstrates that it does not generally persist. Thus exposures via groundwater, when they occur, are likely to be periodic. For example, samples of a drinking water supply in Ohio found no detectable atrazine (<2 ppb) in March, but a strong peak at 36 ppb in mid-April, with levels returning below the detection limit by mid-May (EPA, 2010a). Similar patterns have been seen in other water supplies (EPA, 2010a).

In reviews examining the risk of cancer in general, IARC found atrazine to be not classifiable regarding carcinogenicity in humans (IARC, 1999), and EPA found atrazine unlikely to cause cancer in humans (EPA, 2010a). In 2009, EPA began a reevaluation of the health effects of atrazine; that effort is ongoing (EPA, 2010a).

The few human studies examining atrazine as a potential risk factor for breast cancer have not indicated an association (Sathiakumar et al., 2011). However, most of the studies have been ecological in nature or otherwise would have had difficulty discerning an effect (e.g., studies carried out in occupational populations with few women).

Atrazine does not have direct estrogenic activity, but may indirectly modulate sex hormone levels by affecting steroidogenesis (Fan et al., 2007; Higley et al., 2010; Tinfo et al., 2011). Results of studies in animals have been complicated by findings that atrazine administered to Sprague Dawley female rats affects neuroendocrine pathways to accelerate reproductive senescence and cause mammary tumors not observed in mice or other rat strains (IARC, 1999; Rayner et al., 2005; Enoch et al., 2007; EPA, 2009a; Davis et al., 2011; Hovey et al., 2011). The hormonal manifestations of reproductive aging in humans are very different from those of Sprague Dawley rats, so this mechanism is not thought to be relevant to humans (IARC, 1999). Similar conclusions were drawn regarding a chloro-S-triazine herbicide, cyanazine, from a 2-year bioassay in Sprague Dawley rats (Bogdanffy et al., 2000).

No epidemiologic studies have examined the effects of timing of exposure to atrazine. There are conflicting data from animal studies regarding whether low-dose atrazine exposures in utero can contribute to developmental abnormalities of mammary tissue in offspring (IARC, 1999; EPA, 2009a). Collectively, these data indicate that maternal atrazine exposure has no long-term effects on mammary gland development in female offspring beyond a transitory response to high doses. However, the degree to which atrazine may have effects by modulating steroidogenesis remains an area for further study.

PAHs

Polycyclic aromatic hydrocarbons, or PAHs, exist in more than 100 forms. They are formed from incomplete burning of coal, oil, gas, wood, tobacco, and other organic substances. They are also produced by high-temperature cooking. Humans can be exposed to PAHs through industrial and urban air pollution, tobacco smoke, and diet.

Evaluation of the carcinogenicity of PAHs is complicated by the hundreds of forms of PAHs with differing compositions and properties. An IARC review evaluated evidence through 2005 on 60 PAH compounds, with separate classifications for individual PAH compounds (IARC, 2010d). Benzo(a)pyrene (BaP) was declared carcinogenic to humans (Group 1) "based on sufficient evidence in animals and strong evidence that the mechanisms of carcinogenesis in animals also operate in exposed human beings" (IARC, 2005, p. 23). Cyclopenta[cd]pyrene, dibenz[a,h]anthracene, and dibenzo[a,l]pyrene were classified as probably carcinogenic to humans (Group 2A) based on sufficient evidence in animals and compelling genotoxicity evidence. IARC also found sufficient evidence in humans for the carcinogenicity of a variety of occupational exposures involving PAHs (i.e., during coal gasification, coke production, coal-tar distillation, paving and roofing, and chimney sweeping). The main epidemiologic findings were of increased risk of lung or skin cancer but not breast cancer. Typically, however, such studies are dominated by men and inhalation or dermal exposure.

PAHs' effects on breast cancer risk have been evaluated in a number of noteworthy epidemiologic studies published since 2005, but the results have been inconsistent. A meta-analysis of 10 dietary studies as well as a large prospective cohort study with 8 years of follow-up and 3,818 cases of invasive breast cancer found no correlation between darkly cooked meats and breast cancer (Steck et al., 2007; Kabat et al., 2009). A few studies have attempted to elucidate risks from specific time periods of exposure. A case–control study from western New York used historical levels of total suspended particulates (TSPs) in the air as a proxy for PAH exposure. Residential histories were used to link study participants to TSP levels at specific times in their lives (e.g., birth, menarche). In women with postmenopausal breast cancer, potential exposure to high concentrations of TSPs at birth was associated with an elevated risk that was on the border-line of significance (OR = 2.4, 95% CI, 0.97–6.09), although the relation-ship could have been related to unmeasured confounding factors (Bonner et al., 2005). A more recent study from the same group examined exposure to traffic emissions at specific times on the basis of residence (Nie et al., 2007). Higher exposure at the time of menarche was associated with increased risk for premenopausal breast cancer (OR = 2.05, 95% CI, 0.92–4.54, p for trend = .03). Higher exposures at the time a woman had her first birth were associated with a significantly increased risk for postmenopausal breast cancer (OR = 2.57, 95% CI, 1.16–5.69, p for trend = .19) (Nie et al., 2007).

PAHs' effects on DNA have been explored in a series of case–control studies in the Long Island Breast Cancer Study. The presence of PAH-DNA adducts, which form after exposure to PAHs and are measured in lympho-cytes, were associated with a 29 to 35 percent increase in the risk of breast cancer, with no dose–response relationship (Gammon et al., 2002, 2004). A later analysis in the same study confirmed slightly elevated risks (HR = 1.2, 95% CI, 0.63–2.28) for breast cancer–specific mortality associated with PAH-DNA adducts (Sagiv et al., 2009). In contrast to the generally positive studies from Long Island (Gammon and Santella, 2008), results from the Shanghai Women's Health Study (354 cases, 708 controls) found no asso-ciation between PAH metabolites and oxidative stress markers and breast cancer (Lee et al., 2010). Thus, overall results of epidemiologic studies of PAHs and breast cancer have relied on indirect measures of exposure and been inconsistent.

Inconsistencies in the results from epidemiologic findings on PAHs fol-low from a number of limitations. Case–control designs depend on respon-dent recall of information on diet, smoking, and environmental exposures from the past, proxy measures of exposure, or assays of measures of PAH exposure (PAH-DNA adducts) after the diagnosis of breast cancer. In addi-tion, PAH-DNA adducts may be a measure of exposure rather than of the host's biologic response to PAH. Although the studies from western New York have been generally consistent in their levels of risk estimates, and the studies from the Long Island Breast Cancer Study and western New York have linked PAH-DNA adducts to breast cancer and suggested a number of molecular mechanisms, including gene–environment interactions (Gammon and Santella, 2008), epidemiologic studies of PAHs in breast cancer etiology provide modest support for their carcinogenicity in human breast cancer.

Biologic mechanisms by which PAHs may affect breast cancer risk have been explored rather extensively. PAHs have often been implicated as inducers of mammary tumors in rodents (Tannheimer et al., 1997). However, some of the earlier, most cited evidence of rodent carcinogenicity involved direct applications of PAHs to the mammary gland (Cavalieri et al., 1988; IARC, 2010d). Studies have shown that PAHs are aryl hydrocarbon receptor (AhR) agonists that bind and activate AhR, a receptor that regulates xenobiotic metabolism and initiates homeostatic responses. The nature of the response to AhR binding is specific to the compound bound. AhR affects the expression of CYP 1 enzymes involved in the metabolism of PAHs (IARC, 2010d), and this is hypothesized to lead to greater formation of active metabolites and ultimately DNA mutations (Kemp et al., 2006). Cross talk of AhR with steroid and nuclear receptors can affect many estrogen-dependent pathways, and this cross talk can be influenced by an AhR ligand to PAHs (Hockings et al., 2006). PAHs exhibit weak estrogenic and antiestrogenic activity (Santodonato, 1997), and BaP weakly binds to estrogen receptor α (Pliskova et al., 2005). It is difficult to extrapolate such findings to the potential for breast cancer following human systemic exposure.

However, it is clear that, following oral exposure, various carcinogenic PAHs, including BaP, are absorbed and widely distributed to most tissues, and that PAHs are gradually taken up and also released by fatty tissues (IARC, 2010d), such as mammary tissue. Various enzymes involved in metabolizing carcinogenic PAHs such as BaP to epoxides (e.g., CYP1A1, CYP1B1, and epoxide hydrolase involved in forming diol epoxides [IARC, 2010d]) are present in human breast (Williams and Phillips, 2000). Numerous studies demonstrate and characterize covalent DNA-adducts formed in human mammary tissues from donors or various established cell lines exposed to certain carcinogenic PAHs. Mechanistic and in vitro studies are difficult to interpret due to the complexity of the carcinogenesis. There is a strong chain of mechanistic evidence linking BaP exposure to the cause of a specific mutation in human lung cancer, which, together with numerous studies demonstrating animal carcinogenesis, led IARC (2010d) to declare BaP to be carcinogenic to humans. While overall the mechanistic evidence on various PAHs support the biological plausibility that they may influence breast cancer risk, all the elements of such a chain are not present for any of the PAHs and breast cancer. Animal studies on PAHs have not sufficiently addressed breast cancer endpoints or mammary tumors, and further investigation is required to specifically address carcinogenicity in the mammary gland. Future epidemiologic, in vivo, and in vitro research is needed to further assess the role of PAHs in breast cancer etiology.

Dioxins

The dioxins are a family of highly persistent, lipophilic, and toxic by-products of industrial processes and incineration. The dioxin-like compounds include various furans and coplanar PCBs, but the congener 2,3,7,8-tetrachlorodibenzo-*para*-dioxin (TCDD) is considered the most potent of the dioxins and dioxin-like chemicals (CDC, 2009b) and has been a major focus of concerns about carcinogenicity. The release of dioxins into the environment has declined since the 1970s, and average tissue concentrations in U.S. adults also appear to have declined (CDC, 2009b). In NHANES data from 2003–2004, mean TCDD levels were below the limit of detection (CDC, 2009b). TCDD has been classified by IARC (1997; Baan et al., 2009) as a human carcinogen (Group 1) and by EPA as carcinogenic to humans (EPA, 2000), although the classification of dioxins as "known human carcinogens" by IARC and EPA remains controversial (NRC, 2006). Evidence implicating TCDD and related dioxins as human carcinogens has primarily been based on overall excess cancer mortality in highly exposed occupational cohorts of men and on elevated incidence of some cancers among residents of Seveso, Italy, who experienced high levels of exposure from a major 1976 industrial accident.

Evidence regarding an association between dioxin exposure and breast cancer is more limited. Repeated reviews have found the epidemiologic evidence on the relation between TCDD exposure and breast cancer, including data on the experience of the Seveso residents, inconclusive (IOM, 2011). Follow-up of the Seveso population over 20 years has not found an excess of breast cancer, although the eight observed cases among the small population living in the most contaminated area were more than the expected number (RR = 1.43, 95% CI, 0.71–2.87) (Pesatori et al., 2009). A significant elevation of breast cancer, based on 15 cases, was initially reported associated with a measured 10-fold increase in TCDD levels in blood samples collected from women enrolled in the Seveso Women's Health Study (crude HR = 2.1, 95% CI, 1.0–4.6), a cohort of 981 young women (ages 0–40) enrolled shortly after the Seveso incident in 1976 (Warner et al., 2002). In a 32-year follow-up of the Seveso cohort, with 33 diagnosed breast cancer cases and with adjustment for other risk factors for breast cancer, the risk association was very similar to that from the population-based study, and no longer statistically significant (HR = 1.44, 95% CI, 0.89–2.33) (Warner et al., 2011). Two small hospital-based case–control studies have found that levels of dioxins measured in adipose samples from women undergoing surgery for breast cancer or for benign breast conditions were not significantly different between the cases and controls (Hardell et al., 1996; Reynolds et al., 2005).

Several large and well-conducted TCDD-related cancer bioassays

(Kociba et al., 1978; NTP, 1982a,b, 2004) have reported induction of several types of cancer in both rats and mice. In all studies in which TCDD elicited an increase in tumors, the increase was site specific, most frequently the liver. Mammary tumors were not increased in any study. In some studies, mammary gland tumors in Sprague Dawley rats were significantly reduced at the highest doses (Kociba et al., 1978; NTP, 2006b). Thus, although evidence is clear that TCDD causes liver tumors in experimental animals, none of the standard 2-year in vivo animal oncogenicity bioassays have identified the mammary gland as a target for carcinogenesis from dioxins alone. In all these studies, exposure began when the animals were weaned. However, a single dose of 1 μg/kg TCDD on day 15 of gestation produced alterations in terminal end buds and fewer lobules in 50-day-old offspring. Although this prenatal TCDD treatment did not alter the labeling index in the mammary terminal ductal structures of 21- and 50-day-old rats, it did result in an increase in the number of chemically induced mammary adenocarcinomas in rats (Brown et al., 1998). Other studies have also shown that early-life exposure to TCDD can alter mammary gland development (Brown and Lamartiniere, 1995; Vorderstrasse et al., 2004; Wang et al., 2011). Thus, the potential for exposure to TCDD and other dioxins to alter mammary gland development early in life cannot be excluded.

TCDD and other dioxins are generally not mutagenic and do not bind to the estrogen receptor, although one study found that TCDD can induce oxidative stress and subsequent DNA strand breaks in MCF7 breast cancer cells (Lin et al., 2007).

Dioxins' mode of action as a putative hepatocarcinogen requires binding and activation of the AhR, which causes a cascade of downstream effects on gene expression for genes involved in a variety of biological processes. Whether such changes in AhR-mediated gene expression might alter mammary tumor development later in life has been studied in animals. Two in vivo studies examining whether early-life exposure to dioxins can increase the incidence of carcinogen-initiated mammary tumors did not provide evidence of such an effect (Desaulniers et al., 2004; Wang et al., 2011). It has also been hypothesized that through interactions with other factors, early-life exposure to dioxins may modify mammary gland development and eventually tumorigenesis. A novel mouse experiment that combined maternal TCDD exposure and a high-fat diet in mothers and offspring found that this combined exposure increased mammary cancer incidence in the offspring by two-fold after oral administration of a standard cancer-inducing agent (La Merrill et al., 2010). The maternal oral TCDD dose was high relative to human intakes, resulting in less sequestration by the maternal liver and proportionally more fetal exposure than would be seen at lower doses or from a similar cumulative dose from chronic repeated exposure at lower doses (Bell et al., 2007). These data are indicative of a potential hazard at sufficient doses in combination with a high-fat diet, but the animal exposure experience is not directly equivalent to typical human exposures.

Although 3-methylcholanthrene has been shown to stimulate estrogen receptor alpha in several different ER response assays (Shipley and Waxman, 2006), TCDD and other dioxin analogs induced tissue-specific inhibition of estrogen-induced genes and pathways (Safe and Wormke, 2003; Safe, 2005). Indeed, several structural analogs of chlorinated dioxins have been proposed as tamoxifen-like antiestrogens for treatment of ER-negative breast cancer (Zhang et al., 2009).

Neither human nor animal evidence suggests that exposure to TCDD or other dioxin-like chemicals is directly associated with an increased risk for breast cancer. Some intriguing animal evidence suggests the possibility that early exposure to TCDD may interact with other factors, such as a high-fat diet, to alter breast cancer risks. Although human exposure to TCDD may have declined from peak levels, TCDD persists in the body, and further research may help clarify the nature of its potential interactions with other exposures.

SUMMARY

From the committee's qualitative review of relevant literature on the factors it selected, it found that the factors with the clearest evidence from epidemiologic studies of increased risk of breast cancer were combination HT products, current use of oral contraceptives, overweight and obesity among postmenopausal women, alcohol consumption, and exposure to ionizing radiation. Greater physical activity is associated with decreased risk. Some major reviews have concluded that the evidence on active smoking is consistent with a causal association with breast cancer, and other large-scale reviews describe the evidence as limited. For several other factors reviewed by the committee, the available epidemiologic evidence is less strong but suggests a possible association with increased risk: passive smoking, shift work involving night work, benzene, 1,3-butadiene, and ethylene oxide. In some cases—for example, BPA, zearalenone, vinyl chloride, and alkylphenols—human epidemiologic evidence regarding breast cancer is not available or inconclusive, but findings from animal or mechanistic studies suggest some basis for biological plausibility of an association. A few factors, such as non-ionizing radiation and personal use of hair dyes, have not been associated with breast cancer risk in multiple, well-designed human studies. For several other factors, evidence was too limited or inconsistent to reach a conclusion (e.g., nail products, phthalates). In all cases, these conclusions are based on assessments of the currently available evidence. It is always possible for new evidence to point to different conclusions, as science evolves, new methodologies are applied, and research strategies to examine timing of exposure are developed.

For this review, the evidence was typically considered singly for each chemical or mixture addressed. As discussed further in the next chapter, effects attributed to any one factor evaluated in a study may in fact be due, or due in part, to other factors that might co-occur.

For most of the factors examined, the committee's review found information on the potential for exposure at different life stages to affect risk to be limited or nonexistent. Similarly, the evidence available rarely reported on types of tumors grouped on the basis of characteristics such as hormone receptor status. The committee sees a need for future research to better reflect the growing understanding of a life course perspective whereby the potential for influencing breast cancer risk may depend exquisitely on the timing of exposure, and an appreciation of the potential for different factors to play a role in specific, etiologically distinct varieties of breast cancer based on histologic or molecular subtype.

REFERENCES

Acevedo, R., P. G. Parnell, H. Villanueva, L. M. Chapman, T. Gimenez, S. L. Gray, and W. S. Baldwin. 2005. The contribution of hepatic steroid metabolism to serum estradiol and estriol concentrations in nonylphenol treated MMTVneu mice and its potential effects on breast cancer incidence and latency. *J Appl Toxicol* 25(5):339–353.

Ahn, J., A. Schatzkin, J. V. Lacey, Jr., D. Albanes, R. Ballard-Barbash, K. F. Adams, V. Kipnis, T. Mouw, et al. 2007. Adiposity, adult weight change, and postmenopausal breast cancer risk. *Arch Intern Med* 167(19):2091–2102.

AHRQ. (Agency for Healthcare Research and Quality). 2010. *Alcohol consumption and cancer risk: Understanding possible causal mechanisms for breast and colorectal cancers.* Prepared by O. Oyesanmi, D. Snyder, N. Sullivan, J. Reston, J. Treadwell, and K. M. Schoelles. AHRQ Pub. No. 11-E003. Rockville, MD: AHRQ. http://www.ahrq.gov/downloads/pub/evidence/pdf/alccan/alccan.pdf (accessed October 31, 2011).

Ambrosone, C. B., S. Kropp, J. Yang, S. Yao, P. G. Shields, and J. Chang-Claude. 2008. Cigarette smoking, N-acetyltransferase 2 genotypes, and breast cancer risk: Pooled analysis and meta-analysis. *Cancer Epidemiol Biomarkers Prev* 17(1):15–26.

Andersen, H. R., A. M. Andersson, S. F. Arnold, H. Autrup, M. Barfoed, N. A. Beresford, P. Bjerregaard, L. B. Christiansen, et al. 1999. Comparison of short-term estrogenicity tests for identification of hormone-disrupting chemicals. *Environ Health Perspect* 107(Suppl 1):89–108.

Anderson, L. E., J. E. Morris, L. B. Sasser, and R. G. Stevens. 2000a. Effect of constant light on DMBA mammary tumorigenesis in rats. *Cancer Lett* 148(2):121–126.

Anderson, L. E., J. E. Morris, L. B. Sasser, and W. Loscher. 2000b. Effects of 50- or 60-hertz, 100 μT magnetic field exposure in the DMBA mammary cancer model in Sprague-Dawley rats: Possible explanations for different results from two laboratories. *Environ Health Perspect* 108(9):797–802.

Anderson, W. A., L. Castle, M. J. Scotter, R. C. Massey, and C. Springall. 2001. A biomarker approach to measuring human dietary exposure to certain phthalate diesters. *Food Addit Contam* 18(12):1068–1074.

Anderson, G. L., M. Limacher, A. R. Assaf, T. Bassford, S. A. Beresford, H. Black, D. Bonds, R. Brunner, et al. 2004. Effects of conjugated equine estrogen in postmenopausal women with hysterectomy: The Women's Health Initiative randomized controlled trial. *JAMA* 291(14):1701–1712.

Anisimov, V. N., D. A. Baturin, I. G. Popovich, M. A. Zabezhinski, K. G. Manton, A. V. Semenchenko, and A. I. Yashin. 2004. Effect of exposure to light-at-night on life span and spontaneous carcinogenesis in female CBA mice. *Int J Cancer* 111(4):475–479.

ATSDR (Agency for Toxic Substances and Disease Registry). 1997. *Toxicological profile for benzene.* Atlanta, GA: U.S. Department of Health and Human Services Public Health Service, ATSDR.

ATSDR. 2001. *ToxFAQs for polychlorinated biphenyls (PCBs).* http://www.atsdr.cdc.gov/toxfaqs/tf.asp?id=140&tid=26 (accessed December 22, 2011).

ATSDR. 2004. *Toxicological profile for polybrominated biphenyls and polybrominated diphenyl ethers.* http://www.atsdr.cdc.gov/ToxProfiles/TP.asp?id=529&tid=94 (accessed October 24, 2011).

ATSDR. 2005. *Toxicological profile for nickel.* http://www.atsdr.cdc.gov/ToxProfiles/tp.asp?id=245&tid=44 (accessed November 2, 2011).

ATSDR. 2006. *Toxicological profile for vinyl chloride.* http://www.atsdr.cdc.gov/toxprofiles/tp.asp?id=282&tid=51 (accessed October 24, 2011).

ATSDR. 2007. *Toxicological profile for arsenic.* http://www.atsdr.cdc.gov/ToxProfiles/tp.asp?id=22&tid=3 (accessed November 2, 2011).

ATSDR. 2008a. *Toxicological profile for cadmium.* http://www.atsdr.cdc.gov/ToxProfiles/tp.asp?id=48&tid=15 (accessed November 2, 2011).

ATSDR. 2008b. *Toxicological profile for chromium.* http://www.atsdr.cdc.gov/ToxProfiles/tp.asp?id=62&tid=17 (accessed November 2, 2011).

ATSDR. 2009. *ToxFAQs for 1,3-butadiene.* http://www.atsdr.cdc.gov/toxfaqs/tf.asp?id=458&tid=81 (accessed November 2, 2011).

ATSDR. 2011a. *ToxFAQs for benzene.* http://www.atsdr.cdc.gov/toxfaqs/tf.asp?id=38&tid=14 (accessed October 24, 2011).

ATSDR. 2011b. *Toxicological profile for aldrin/dieldrin.* http://www.atsdr.cdc.gov/toxprofiles/tp.asp?id=317&tid=56 (accessed October 24, 2011).

Baan, R., K. Straif, Y. Grosse, B. Secretan, F. El Ghissassi, V. Bouvard, A. Altieri, and V. Cogliano. 2007. Carcinogenicity of alcoholic beverages. *Lancet Oncol* Apr(8):292–293.

Baan, R., K. Straif, Y. Grosse, B. Secretan, F. El Ghissassi, V. Bouvard, L. Benbrahim-Tallaa, and V. Cogliano. 2008. Carcinogenicity of some aromatic amines, organic dyes, and related exposures. *Lancet Oncol* 9(4):322–323.

Baan, R., Y. Grosse, K. Straif, B. Secretan, F. El Ghissassi, V. Bouvard, L. Benbrahim-Tallaa, N. Guha, et al. 2009. A review of human carcinogens—Part F: Chemical agents and related occupations. *Lancet Oncol* 10(12):1143–1144.

Barnes, B. B., K. Steindorf, R. Hein, D. Flesch-Janys, and J. Chang-Claude. 2010. Population attributable risk of invasive postmenopausal breast cancer and breast cancer subtypes for modifiable and non-modifiable risk factors. *Cancer Epidemiol* 35(4):345–352.

Baumgartner, K. B., T. J. Schlierf, D. Yang, M. A. Doll, and D. W. Hein. 2009. N-acetyltransferase 2 genotype modification of active cigarette smoking on breast cancer risk among Hispanic and non-Hispanic white women. *Toxicol Sci* 112(1):211–220.

Bell, D. R., S. Clode, M. Q. Fan, A. Fernandes, P. M. Foster, T. Jiang, G. Loizou, A. MacNicoll, et al. 2007. Relationships between tissue levels of 2,3,7,8-tetrachlorodibenzo-p-dioxin (TCDD), mRNAs, and toxicity in the developing male Wistar(Han) rat. *Toxicol Sci* 99(2):591–604.

Belli, P., C. Bellaton, J. Durand, S. Balleydier, N. Milhau, M. Mure, J. F. Mornex, M. Benahmed, et al. 2010. Fetal and neonatal exposure to the mycotoxin zearalenone induces phenotypic alterations in adult rat mammary gland. *Food Chem Toxicol* 48(10):2818–2826.

Benbrahim-Tallaa, L., E. J. Tokar, B. A. Diwan, A. L. Dill, J.-F. Coppin, and M. P. Waalkes. 2009. Cadmium malignantly transforms normal human breast epithelial cells into a basal-like phenotype. *Environ Health Perspect* 117(12):1847–1852.

Benzon Larsen, S., U. Vogel, J. Christensen, R. D. Hansen, H. Wallin, K. Overvad, A. Tjonneland, and J. Tolstrup. 2010. Interaction between ADH1C Arg(272)Gln and alcohol intake in relation to breast cancer risk suggests that ethanol is the causal factor in alcohol related breast cancer. *Cancer Lett* 295(2):191–197.

Beral, V., G. Reeves, D. Bull, and J. Green. 2011. Breast cancer risk in relation to the interval between menopause and starting hormone therapy. *J Natl Cancer Inst* 103(4):296–305.

Berkey, C. S., W. C. Willett, A. L. Frazier, B. Rosner, R. M. Tamimi, H. R. Rockett, and G. A. Colditz. 2010. Prospective study of adolescent alcohol consumption and risk of benign breast disease in young women. *Pediatrics* 125(5):e1081–e1087.

Berrington de Gonzalez, A., M. Mahesh, K. P. Kim, M. Bhargavan, R. Lewis, F. Mettler, and C. Land. 2009. Projected cancer risks from computed tomographic scans performed in the United States in 2007. *Arch Intern Med* 169(22):2071–2077.

Berstad, P., R. J. Coates, L. Bernstein, S. G. Folger, K. E. Malone, P. A. Marchbanks, L. K. Weiss, J. M. Liff, et al. 2010. A case–control study of body mass index and breast cancer risk in white and African-American women. *Cancer Epidemiol Biomarkers Prev* 19(6):1532–1544.

Betts, K. S. 2008. Unwelcome guest: PBDEs in indoor dust. *Environ Health Perspect* 116(5): A202–A208.

Biedermann, S., P. Tschudin, and K. Grob. 2010. Transfer of bisphenol A from thermal printer paper to the skin. *Anal Bioanal Chem* 398(1):571–576.

Biro, F. M., M. P. Galvez, L. C. Greenspan, P. A. Succop, N. Vangeepuram, S. M. Pinney, S. Teitelbaum, G. C. Windham, et al. 2010. Pubertal assessment method and baseline characteristics in a mixed longitudinal study of girls. *Pediatrics* 126(3):e583–e590.

Blask, D. E., R. T. Dauchy, L. A. Sauer, J. A. Krause, and G. C. Brainard. 2002. Light during darkness, melatonin suppression and cancer progression. *Neuro Endocrinol Lett* 23(Suppl 2):52–56.

Bogdanffy, M. S., J. C. O'Connor, J. F. Hansen, V. Gaddamidi, C. S. Van Pelt, J. W. Green, and J. C. Cook. 2000. Chronic toxicity and oncogenicity bioassay in rats with the chloro-s-triazine herbicide cyanazine. *J Toxicol Environ Health A* 60(8):567–586.

Bolt, H. M., and K. Golka. 2007. The debate on carcinogenicity of permanent hair dyes: New insights. *Crit Rev Toxicol* 37(6):521–536.

Bonner, M. R., D. Han, J. Nie, P. Rogerson, J. E. Vena, P. Muti, M. Trevisan, S. B. Edge, et al. 2005. Breast cancer risk and exposure in early life to polycyclic aromatic hydrocarbons using total suspended particulates as a proxy measure. *Cancer Epidemiol Biomarkers Prev* 14(1):53–60.

Brody, J. G., and R. A. Rudel. 2003. Environmental pollutants and breast cancer. *Environ Health Perspect* 111(8):1007–1019.

Brody, J. G., K. B. Moysich, O. Humblet, K. R. Attfield, G. P. Beehler, and R. A. Rudel. 2007. Environmental pollutants and breast cancer: Epidemiologic studies. *Cancer* 109(12 Suppl):2667–2711.

Brown, N. M., and C. A. Lamartiniere. 1995. Xenoestrogens alter mammary gland differentiation and cell proliferation in the rat. *Environ Health Perspect* 103(7–8):708–713.

Brown, N. M., P. A. Manzolillo, J. X. Zhang, J. Wang, and C. A. Lamartiniere. 1998. Prenatal TCDD and predisposition to mammary cancer in the rat. *Carcinogenesis* 19(9): 1623–1629.

Brown, L. M., G. Gridley, A. H. Wu, R. T. Falk, M. Hauptmann, L. N. Kolonel, D. W. West, A. M. Nomura, et al. 2010. Low level alcohol intake, cigarette smoking and risk of breast cancer in Asian-American women. *Breast Cancer Res Treat* 120(1):203–210.

Calafat, A. M., L. Y. Wong, Z. Kuklenyik, J. A. Reidy, and L. L. Needham. 2007. Polyfluoroalkyl chemicals in the U.S. population: Data from the National Health and Nutrition Examination Survey (NHANES) 2003–2004 and comparisons with NHANES 1999–2000. *Environ Health Perspect* 115(11):1596–1602.

Calafat, A. M., X. Ye, L. Y. Wong, J. A. Reidy, and L. L. Needham. 2008. Exposure of the U.S. population to bisphenol A and 4-tertiary-octylphenol: 2003–2004. *Environ Health Perspect* 116(1):39–44.

CalEPA (California Environmental Protection Agency). 2001. *Public health goal for benzene in drinking water*. Office of Environmental Health Hazard Assessment (OEHHA). http://oehha.ca.gov/water/phg/pdf/BenzeneFinPHG.pdf (accessed July 25, 2011).

CalEPA. 2005. *Proposed identification of environmental tobacco smoke as a toxic air contaminant*. http://www.arb.ca.gov/regact/ets2006/ets2006.htm (accessed October 24, 2011).

California Department of Health Services. Hazard Evaluation System and Information Service (HESIS). 1999. *Artificial fingernail products: A HESIS guide to chemical exposures in the nail salon*. http://www.cdph.ca.gov/programs/hesis/documents/artnails.pdf (accessed October 24, 2011).

Calle, E. E., H. Frumkin, S. J. Henley, D. A. Savitz, and M. J. Thun. 2002. Organochlorines and breast cancer risk. *CA Cancer J Clin* 52(5):301–309.

Cantón, R. F., D. E. Scholten, G. Marsh, P. C. de Jong, and M. van den Berg. 2008. Inhibition of human placental aromatase activity by hydroxylated polybrominated diphenyl ethers (OH-PBDEs). *Toxicol Appl Pharmacol* 227(1):68–75.

Cantor, K. P., M. Dosemeci, L. A. Brinton, and P. A. Stewart. 1995. Re: Breast cancer mortality among female electrical workers in the United States. *J Natl Cancer Inst* 87(3):227–228.

Carmichael, A. R., A. S. Sami, and J. M. Dixon. 2003. Breast cancer risk among the survivors of atomic bomb and patients exposed to therapeutic ionising radiation. *Eur J Surg Oncol* 29(5):475–479.

Cavalieri, E., E. Rogan, and D. Sinha. 1988. Carcinogenicity of aromatic hydrocarbons directly applied to rat mammary gland. *J Cancer Res Clin Oncol* 114(1):3–9.

CDC (Centers for Disease Control and Prevention). 2003. *Second national report on human exposure to environmental chemicals*. NCEH Pub. No. 02-0716. Atlanta, GA: CDC. http://www.jhsph.edu/ephtcenter/Second%20Report.pdf (accessed October 24, 2011).

CDC. 2005. *Third national report on human exposure to environmental chemicals*. NCEH Pub. No. 05-0570. Atlanta, GA: CDC. http://www.cphfoundation.org/documents/third report_001.pdf (accessed December 22, 2011).

CDC. 2008. *DDT, DDE, DDD*. http://www.atsdr.cdc.gov/substances/toxsubstance.asp? toxid=20 (accessed October 24, 2011).

CDC. 2009a. *4-tert-Octylphenol CAS No. 140-66-9*. http://www.cdc.gov/exposurereport/ data_tables/Octylphenol_ChemicalInformation.html (accessed October 24, 2011).

CDC. 2009b. *National report on human exposure to environmental chemicals*. Fourth Rpt. http://www.cdc.gov/exposurereport/ (accessed October 24, 2011).

CDC. 2011a. *Alcohol and public health—frequently asked questions*. http://www.cdc.gov/ alcohol/faqs.htm#heavyDrinking (accessed October 24, 2011).

CDC. 2011b. *Phthalates—general information*. http://www.cdc.gov/exposurereport/data_ tables/chemical_group_12.html (accessed September 12, 2011)

Chen, W. Y., G. A. Colditz, B. Rosner, S. E. Hankinson, D. J. Hunter, J. E. Manson, M. J. Stampfer, W. C. Willett, et al. 2002. Use of postmenopausal hormones, alcohol, and risk for invasive breast cancer. *Ann Intern Med* 137(10):798–804.

Chen, C., X. Ma, M. Zhong, and Z. Yu. 2010. Extremely low-frequency electromagnetic fields exposure and female breast cancer risk: A meta-analysis based on 24,338 cases and 60,628 controls. *Breast Cancer Res Treat* 123(2):569–576.

Chlebowski, R. T., L. H. Kuller, R. L. Prentice, M. L. Stefanick, J. E. Manson, M. Gass, A. K. Aragaki, J. K. Ockene, et al. 2009. Breast cancer after use of estrogen plus progestin in postmenopausal women. *N Engl J Med* 360(6):573–587.

Choe, S.-Y., S.-J. Kim, H.-G. Kim, J. H. Lee, Y. Choi, H. Lee, and Y. Kim. 2003. Evaluation of estrogenicity of major heavy metals. *Sci Total Environ* 312(1–3):15–21.

Chou, Y. Y., P. C. Huang, C. C. Lee, M. H. Wu, and S. J. Lin. 2009. Phthalate exposure in girls during early puberty. *J Pediatr Endocrinol Metab* 22(1):69–77.

Chuang, S. C., V. Gallo, D. Michaud, K. Overvad, A. Tjonneland, F. Clavel-Chapelon, I. Romieu, K. Straif, et al. 2011. Exposure to environmental tobacco smoke in childhood and incidence of cancer in adulthood in never smokers in the European Prospective Investigation into Cancer and Nutrition. *Cancer Causes Control* 22(3):487–494.

Cini, G., B. Neri, A. Pacini, V. Cesati, C. Sassoli, S. Quattrone, M. D'Apolito, A. Fazio, et al. 2005. Antiproliferative activity of melatonin by transcriptional inhibition of cyclin D1 expression: A molecular basis for melatonin-induced oncostatic effects. *J Pineal Res* 39(1):12–20.

Cohn, B. A., M. S. Wolff, P. M. Cirillo, and R. I. Scholtz. 2007. DDT and breast cancer in young women: New data on the significance of age at exposure. *Environ Health Perspect* 115(10):1406–1414.

Collaborative Group on Hormonal Factors in Breast Cancer. 2002. Alcohol, tobacco and breast cancer—collaborative reanalysis of individual data from 53 epidemiological studies, including 58,515 women with breast cancer and 95,067 women without the disease. *Br J Cancer* 87(11):1234–1245.

Collishaw, N. C., N. F. Boyd, K. P. Cantor, S. K. Hammond, K. C. Johnson, J. Millar, A. B. Miller, M. Miller, J. R. Palmer, A. G. Salmon, and F. Turcotte. 2009. *Canadian expert panel on tobacco smoke and breast cancer risk.* OTRU Special Report Series. Toronto, Canada: Ontario Tobacco Research Unit. http://www.otru.org/pdf/special/expert_panel_tobacco_breast_cancer.pdf (accessed October 29, 2011).

Conlon, M. S., K. C. Johnson, M. A. Bewick, R. M. Lafrenie, and A. Donner. 2010. Smoking (active and passive), N-acetyltransferase 2, and risk of breast cancer. *Cancer Epidemiol* 34(2):142–149.

Cos, S., D. Mediavilla, C. Martinez-Campa, A. Gonzalez, C. Alonso-Gonzalez, and E. J. Sanchez-Barcelo. 2006. Exposure to light-at-night increases the growth of DMBA-induced mammary adenocarcinomas in rats. *Cancer Lett* 235(2):266–271.

Cosmetic Ingredient Review. 2008. Annual review of cosmetic ingredient safety assessments: 2005/2006. *Int J Toxicol* 27(Suppl 1):77–142.

Costa, L. G., G. Giordano, S. Tagliaferri, A. Caglieri, and A. Mutti. 2008. Polybrominated diphenyl ether (PBDE) flame retardants: Environmental contamination, human body burden and potential adverse health effects. *Acta Biomed* 79(3):172–183.

Costantini, A. S., G. Gorini, D. Consonni, L. Miligi, L. Giovannetti, and M. Quinn. 2009. Exposure to benzene and risk of breast cancer among shoe factory workers in Italy. *Tumori* 95(1):8–12.

Coutelle, C., B. Hohn, M. Benesova, C. M. Oneta, P. Quattrochi, H. J. Roth, H. Schmidt-Gayk, A. Schneeweiss, et al. 2004. Risk factors in alcohol associated breast cancer: Alcohol dehydrogenase polymorphism and estrogens. *Int J Oncol* 25(4):1127–1132.

Cowan-Ellsberry, C. E., and S. H. Robison. 2009. Refining aggregate exposure: Example using parabens. *Regul Toxicol Pharmacol* 55(3):321–329.

CPSC (Consumer Product Safety Commission). 2001. *Chronic hazard advisory panel on diisonyl phthalate (DINP)*. http://www.cpsc.gov/library/foia/foia01/os/dinp.pdf (accessed November 2, 2011).

Cronkite, E. P., J. Bullis, T. Inoue, and R. T. Drew. 1984. Benzene inhalation produces leukemia in mice. *Toxicol Appl Pharmacol* 75(2):358–361.

Darbre, P. D., and P. W. Harvey. 2008. Paraben esters: Review of recent studies of endocrine toxicity, absorption, esterase and human exposure, and discussion of potential human health risks. *J Appl Toxicol* 28(5):561–578.

Darnerud, P. O. 2008. Brominated flame retardants as possible endocrine disrupters. *Int J Androl* 31(2):152–160.

Darnerud, P. O., G. S. Eriksen, T. Johannesson, P. B. Larsen, and M. Viluksela. 2001. Polybrominated diphenyl ethers: Occurrence, dietary exposure, and toxicology. *Environ Health Perspect* 109(Suppl 1):49–68.

Dauchy, R. T., L. A. Sauer, D. E. Blask, and G. M. Vaughan. 1997. Light contamination during the dark phase in "photoperiodically controlled" animal rooms: Effect on tumor growth and metabolism in rats. *Lab Anim Sci* 47(5):511–518.

Dauchy, R. T., D. E. Blask, L. A. Sauer, G. C. Brainard, and J. A. Krause. 1999. Dim light during darkness stimulates tumor progression by enhancing tumor fatty acid uptake and metabolism. *Cancer Lett* 144(2):131–136.

Davidson, T., K. E. Qingdong, and M. Costa. 2007. Selected molecular mechanisms of metal toxicity and carcinogenicity. In *Handbook on the Toxicology of Metals*, 3rd ed. Edited by G. F. Nordberg, B. A. Fowler, M. Nordberg, and L. T. Friberg. Boston, MA: Academic Press.

Davis, S., D. K. Mirick, and R. G. Stevens. 2001. Night shift work, light at night, and risk of breast cancer. *J Natl Cancer Inst* 93(20):1557–1562.

Davis, S., D. K. Mirick, and R. G. Stevens. 2002. Residential magnetic fields and the risk of breast cancer. *American Journal of Epidemiology* 155(5):446–454.

Davis, L. K., A. S. Murr, D. S. Best, M. J. Fraites, L. M. Zorrilla, M. G. Narotsky, T. E. Stoker, J. M. Goldman, et al. 2011. The effects of prenatal exposure to atrazine on pubertal and postnatal reproductive indices in the female rat. *Reprod Toxicol* 32(1):43–51.

Dean, B. J. 1978. Genetic toxicology of benzene, toluene, xylenes and phenols. *Mutat Res* 47(2):75–97.

Dean, B. J. 1985. Recent findings on the genetic toxicology of benzene, toluene, xylenes and phenols. *Mutat Res* 154(3):153–181.

DeKeyser, J. G., M. C. Stagliano, S. S. Auerbach, K. S. Prabhu, A. D. Jones, and C. J. Omiecinski. 2009. Di(2-ethylhexyl) phthalate is a highly potent agonist for the human constitutive androstane receptor splice variant CAR2. *Mol Pharmacol* 75(5):1005–1013.

DeKeyser, J. G., E. M. Laurenzana, E. C. Peterson, T. Chen, and C. J. Omiecinski. 2011. Selective phthalate activation of naturally occurring human constitutive androstane receptor splice variants and the pregnane X receptor. *Toxicol Sci* 120(2):381–391.

del Rio, B., J. M. Garcia Pedrero, C. Martinez-Campa, P. Zuazua, P. S. Lazo, and S. Ramos. 2004. Melatonin, an endogenous-specific inhibitor of estrogen receptor alpha via calmodulin. *J Biol Chem* 279(37):38294–38302.

Demers, P. A., D. B. Thomas, K. A. Rosenblatt, L. M. Jimenez, A. McTiernan, H. Stalsberg, A. Stemhagen, W. D. Thompson, et al. 1991. Occupational exposure to electromagnetic fields and breast cancer in men. *Am J Epidemiol* 134(4):340–347.

Den Hond, E., H. A. Roels, K. Hoppenbrouwers, T. Nawrot, L. Thijs, C. Vandermeulen, G. Winneke, D. Vanderschueren, et al. 2002. Sexual maturation in relation to polychlorinated aromatic hydrocarbons: Sharpe and Skakkebaek's hypothesis revisited. *Environ Health Perspect* 110(8):771–776.

DeRoo, L. A., P. Cummings, and B. A. Mueller. 2011. Smoking before the first pregnancy and the risk of breast cancer: A meta-analysis. *Am J Epidemiol* 174(4):390–402.

DeSantis, C., N. Howlader, K. A. Cronin, and A. Jemal. 2011. Breast cancer incidence rates in U.S. women are no longer declining. *Cancer Epidemiol Biomarkers Prev* 20(5):733–739.

Desaulniers, D., K. Leingartner, B. Musicki, J. Cole, M. Li, M. Charbonneau, and B. K. Tsang. 2004. Lack of effects of postnatal exposure to a mixture of aryl hydrocarbon-receptor agonists on the development of methylnitrosourea-induced mammary tumors in Sprague-Dawley rats. *J Toxicol Environ Health A* 67(18):1457–1475.

Doherty, L., J. G. Bromer, Y. Zhou, T. Aldad, and H. S. Taylor. 2010. In utero exposure to diethylstilbestrol (DES) or bisphenol-A (BPA) increases EZH2 expression in the mammary gland: An epigenetic mechanism linking endocrine disruptors to breast cancer. *Hormones and Cancer* 1:146–155.

Dorgan, J. F., J. W. Brock, N. Rothman, L. L. Needham, R. Miller, H. E. Stephenson, Jr., N. Schussler, and P. R. Taylor. 1999. Serum organochlorine pesticides and PCBs and breast cancer risk: Results from a prospective analysis (USA). *Cancer Causes Control* 10(1):1–11.

Dostal, L. A., R. P. Weaver, and B. A. Schwetz. 1987. Transfer of di(2-ethylhexyl) phthalate through rat milk and effects on milk composition and the mammary gland. *Toxicol Appl Pharmacol* 91(3):315–325.

Durando, M., L. Kass, J. Piva, C. Sonnenschein, A. M. Soto, E. H. Luque, and M. Muñoz-de-Toro. 2007. Prenatal bisphenol A exposure induces preneoplastic lesions in the mammary gland in Wistar rats. *Environ Health Perspect* 115(1):80–86.

EFSA (European Food Safety Authority). 2006. *Opinion of the Scientific Panel on Food Additives, Flavourings, Processing Aids and Materials in Contact with Food on a request from the Commission related to 2,2-bis(4-hydroxyphenyl)propane (Bisphenol A).* http://www.efsa.europa.eu/de/scdocs/doc/428.pdf (accessed November 18, 2011).

EFSA. 2008. *Toxicokinetics of bisphenol A: Scientific opinion of the Panel on Food Additives, Flavourings, Processing Aids and Materials in Contact with Food (AFC) (Question No EFSA-Q-2008-382).* http://www.efsa.europa.eu/fr/scdocs/doc/759.pdf (accessed November 18, 2011).

EFSA. 2010. *Results of the monitoring of non dioxin-like PCBs in food and feed.* http://www.efsa.europa.eu/en/efsajournal/pub/1701.htm (accessed October 24, 2011).

Eliassen, A. H., G. A. Colditz, B. Rosner, W. C. Willett, and S. E. Hankinson. 2006. Adult weight change and risk of postmenopausal breast cancer. *JAMA* 296(2):193–201.

Enoch, R. R., J. P. Stanko, S. N. Greiner, G. L. Youngblood, J. L. Rayner, and S. E. Fenton. 2007. Mammary gland development as a sensitive end point after acute prenatal exposure to an atrazine metabolite mixture in female Long-Evans rats. *Environ Health Perspect* 115(4):541–547.

EPA (Environmental Protection Agency). 1997. *Di(2-ethylhexyl)phthalate (DEHP) (CASRN 117-81-7).* http://www.epa.gov/iris/subst/0014.htm (accessed December 22, 2011).

EPA. 2000. *Exposure and human health reassessment of 2,3,7,8-tetrachlorodibenzo-p-dioxin (TCDD) and related compounds.* Part II: Health assessment for 2,3,7,8-tetrachlorodibenzo-p-dioxin (TCDD) and related compounds. http://www.epa.gov/ncea/pdfs/dioxin/nas-review/pdfs/part2/dioxin_pt2_ch08_dec2003.pdf (accessed October 24, 2011).

EPA. 2004. *Protecting the health of nail salon workers.* http://www.epa.gov/dfe/pubs/projects/salon/nailsalonguide.pdf (accessed December 22, 2011).

EPA. 2005a. *Draft risk assessment of the potential human health effects associated with exposure to perfluorooctanoic acid and its salts (PFOA).* http://www.epa.gov/opptintr/pfoa/pubs/pfoarisk.pdf (accessed November 2, 2011).

EPA. 2005b. *Guidelines for carcinogen risk assessment (2005).* http://www.epa.gov/cancerguidelines/ (accessed October 24, 2011).

EPA. 2008. *Toxicological review of 2,2',4,4',5-pentabromodiphenyl ether (BDE-99) in support of summary information on the Integrated Risk Information System (IRIS).* http://www. epa.gov/iris/toxreviews/1008tr.pdf (accessed October 18, 2011).

EPA. 2009a. *Atrazine science reevaluation: Potential health impacts document.* http://www. epa.gov/pesticides/reregistration/atrazine/atrazine_update.htm (accessed October 24, 2011).

EPA. 2009b. Final list of initial pesticide active ingredients and pesticide inert ingredients to be screened under the Federal Food, Drug, and Cosmetic Act. *Federal Register* 74(71): 17579–17585. http://www.epa.gov/scipoly/oscpendo/pubs/final_list_frn_041509.pdf (accessed October 25, 2011).

EPA. 2010a. *Atrazine updates.* http://www.epa.gov/opp00001/reregistration/atrazine/ atrazine_update.htm#cancer (accessed October 24, 2011).

EPA. 2010b. *Bisphenol A action plan (CASRN 80-05-7).* http://www.epa.gov/opptintr/ existingchemicals/pubs/actionplans/bpa_action_plan.pdf (accessed October 24, 2011).

EPA. 2011a. *Bisphenol A.* http://www.epa.gov/iris/subst/0356.htm (accessed October 24, 2011).

EPA. 2011b. *Integrated risk information system (IRIS).* http://www.epa.gov/IRIS/ (accessed December 21, 2011).

EPA. 2011c. *p,p'-Dichlorodiphenyltrichloroethane (DDT) (CASRN 50-29-3).* http://www.epa. gov/iris/subst/0147.htm (accessed December 21, 2011).

Erren, T. C. 2001. A meta-analysis of epidemiologic studies of electric and magnetic fields and breast cancer in women and men. *Bioelectromagnetics* Suppl 5:S105–S119.

European Commission. 2000. *Opinion of the scientific committee on food on Fusarium toxins, Part 21: Zearalenone (ZEA).* http://ec.europa.eu/food/fs/sc/scf/out65_en.pdf (accessed December 22, 2011).

European Commission. 2007. *Hormones and meat.* http://ec.europa.eu/food/food/chemical- safety/contaminants/hormones/index_en.htm (accessed October 24, 2011).

Fan, W., T. Yanase, H. Morinaga, S. Gondo, T. Okabe, M. Nomura, T. Komatsu, K. Morohashi, et al. 2007. Atrazine-induced aromatase expression is SF-1 dependent: Implications for endocrine disruption in wildlife and reproductive cancers in humans. *Environ Health Perspect* 115(5):720–727.

FAO/WHO (Food and Agriculture Organization of the United Nations and World Health Orga- nization Expert Committee on Food Additives). 2005. *Evaluation of certain food additives.* WHO Technical Report Series; No. 928. http://whqlibdoc.who.int/trs/WHO_TRS_928.pdf (accessed October 24, 2011)

FAO/WHO. 2007. *Evaluation of certain food additives and contaminants.* WHO Technical Report Series; No. 940. http://whqlibdoc.who.int/trs/WHO_TRS_940_eng.pdf (accessed December 21, 2011).

Farquhar, C., J. Marjoribanks, A. Lethaby, J. A. Suckling, and Q. Lamberts. 2009. Long term hormone therapy for perimenopausal and postmenopausal women. *Cochrane Database Syst Rev* 2009(2):CD004143.

FDA (Food and Drug Administration). 2007. *Product and ingredient safety: Parabens.* http://www.fda.gov/cosmetics/productandingredientsafety/selectedcosmeticingredients/ ucm128042.htm (accessed October 24, 2011).

FDA. 2008. *Draft assessment of bisphenol A for use in food contact applications.* http://www. fda.gov/ohrms/dockets/ac/08/briefing/2008-0038b1_01_02_FDA%20BPA%20Draft%20 Assessment.pdf (accessed November 18, 2011).

FDA. 2010. *Update on bisphenol A for use in food contact applications: January 2010.* http://www.fda.gov/NewsEvents/PublicHealthFocus/ucm197739.htm (accessed October 31, 2011).

Fedrowitz, M., K. Kamino, and W. Löscher. 2004. Significant differences in the effects of magnetic field exposure on 7,12-dimethylbenz(a)anthracene-induced mammary carcinogenesis in two substrains of Sprague-Dawley rats. *Cancer Research* 64(1):243–251.

Feigelson, H. S., C. R. Jonas, A. S. Robertson, M. L. McCullough, M. J. Thun, and E. E. Calle. 2003. Alcohol, folate, methionine, and risk of incident breast cancer in the American Cancer Society Cancer Prevention Study II Nutrition Cohort. *Cancer Epidemiol Biomarkers Prev* 12(2):161–164.

Fenet, H., E. Gomez, A. Pillon, D. Rosain, J. C. Nicolas, C. Casellas, and P. Balaguer. 2003. Estrogenic activity in water and sediments of a French river: Contribution of alkylphenols. *Arch Environ Contam Toxicol* 44(1):1–6.

Flegal, K. M., M. D. Carroll, C. L. Ogden, and L. R. Curtin. 2010. Prevalence and trends in obesity among U.S. adults, 1999–2008. *JAMA* 303(3):235–241.

Floderus, B., S. Tornqvist, and C. Stenlund. 1994. Incidence of selected cancers in Swedish railway workers, 1961–79. *Cancer Causes Control* 5(2):189–194.

Forssen, U. M., M. Feychting, L. E. Rutqvist, B. Floderus, and A. Ahlbom. 2000. Occupational and residential magnetic field exposure and breast cancer in females. *Epidemiology* 11(1):24–29.

Forssen, U. M., L. E. Rutqvist, A. Ahlbom, and M. Feychting. 2005. Occupational magnetic fields and female breast cancer: A case–control study using Swedish population registers and new exposure data. *Am J Epidemiol* 161(3):250–259.

Freudenheim, J. L., J. R. Marshall, S. Graham, R. Laughlin, J. E. Vena, M. Swanson, C. Ambrosone, and T. Nemoto. 1995. Lifetime alcohol consumption and risk of breast cancer. *Nutr Cancer* 23(1):1–11.

Freudenheim, J. L., C. B. Ambrosone, K. B. Moysich, J. E. Vena, S. Graham, J. R. Marshall, P. Muti, R. Laughlin, et al. 1999. Alcohol dehydrogenase 3 genotype modification of the association of alcohol consumption with breast cancer risk. *Cancer Causes Control* 10(5):369–377.

Gajecka, M., L. Rybarczyk, E. Jakimiuk, L. Zielonka, K. Obremski, W. Zwierzchowski, and M. Gajecki. 2011. The effect of experimental long-term exposure to low-dose zearalenone on uterine histology in sexually immature gilts. *Exp Toxicol Pathol* Jan 10 (Epub ahead of print).

Galloway, S. M., J. E. Miller, M. J. Armstrong, C. L. Bean, T. R. Skopek, and W. W. Nichols. 1998. DNA synthesis inhibition as an indirect mechanism of chromosome aberrations: Comparison of DNA-reactive and non-DNA-reactive clastogens. *Mutat Res* 400(1–2):169–186.

Gammon, M. D., and R. M. Santella. 2008. PAH, genetic susceptibility and breast cancer risk: An update from the Long Island Breast Cancer Study Project. *Eur J Cancer* 44(5):636–640.

Gammon, M. D., M. S. Wolff, A. I. Neugut, S. M. Eng, S. L. Teitelbaum, J. A. Britton, M. B. Terry, B. Levin, et al. 2002. Environmental toxins and breast cancer on Long Island. Organochlorine compound levels in blood. *Cancer Epidemiol Biomarkers Prev* 11(8):686–697.

Gammon, M. D., S. K. Sagiv, S. M. Eng, S. Shantakumar, M. M. Gaudet, S. L. Teitelbaum, J. A. Britton, M. B. Terry, et al. 2004. Polycyclic aromatic hydrocarbon-DNA adducts and breast cancer: A pooled analysis. *Arch Environ Health* 59(12):640–649.

Gapstur, S. M., J. D. Potter, T. A. Sellers, and A. R. Folsom. 1992. Increased risk of breast cancer with alcohol consumption in postmenopausal women. *Am J Epidemiol* 136(10):1221–1231.

Gapstur, S. M., J. D. Potter, C. Drinkard, and A. R. Folsom. 1995. Synergistic effect between alcohol and estrogen replacement therapy on risk of breast cancer differs by estrogen/progesterone receptor status in the Iowa Women's Health Study. *Cancer Epidemiol Biomarkers Prev* 4(4):313–318.

Garland, M., D. J. Hunter, G. A. Colditz, D. L. Spiegelman, J. E. Manson, M. J. Stampfer, and W. C. Willett. 1999. Alcohol consumption in relation to breast cancer risk in a cohort of United States women 25–42 years of age. *Cancer Epidemiol Biomarkers Prev* 8(11):1017–1021.

Gatto, N. M., M. P. Longnecker, M. F. Press, J. Sullivan-Halley, R. McKean-Cowdin, and L. Bernstein. 2007. Serum organochlorines and breast cancer: A case–control study among African-American women. *Cancer Causes Control* 18(1):29–39.

Gladen, B. C., N. B. Ragan, and W. J. Rogan. 2000. Pubertal growth and development and prenatal and lactational exposure to polychlorinated biphenyls and dichlorodiphenyl dichloroethene. *J Pediatr* 136(4):490–496.

Golden, R., and R. Kimbrough. 2009. Weight of evidence evaluation of potential human cancer risks from exposure to polychlorinated biphenyls: An update based on studies published since 2003. *Crit Rev Toxicol* 39(4):299–331.

Golden, R., J. Gandy, and G. Vollmer. 2005. A review of the endocrine activity of parabens and implications for potential risks to human health. *Crit Rev Toxicol* 35(5):435–458.

Gollenberg, A. L., M. L. Hediger, P. A. Lee, J. H. Himes, and G. M. Louis. 2010. Association between lead and cadmium and reproductive hormones in peripubertal U.S. girls. *Environ Health Perspect* 118(12):1782–1787.

Gordon, G. B., T. L. Bush, K. J. Helzlsouer, S. R. Miller, and G. W. Comstock. 1990. Relationship of serum levels of dehydroepiandrosterone and dehydroepiandrosterone sulfate to the risk of developing postmenopausal breast cancer. *Cancer Res* 50(13):3859–3862.

Graham, C., M. R. Cook, M. M. Gerkovich, and A. Sastre. 2001. Examination of the melatonin hypothesis in women exposed at night to EMF or bright light. *Environ Health Perspect* 109(5):501–507.

Greenfield, T. K., L. T. Midanik, and J. D. Rogers. 2000. A 10-year national trend study of alcohol consumption, 1984–1995: Is the period of declining drinking over? *Am J Public Health* 90(1):47–52.

Grice, M. M., B. H. Alexander, R. Hoffbeck, and D. M. Kampa. 2007. Self-reported medical conditions in perfluorooctanesulfonyl fluoride manufacturing workers. *J Occup Environ Med* 49(7):722–729.

Gross, B., J. Montgomery-Brown, A. Naumann, and M. Reinhard. 2004. Occurrence and fate of pharmaceuticals and alkylphenol ethoxylate metabolites in an effluent-dominated river and wetland. *Environ Toxicol Chem* 23(9):2074–2083.

Grosse, Y., R. Baan, B. Secretan-Lauby, F. El Ghissassi, V. Bouvard, L. Benbrahim-Tallaa, N. Guha, F. Islami, et al. 2011. Carcinogenicity of chemicals in industrial and consumer products, food contaminants, and flavourings, and water chlorination byproducts. *Lancet Oncol* 12(4):328–329.

Guenel, P., P. Raskmark, J. B. Andersen, and E. Lynge. 1993. Incidence of cancer in persons with occupational exposure to electromagnetic fields in Denmark. *Br J Ind Med* 50(8):758–764.

Haas, J. S., C. P. Kaplan, E. P. Gerstenberger, and K. Kerlikowske. 2004. Changes in the use of postmenopausal hormone therapy after the publication of clinical trial results. *Ann Intern Med* 140(3):184–188.

Hansen, J. 2000. Elevated risk for male breast cancer after occupational exposure to gasoline and vehicular combustion products. *Am J Ind Med* 37(4):349–352.

Hansen, J. 2001. Increased breast cancer risk among women who work predominantly at night. *Epidemiology* 12(1):74–77.

Hansen, J. 2006. Risk of breast cancer after night- and shift work: Current evidence and ongoing studies in Denmark. *Cancer Causes Control* 17(4):531–537.

Hardell, L., G. Lindstrom, G. Liljegren, P. Dahl, and A. Magnuson. 1996. Increased concentrations of octachlorodibenzo-p-dioxin in cases with breast cancer—results from a case–control study. *Eur J Cancer Prev* 5(5):351–357.

Hardell, L., G. Lindstrom, B. van Bavel, H. Wingfors, E. Sundelin, and G. Liljegren. 1998. Concentrations of the flame retardant 2,2',4,4'-tetrabrominated diphenyl ether in human adipose tissue in Swedish persons and the risk for non-Hodgkin's lymphoma. *Oncol Res* 10(8):429–432.

Hardell, L., B. Bavel, G. Lindstrom, M. Eriksson, and M. Carlberg. 2006. In utero exposure to persistent organic pollutants in relation to testicular cancer risk. *Int J Androl* 29(1):228–234.

Harris, H. R., R. M. Tamimi, W. C. Willett, S. E. Hankinson, and K. B. Michels. 2011. Body size across the life course, mammographic density, and risk of breast cancer. *Am J Epidemiol* 174(8):909–918.

Harvey, E. B., C. Schairer, L. A. Brinton, R. N. Hoover, and J. F. Fraumeni, Jr. 1987. Alcohol consumption and breast cancer. *J Natl Cancer Inst* 78(4):657–661.

Harvie, M., A. Howell, R. A. Vierkant, N. Kumar, J. R. Cerhan, L. E. Kelemen, A. R. Folsom, and T. A. Sellers. 2005. Association of gain and loss of weight before and after menopause with risk of postmenopausal breast cancer in the Iowa Women's Health Study. *Cancer Epidemiol Biomarkers Prev* 14(3):656–661.

Helzlsouer, K. J., G. B. Gordon, A. J. Alberg, T. L. Bush, and G. W. Comstock. 1992. Relationship of prediagnostic serum levels of dehydroepiandrosterone and dehydroepiandrosterone sulfate to the risk of developing premenopausal breast cancer. *Cancer Res* 52(1):1–4.

Helzlsouer, K. J., A. J. Alberg, H. Y. Huang, S. C. Hoffman, P. T. Strickland, J. W. Brock, V. W. Burse, L. L. Needham, et al. 1999. Serum concentrations of organochlorine compounds and the subsequent development of breast cancer. *Cancer Epidemiol Biomarkers Prev* 8(6):525–532.

Henderson, T. O., A. Amsterdam, S. Bhatia, M. M. Hudson, A. T. Meadows, J. P. Neglia, L. R. Diller, L. S. Constine, et al. 2010. Systematic review: Surveillance for breast cancer in women treated with chest radiation for childhood, adolescent, or young adult cancer. *Ann Intern Med* 152(7):444–455.

Hercberg, S., P. Galan, P. Preziosi, S. Bertrais, L. Mennen, D. Malvy, A. M. Roussel, A. Favier, et al. 2004. The SU.VI.MAX Study: A randomized, placebo-controlled trial of the health effects of antioxidant vitamins and minerals. *Arch Intern Med* 164(21):2335–2342.

Hersh, A. L., M. L. Stefanick, and R. S. Stafford. 2004. National use of postmenopausal hormone therapy: Annual trends and response to recent evidence. *JAMA* 291(1):47–53.

HHS (U.S. Department of Health and Human Services). 1996. *Physical activity and health: A report of the Surgeon General.* Atlanta, GA: U.S. Department of Health and Human Services, Centers for Disease Control and Prevention, National Center for Chronic Disease Prevention and Health Promotion. http://www.cdc.gov/nccdphp/sgr/index.htm (accessed November 29, 2011).

HHS. 2000. *10th special report to the U.S. Congress on alcohol and health.* http://pubs.niaaa.nih.gov/publications/10report/intro.pdf (accessed November 18, 2011).

HHS. 2006. *The health consequences of involuntary exposure to tobacco smoke: A report of the Surgeon General.* Atlanta, GA: Department of Health and Human Services, Centers for Disease Control and Prevention, Coordinating Center for Health Promotion, National Center for Chronic Disease Prevention and Health Promotion, Office on Smoking and Health. http://www.surgeongeneral.gov/library/secondhandsmoke/index.html (accessed December 22, 2011).

HHS. 2010a. *Household products database: Nonylphenol.* http://householdproducts.nlm.nih. gov/cgi-bin/household/brands?tbl=chem&id=2591 (accessed October 24, 2011).

HHS. 2010b. *Household products database: Nonylphenyl polyethoxylate.* http://household products.nlm.nih.gov/cgi-bin/household/brands?tbl=chem&id=960 (accessed October 24, 2011).

Hiatt, R. A., and R. D. Bawol. 1984. Alcoholic beverage consumption and breast cancer incidence. *Am J Epidemiol* 120(5):676–683.

Hiatt, R. A., A. Klatsky, and M. A. Armstrong. 1988. Alcohol and breast cancer. *Prev Med* 17(6):683–685.

Hiatt, R. A., S. Z. Haslam, and J. Osuch. 2009. The breast cancer and the environment research centers: Transdisciplinary research on the role of the environment in breast cancer etiology. *Environ Health Perspect* 117(12):1814–1822.

Higley, E. B., J. L. Newsted, X. Zhang, J. P. Giesy, and M. Hecker. 2010. Assessment of chemical effects on aromatase activity using the H295R cell line. *Environ Sci Pollut Res Int* 17(5):1137–1148.

Hilakivi-Clarke, L., I. Onojafe, M. Raygada, E. Cho, T. Skaar, I. Russo, and R. Clarke. 1999. Prepubertal exposure to zearalenone or genistein reduces mammary tumorigenesis. *Br J Cancer* 80(11):1682–1688.

Hilakivi-Clarke, L., A. Cabanes, S. de Assis, M. Wang, G. Khan, W. J. Shoemaker, and R. G. Stevens. 2004. In utero alcohol exposure increases mammary tumorigenesis in rats. *Br J Cancer* 90(11):2225–2231.

Hilliard, C. A., M. J. Armstrong, C. I. Bradt, R. B. Hill, S. K. Greenwood, and S. M. Galloway. 1998. Chromosome aberrations in vitro related to cytotoxicity of nonmutagenic chemicals and metabolic poisons. *Environ Mol Mutagen* 31(4):316–326.

Hines, L. M., S. E. Hankinson, S. A. Smith-Warner, D. Spiegelman, K. T. Kelsey, G. A. Colditz, W. C. Willett, and D. J. Hunter. 2000. A prospective study of the effect of alcohol consumption and ADH3 genotype on plasma steroid hormone levels and breast cancer risk. *Cancer Epidemiol Biomarkers Prev* 9(10):1099–1105.

Hines, L. M., B. Risendal, M. L. Slattery, K. B. Baumgartner, A. R. Giuliano, C. Sweeney, D. E. Rollison, and T. Byers. 2010. Comparative analysis of breast cancer risk factors among Hispanic and non-Hispanic white women. *Cancer* 116(13):3215–3223.

Hites, R. A. 2004. Polybrominated diphenyl ethers in the environment and in people: A meta-analysis of concentrations. *Environ Sci Technol* 38(4):945–956.

Hockings, J. K., P. A. Thorne, M. Q. Kemp, S. S. Morgan, O. Selmin, and D. F. Romagnolo. 2006. The ligand status of the aromatic hydrocarbon receptor modulates transcriptional activation of BRCA-1 promoter by estrogen. *Cancer Res* 66(4):2224–2232.

Holmberg, L., J. A. Baron, T. Byers, A. Wolk, E. M. Ohlander, M. Zack, and H. O. Adami. 1995. Alcohol intake and breast cancer risk: Effect of exposure from 15 years of age. *Cancer Epidemiol Biomarkers Prev* 4(8):843–847.

Hong, E. J., Y. K. Ji, K. C. Choi, N. Manabe, and E. B. Jeung. 2005. Conflict of estrogenic activity by various phthalates between in vitro and in vivo models related to the expression of Calbindin-D_{9k}. *J Reprod Dev* 51(2):253–263.

Horn-Ross, P. L., A. J. Canchola, D. W. West, S. L. Stewart, L. Bernstein, D. Deapen, R. Pinder, R. K. Ross, et al. 2004. Patterns of alcohol consumption and breast cancer risk in the California Teachers Study cohort. *Cancer Epidemiol Biomarkers Prev* 13(3):405–411.

Houle, C. D., T. V. Ton, N. Clayton, J. Huff, H. H. Hong, and R. C. Sills. 2006. Frequent p53 and H-ras mutations in benzene- and ethylene oxide-induced mammary gland carcinomas from B6C3F1 mice. *Toxicol Pathol* 34 (6):752–762.

Hovey, R. C., P. S. Coder, J. C. Wolf, R. L. Sielken, Jr., M. O. Tisdel, and C. B. Breckenridge. 2011. Quantitative assessment of mammary gland development in female Long Evans rats following in utero exposure to atrazine. *Toxicol Sci* 119(2):380–390.

Hoyer, A. P., P. Grandjean, T. Jorgensen, J. W. Brock, and H. B. Hartvig. 1998. Organochlorine exposure and risk of breast cancer. *Lancet* 352(9143):1816–1820.

Hunter, D. J., S. E. Hankinson, F. Laden, G. A. Colditz, J. E. Manson, W. C. Willett, F. E. Speizer, and M. S. Wolff. 1997. Plasma organochlorine levels and the risk of breast cancer. *N Engl J Med* 337(18):1253–1258.

Hunter, D. J., G. A. Colditz, S. E. Hankinson, S. Malspeis, D. Spiegelman, W. Chen, M. J. Stampfer, and W. C. Willett. 2010. Oral contraceptive use and breast cancer: A prospective study of young women. *Cancer Epidemiol Biomarkers Prev* 19(10):2496–2502.

Hurley, S., P. Reynolds, D. Goldberg, D. O. Nelson, S. S. Jeffrey, and M. Petreas. 2011. Adipose levels of polybrominated diphenyl ethers and risk of breast cancer. *Breast Cancer Res Treat* 129(2):505–511.

IARC (International Agency for Research on Cancer). 1982. *Benzene: Some industrial chemicals and dyestuffs. IARC monographs on the evaluation of carcinogenic risks to humans: Summary of data reported and evaluation.* Vol. 29. Lyon, France: IARC.

IARC. 1987. *Overall evaluations of carcinogenicity. IARC monographs on the evaluation of carcinogenic risks to humans: Summary of data reported and evaluation.* Suppl 7. Lyon, France: IARC.

IARC. 1991. *DDT and associated compounds. Occupational exposures in insecticide application, and some pesticides: IARC monographs on the evaluation of carcinogenic risks to humans.* Vol. 53. Lyon, France: IARC.

IARC. 1993. *Some naturally occurring substances: Food items and constituents, heterocyclic aromatic amines and mycotoxins. IARC monographs on the evaluation of carcinogenic risks to humans.* Vol. 56. Lyon, France: IARC.

IARC. 1997. *Polychlorinated dibenzo-para-dioxins and polychlorinated dibenzofurans. IARC monographs on the evaluation of carcinogenic risks to humans.* Vol. 69. Lyon, France: IARC.

IARC. 1999. *Some chemicals that cause tumours of the kidney or urinary bladder in rodents and some other substances: Summary of data reported and evaluation. IARC monographs on the evaluation of carcinogenic risks to humans.* Vol. 73. Lyon, France: IARC.

IARC. 2000. *Ionizing radiation, Part 1: X- and gamma (γ)-radiation, and neutrons. IARC monographs on the evaluation of carcinogenic risks to humans.* Vol. 75. Lyon, France: IARC.

IARC. 2002a. *Non-ionizing radiation, Part 1: Static and extremely low-frequency (ELF) electric and magnetic fields. IARC monographs on the evaluation of carcinogenic risks to humans.* Vol. 80. Lyon, France: IARC.

IARC. 2002b. *Weight control and physical activity. IARC handbook of cancer prevention.* Vol. 6. Lyon, France: IARC.

IARC. 2004. *Tobacco smoke and involuntary smoking. IARC monographs on the evaluation of carcinogenic risks to humans.* Vol. 83. Lyon, France: IARC.

IARC. 2005. *Some non-heterocyclic polycyclic aromatic hydrocarbons and some related exposures. IARC monographs on the evaluation of carcinogenic risks to humans.* Vol. 92. Lyon, France: IARC.

IARC. 2006a. *Formaldehyde, 2-butoxyethanol and 1-tert-butoxypropan-2-ol. IARC monographs on the evaluation of carcinogenic risks to humans.* Vol. 88. Lyon, France: IARC.

IARC. 2006b. *Preamble to the IARC monographs.* Lyon, France: IARC. http://monographs.iarc.fr/ENG/Preamble/currenta2objective0706.php (accessed October 31, 2011).

IARC. 2007. *Combined estrogen-progestogen contraceptives and combined estrogen-progestogen menopausal therapy. IARC monographs on the evaluation of carcinogenic risks to humans.* Vol. 91. Lyon, France: IARC.

IARC. 2008. *1,3-Butadiene, ethylene oxide and vinyl halides (vinyl fluoride, vinyl chloride and vinyl bromide). IARC monographs on the evaluation of carcinogenic risks to humans.* Vol. 97. Lyon, France: IARC.

IARC. 2010a. *Alcohol consumption and ethyl carbamate. IARC monographs on the evaluation of carcinogenic risks to humans.* Vol. 96. Lyon, France: IARC.

IARC. 2010b. *Painting, firefighting, and shiftwork. IARC monographs on the evaluation of carcinogenic risks to humans.* Vol. 98. Lyon, France: IARC.

IARC. 2010c. *Some aromatic amines, organic dyes, and related exposures. IARC monographs on the evaluation of carcinogenic risks to humans.* Vol. 99. Lyon, France: IARC.

IARC. 2010d. *Some non-heterocyclic polycyclic aromatic hydrocarbons and some related exposures. IARC monographs on the evaluation of carcinogenic risks to humans.* Vol. 92. Lyon, France: IARC.

IARC. 2011. *IARC Monographs on the Evaluation of Carcinogenic Risks to Humans.* http://monographs.iarc.fr/ (accessed November 16, 2011).

Integrated Laboratory Systems. 2005. *Butylparaben [CAS No. 94-26-8]: Review of toxicological literature.* http://ntp.niehs.nih.gov/ntp/htdocs/chem_background/exsumpdf/butylparaben.pdf (accessed November 2, 2011).

IPCS (International Programme on Chemical Safety). 1999. *Environmental Health Criteria 215: Vinyl Chloride.* http://whqlibdoc.who.int/ehc/WHO_EHC_215.pdf (accessed November 18, 2011).

IOM (Institute of Medicine). 1991. *Adverse effects of pertussis and rubella vaccines.* Washington, DC: National Academy Press.

IOM. 2001. *Immunization Safety Review: Measles-Mumps-Rubella Vaccine and Autism.* Washington, DC: National Academy Press

IOM. 2010. *Gulf War and health. Update of Health Effects of Serving in the Gulf War:* Washington, DC: The National Academies Press.

IOM. 2011. *Veterans and Agent Orange: Update 2010.* Washington, DC: The National Academies Press.

Ishitani, K., J. Lin, J. E. Manson, J. E. Buring, and S. M. Zhang. 2008. A prospective study of multivitamin supplement use and risk of breast cancer. *Am J Epidemiol* 167(10):1197–1206.

Iso, T., T. Watanabe, T. Iwamoto, A. Shimamoto, and Y. Furuichi. 2006. DNA damage caused by bisphenol A and estradiol through estrogenic activity. *Biol Pharm Bull* 29(2):206–210.

Itoh, H., M. Iwasaki, T. Hanaoka, Y. Kasuga, S. Yokoyama, H. Onuma, H. Nishimura, R. Kusama, et al. 2009. Serum organochlorines and breast cancer risk in Japanese women: A case–control study. *Cancer Causes & Control* 20(5):567–580.

Iwasaki, M., M. Inoue, S. Sasazuki, N. Kurahashi, H. Itoh, M. Usuda, S. Tsugane, and Japan Public Health Center-based Prospective Study Group. 2008. Plasma organochlorine levels and subsequent risk of breast cancer among Japanese women: A nested case–control study. *Sci Total Environ* 402(2–3):176–183.

Johnson, M. D., N. Kenney, A. Stoica, L. Hilakivi-Clarke, B. Singh, G. Chepko, R. Clarke, P. F. Sholler, et al. 2003. Cadmium mimics the in vivo effects of estrogen in the uterus and mammary gland. *Nat Med* 9(8):1081–1084.

Johnson, K. C., A. B. Miller, N. E. Collishaw, J. R. Palmer, S. K. Hammond, A. G. Salmon, K. P. Cantor, M. D. Miller, et al. 2011. Active smoking and secondhand smoke increase breast cancer risk: The report of the Canadian Expert Panel on Tobacco Smoke and Breast Cancer Risk (2009). *Tob Control* 20(1):e2.

JECFA (Joint FAO/WHO Expert Committee on Food Additives). 2010. *Joint FAO/WHO expert meeting to review toxicological and health aspects of bisphenol A: Final report, including report of stakeholder meeting, Ottawa, Canada.* http://whqlibdoc.who.int/publications/2011/97892141564274_eng.pdf (accessed December 22, 2011).

Kaaks, R., F. Berrino, T. Key, S. Rinaldi, L. Dossus, C. Biessy, G. Secreto, P. Amiano, et al. 2005. Serum sex steroids in premenopausal women and breast cancer risk within the European Prospective Investigation into Cancer and Nutrition (EPIC). *J Natl Cancer Inst* 97(10):755–765.

Kabat, G. C., E. S. O'Leary, E. R. Schoenfeld, J. M. Greene, R. Grimson, K. Henderson, W. T. Kaune, M. D. Gammon, et al. 2003. Electric blanket use and breast cancer on Long Island. *Epidemiology* 14(5):514–520.

Kabat, G. C., A. J. Cross, Y. Park, A. Schatzkin, A. R. Hollenbeck, T. E. Rohan, and R. Sinha. 2009. Meat intake and meat preparation in relation to risk of postmenopausal breast cancer in the NIH–AARP Diet and Health Study. *Int J Cancer* 124(10):2430–2435.

Kabat, G. C., M. Kim, C. Kakani, H. Tindle, J. Wactawski-Wende, J. K. Ockene, J. Luo, S. Wassertheil-Smoller, et al. 2010. Cigarette smoking in relation to risk of ductal carcinoma in situ of the breast in a cohort of postmenopausal women. *Am J Epidemiol* 172(5):591–599.

Kabat, G. C., M. Kim, A. I. Phipps, C. I. Li, C. R. Messina, J. Wactawski-Wende, L. Kuller, M. S. Simon, et al. 2011. Smoking and alcohol consumption in relation to risk of triple-negative breast cancer in a cohort of postmenopausal women. *Cancer Causes Control* 22(5):775–783.

Kaplowitz, P. B., E. J. Slora, R. C. Wasserman, S. E. Pedlow, and M. E. Herman-Giddens. 2001. Earlier onset of puberty in girls: Relation to increased body mass index and race. *Pediatrics* 108(2):347–353.

Kawase, T., K. Matsuo, A. Hiraki, T. Suzuki, M. Watanabe, H. Iwata, H. Tanaka, and K. Tajima. 2009. Interaction of the effects of alcohol drinking and polymorphisms in alcohol-metabolizing enzymes on the risk of female breast cancer in Japan. *J Epidemiol* 19(5):244–250.

Keating, N. L., P. D. Cleary, A. S. Rossi, A. M. Zaslavsky, and J. Z. Ayanian. 1999. Use of hormone replacement therapy by postmenopausal women in the United States. *Ann Intern Med* 130(7):545–553.

Kemp, M. Q., W. Liu, P. A. Thorne, M. D. Kane, O. Selmin, and D. F. Romagnolo. 2006. Induction of the transferrin receptor gene by benzo[a]pyrene in breast cancer MCF-7 cells: Potential as a biomarker of PAH exposure. *Environ Mol Mutagen* 47(7):518–526.

Keri, R. A., S. M. Ho, P. A. Hunt, K. E. Knudsen, A. M. Soto, and G. S. Prins. 2007. An evaluation of evidence for the carcinogenic activity of bisphenol A. *Reprod Toxicol* 24(2):240–252.

Key, T., P. Appleby, I. Barnes, and G. Reeves. 2002. Endogenous sex hormones and breast cancer in postmenopausal women: Reanalysis of nine prospective studies. *J Natl Cancer Inst* 94(8):606–616.

Kheifets, L. I., A. A. Afifi, P. A. Buffler, and Z. W. Zhang. 1995. Occupational electric and magnetic field exposure and brain cancer: A meta-analysis. *J Occup Environ Med* 37(12):1327–1341.

Klatsky, A. L. 2010. Alcohol and cardiovascular health. *Physiol Behav* 100(1):76–81.

Kliukiene, J., T. Tynes, and A. Andersen. 2003. Follow-up of radio and telegraph operators with exposure to electromagnetic fields and risk of breast cancer. *Eur J Cancer Prev* 12(4):301–307.

Kochukov, M. Y., Y. J. Jeng, and C. S. Watson. 2009. Alkylphenol xenoestrogens with varying carbon chain lengths differentially and potently activate signaling and functional responses in GH3/B6/F10 somatomammotropes. *Environ Health Perspect* 117(5):723–730.

Kociba, R. J., D. G. Keyes, J. E. Beyer, R. M. Carreon, C. E. Wade, D. A. Dittenber, R. P. Kalnins, L. E. Frauson, et al. 1978. Results of a two-year chronic toxicity and oncogenicity study of 2,3,7,8-tetrachlorodibenzo-p-dioxin in rats. *Toxicol Appl Pharmacol* 46(2):279–303.

Kolstad, H. A. 2008. Nightshift work and risk of breast cancer and other cancers—a critical review of the epidemiologic evidence. *Scand J Work Environ Health* 34(1):5–22.

Korte, J. E., I. Hertz-Picciotto, M. R. Schulz, L. M. Ball, and E. J. Duell. 2000. The contribution of benzene to smoking-induced leukemia. *Environ Health Perspect* 108(4):333–339.

Krieger, N., M. S. Wolff, R. A. Hiatt, M. Rivera, J. Vogelman, and N. Orentreich. 1994. Breast cancer and serum organochlorines: A prospective study among white, black, and Asian women. *J Natl Cancer Inst* 86(8):589–599.

Kuczmarski, R. J., C. L. Ogden, S. S. Guo, L. M. Grummer-Strawn, K. M. Flegal, Z. Mei, R. Wei, L. R. Curtin, et al. 2002. 2000 CDC growth charts for the United States: Methods and development. *Vital Health Stat* 11(246):1–190.

Kuiper, G. G., B. Carlsson, K. Grandien, E. Enmark, J. Haggblad, S. Nilsson, and J. A. Gustafsson. 1997. Comparison of the ligand binding specificity and transcript tissue distribution of estrogen receptors alpha and beta. *Endocrinology* 138(3):863–870.

La Merrill, M., R. Harper, L. S. Birnbaum, R. D. Cardiff, and D. W. Threadgill. 2010. Maternal dioxin exposure combined with a diet high in fat increases mammary cancer incidence in mice. *Environ Health Perspect* 118(5):596–601.

La Vecchia, C., E. Negri, F. Parazzini, P. Boyle, P. Fasoli, A. Gentile, and S. Franceschi. 1989. Alcohol and breast cancer: Update from an Italian case–control study. *Eur J Cancer Clin Oncol* 25(12):1711–1717.

Labreche, F., M. S. Goldberg, M.-F. Valois, L. Nadon, L. Richardson, R. Lakhani, and B. Latreille. 2003. Occupational exposures to extremely low frequency magnetic fields and postmenopausal breast cancer. *Am J Ind Med* 44(6):643–652.

LaCroix, A. Z., R. T. Chlebowski, J. E. Manson, A. K. Aragaki, K. C. Johnson, L. Martin, K. L. Margolis, M. L. Stefanick, et al. 2011. Health outcomes after stopping conjugated equine estrogens among postmenopausal women with prior hysterectomy: A randomized controlled trial. *JAMA* 305(13):1305–1314.

Laden, F., L. M. Neas, P. E. Tolbert, M. D. Holmes, S. E. Hankinson, D. Spiegelman, F. E. Speizer, and D. J. Hunter. 2000. Electric blanket use and breast cancer in the Nurses' Health Study. *Am J Epidemiol* 152(1):41–49.

Laden, F., N. Ishibe, S. E. Hankinson, M. S. Wolff, D. M. Gertig, D. J. Hunter, and K. T. Kelsey. 2002. Polychlorinated biphenyls, cytochrome P450 1A1, and breast cancer risk in the Nurses' Health Study. *Cancer Epidemiol Biomarkers Prev* 11(12):1560–1565.

Lang, I. A., T. S. Galloway, A. Scarlett, W. E. Henley, M. Depledge, R. B. Wallace, and D. Melzer. 2008. Association of urinary bisphenol A concentration with medical disorders and laboratory abnormalities in adults. *JAMA* 300(11):1303–1310.

Lani, A. 2010. Basis statement for Chapter 883, Designation of the chemical class nonylphenol and nonylphenol ethoxylates as a priority chemical, and Safer Chemicals Program support document for the designation as a priority chemical of nonylphenol and nonylphenol ethoxylates. Bureau of Remediation and Waste Management, Maine Department of Environmental Protection.

Larsson, S. C., A. Akesson, L. Bergkvist, and A. Wolk. 2010. Multivitamin use and breast cancer incidence in a prospective cohort of Swedish women. *Am J Clin Nutr* 91(5):1268–1272.

Lee, K. Y., M. Shibutani, H. Takagi, N. Kato, S. Takigami, C. Uneyama, and M. Hirose. 2004. Diverse developmental toxicity of di-n-butyl phthalate in both sexes of rat offspring after maternal exposure during the period from late gestation through lactation. *Toxicology* 203(1–3):221–238.

Lee, J. M., D. Appugliese, N. Kaciroti, R. F. Corwyn, R. H. Bradley, and J. C. Lumeng. 2007. Weight status in young girls and the onset of puberty. *Pediatrics* 119(3):e624–e630.

Lee, K. H., X. O. Shu, Y. T. Gao, B. T. Ji, G. Yang, A. Blair, N. Rothman, W. Zheng, et al. 2010. Breast cancer and urinary biomarkers of polycyclic aromatic hydrocarbon and oxidative stress in the Shanghai Women's Health Study. *Cancer Epidemiol Biomarkers Prev* 19(3):877–883.

Legler, J. 2008. New insights into the endocrine disrupting effects of brominated flame retardants. *Chemosphere* 73(2):216–222.

Lenz, S. K., M. S. Goldberg, F. Labreche, M. E. Parent, and M. F. Valois. 2002. Association between alcohol consumption and postmenopausal breast cancer: Results of a case–control study in Montreal, Quebec, Canada. *Cancer Causes Control* 13(8):701–710.

Li, Y., R. C. Millikan, D. A. Bell, L. Cui, C.-K. J. Tse, B. Newman, and K. Conway. 2005. Polychlorinated biphenyls, cytochrome P450 1A1 (CYP1A1) polymorphisms, and breast cancer risk among African American women and white women in North Carolina: A population-based case–control study. *Breast Cancer Research* 7(1):R12–R18.

Li, J., N. Li, M. Ma, J. P. Giesy, and Z. Wang. 2008. In vitro profiling of the endocrine disrupting potency of organochlorine pesticides. *Toxicol Lett* 183(1–3):65–71.

Lie, J. A., J. Roessink, and K. Kjaerheim. 2006. Breast cancer and night work among Norwegian nurses. *Cancer Causes Control* 17(1):39–44.

Lin, Y., S. Kikuchi, K. Tamakoshi, K. Wakai, T. Kondo, Y. Niwa, H. Yatsuya, K. Nishio, et al. 2005. Prospective study of alcohol consumption and breast cancer risk in Japanese women. *Int J Cancer* 116(5):779–783.

Lin, P. H., C. H. Lin, C. C. Huang, M. C. Chuang, and P. Lin. 2007. 2,3,7,8-Tetrachlorodibenzo-p-dioxin (TCDD) induces oxidative stress, DNA strand breaks, and poly(ADP-ribose) polymerase-1 activation in human breast carcinoma cell lines. *Toxicol Lett* 172(3):146–158.

London, S. J., J. M. Pogoda, K. L. Hwang, B. Langholz, K. R. Monroe, L. N. Kolonel, W. T. Kaune, J. M. Peters, et al. 2003. Residential magnetic field exposure and breast cancer risk: A nested case–control study from a multiethnic cohort in Los Angeles County, California. *Am J Epidemiol* 158(10):969–980.

Longnecker, M. P. 1994. Alcoholic beverage consumption in relation to risk of breast cancer: Meta-analysis and review. *Cancer Causes Control* 5(1):73–82.

Longnecker, M. P., P. A. Newcomb, R. Mittendorf, E. R. Greenberg, R. W. Clapp, G. Bogdan, W. C. Willett, and B. MacMahon. 1992. The reliability of self-reported alcohol consumption in the remote past. *Epidemiology* 3(6):535–539.

Loomis, D. P. 1992. Cancer of breast among men in electrical occupations. *Lancet* 339(8807):1482–1483.

Lopez-Carrillo, L., R. U. Hernandez-Ramirez, A. M. Calafat, L. Torres-Sanchez, M. Galvan-Portillo, L. L. Needham, R. Ruiz-Ramos, and M. E. Cebrian. 2010. Exposure to phthalates and breast cancer risk in northern Mexico. *Environ Health Perspect* 118(4):539–544.

Lopez-Cervantes, M., L. Torres-Sanchez, A. Tobias, and L. Lopez-Carrillo. 2004. Dichlorodiphenyldichloroethane burden and breast cancer risk: A meta-analysis of the epidemiologic evidence. *Environ Health Perspect* 112(2):207–214.

Lorber, M. 2008. Exposure of Americans to polybrominated diphenyl ethers. *J Expo Sci Environ Epidemiol* 18(1):2–19.

Luo, J., K. Horn, J. K. Ockene, M. S. Simon, M. L. Stefanick, E. Tong, and K. L. Margolis. 2011a. Interaction between smoking and obesity and the risk of developing breast cancer among postmenopausal women: The Women's Health Initiative observational study. *Am J Epidemiol* 174(8):919–928.

Luo, J., K. L. Margolis, J. Wactawski-Wende, K. Horn, C. Messina, M. L. Stefanick, H. A. Tindle, E. Tong, and T. E. Rohan. 2011b. Association of active and passive smoking with risk of breast cancer among postmenopausal women: A prospective cohort study. *BMJ* 342:d1016.

Ma, H., L. Bernstein, M. C. Pike, and G. Ursin. 2006. Reproductive factors and breast cancer risk according to joint estrogen and progesterone receptor status: A meta-analysis of epidemiological studies. *Breast Cancer Res* 8(4):R43.

Macon, M. B., L. R. Villanueva, K. Tatum-Gibbs, R. D. Zehr, M. J. Strynar, J. P. Stanko, S. S. White, L. Helfant, et al. 2011. Prenatal perfluorooctanoic acid exposure in CD-1 mice: Low dose developmental effects and internal dosimetry. *Toxicol Sci* 122(1):134–145.

Makela, S., V. L. Davis, W. C. Tally, J. Korkman, L. Salo, R. Vihko, R. Santti, and K. S. Korach. 1994. Dietary estrogens act through estrogen receptor-mediated processes and show no antiestrogenicity in cultured breast cancer cells. *Environ Health Perspect* 102(6–7):572–578.

Malekinejad, H., R. Maas-Bakker, and J. Fink-Gremmels. 2006. Species differences in the hepatic biotransformation of zearalenone. *Vet J* 172(1):96–102.

Maltoni, C., A. Ciliberti, G. Cotti, B. Conti, and F. Belpoggi. 1989. Benzene, an experimental multipotential carcinogen: Results of the long-term bioassays performed at the Bologna Institute of Oncology. *Environ Health Perspect* 82:109–124.

Maruti, S. S., C. M. Ulrich, and E. White. 2009. Folate and one-carbon metabolism nutrients from supplements and diet in relation to breast cancer risk. *Am J Clin Nutr* 89(2):624–633.

Matanoski, G. M., P. N. Breysse, and E. A. Elliott. 1991. Electromagnetic field exposure and male breast cancer. *Lancet* 337(8743):737.

Matthews, J. B., K. Twomey, and T. R. Zacharewski. 2001. In vitro and in vivo interactions of bisphenol A and its metabolite, bisphenol A glucuronide, with estrogen receptors alpha and beta. *Chem Res Toxicol* 14(2):149–157.

McElroy, J. A., P. A. Newcomb, P. L. Remington, K. M. Egan, L. Titus-Ernstoff, A. Trentham-Dietz, J. M. Hampton, J. A. Baron, et al. 2001. Electric blanket or mattress cover use and breast cancer incidence in women 50–79 years of age. *Epidemiology* 12(6):613–617.

McElroy, J. A., M. M. Shafer, A. Trentham-Dietz, J. M. Hampton, and P. A. Newcomb. 2006. Cadmium exposure and breast cancer risk. *J Natl Cancer Inst* 98(12):869–873.

McGrath, K. G. 2003. An earlier age of breast cancer diagnosis related to more frequent use of antiperspirants/deodorants and underarm shaving. *Eur J Cancer Prev* 12(6):479–485.

McGregor, D. B., R. A. Baan, C. Partensky, J. M. Rice, and J. D. Wilbourn. 2000. Evaluation of the carcinogenic risks to humans associated with surgical implants and other foreign bodies—a report of an IARC Monographs Programme Meeting. International Agency for Research on Cancer. *Eur J Cancer* 36(3):307–313.

Meerts, I. A., R. J. Letcher, S. Hoving, G. Marsh, A. Bergman, J. G. Lemmen, B. van der Burg, and A. Brouwer. 2001. In vitro estrogenicity of polybrominated diphenyl ethers, hydroxylated PDBEs, and polybrominated bisphenol A compounds. *Environ Health Perspect* 109(4):399–407.

Megdal, S. P., C. H. Kroenke, F. Laden, E. Pukkala, and E. S. Schernhammer. 2005. Night work and breast cancer risk: A systematic review and meta-analysis. *Eur J Cancer* 41(13):2023–2032.

Melnick, R. L., and R. C. Sills. 2001. Comparative carcinogenicity of 1,3-butadiene, isoprene, and chloroprene in rats and mice. *Chem Biol Interact* 135–136:27–42.

Mercado-Feliciano, M., and R. M. Bigsby. 2008a. Hydroxylated metabolites of the polybrominated diphenyl ether mixture DE-71 are weak estrogen receptor-alpha ligands. *Environ Health Perspect* 116(10):1315–1321.

Mercado-Feliciano, M., and R. M. Bigsby. 2008b. The polybrominated diphenyl ether mixture DE-71 is mildly estrogenic. *Environ Health Perspect* 116(5):605–611.

Midanik, L. T., and W. B. Clark. 1994. The demographic distribution of US drinking patterns in 1990: Description and trends from 1984. *Am J Public Health* 84(8):1218–1222.

Miller, M. D., M. A. Marty, R. Broadwin, K. C. Johnson, A. G. Salmon, B. Winder, and C. Steinmaus. 2007. The association between exposure to environmental tobacco smoke and breast cancer: A review by the California Environmental Protection Agency. *Prev Med* 44(2):93–106.

Mirick, D. K., S. Davis, and D. B. Thomas. 2002. Antiperspirant use and the risk of breast cancer. *J Natl Cancer Inst* 94(20):1578–1580.

Mirocha, C. J., B. Schauerhamer, C. M. Christensen, M. L. Niku-Paavola, and M. Nummi. 1979. Incidence of zearalenol (*Fusarium* mycotoxin) in animal feed. *Appl Environ Microbiol* 38(4):749–750.

Molis, T. M., L. L. Spriggs, and S. M. Hill. 1994. Modulation of estrogen receptor mRNA expression by melatonin in MCF-7 human breast cancer cells. *Mol Endocrinol* 8(12):1681–1690.

Monteiro, R., C. Calhau, A. O. Silva, S. Pinheiro-Silva, S. Guerreiro, F. Gartner, I. Azevedo, and R. Soares. 2008. Xanthohumol inhibits inflammatory factor production and angiogenesis in breast cancer xenografts. *J Cell Biochem* 104(5):1699–1707.

Moon, H. J., S. Y. Han, J. H. Shin, I. H. Kang, T. S. Kim, J. H. Hong, S. H. Kim, and S. E. Fenton. 2007. Gestational exposure to nonylphenol causes precocious mammary gland development in female rat offspring. *J Reprod Dev* 53(2):333–344.

Morabia, A. 2002. Smoking (active and passive) and breast cancer: Epidemiologic evidence up to June 2001. *Environ Mol Mutagen* 39(2–3):89–95.

Moral, R., R. Wang, I. H. Russo, D. A. Mailo, C. A. Lamartiniere, and J. Russo. 2007. The plasticizer butyl benzyl phthalate induces genomic changes in rat mammary gland after neonatal/prepubertal exposure. *BMC Genomics* 8:453.

Moral, R., R. Wang, I. H. Russo, C. A. Lamartiniere, J. Pereira, and J. Russo. 2008. Effect of prenatal exposure to the endocrine disruptor bisphenol A on mammary gland morphology and gene expression signature. *J Endocrinol* 196(1):101–112.

Moral, R., J. Santucci-Pereira, R. Wang, I. H. Russo, C. A. Lamartiniere, and J. Russo. 2011. In utero exposure to butyl benzyl phthalate induces modifications in the morphology and the gene expression profile of the mammary gland: An experimental study in rats. *Environ Health* 10(1):5.

Mosher, W. D., and J. Jones. 2010. Use of contraception in the United States: 1982–2008. *Vital Health Stat* 23(29):1–44.

Moysich, K. B., P. G. Shields, J. L. Freudenheim, E. F. Schisterman, J. E. Vena, P. Kostyniak, H. Greizerstein, J. R. Marshall, et al. 1999. Polychlorinated biphenyls, cytochrome P4501A1 polymorphism, and postmenopausal breast cancer risk. *Cancer Epidemiol Biomarkers Prev* 8(1):41–44.

Moysich, K. B., R. J. Menezes, J. A. Baker, and K. L. Falkner. 2002. Environmental exposure to polychlorinated biphenyls and breast cancer risk. *Rev Environ Health* 17(4):263–277.

Muñoz-de-Toro, M., C. M. Markey, P. R. Wadia, E. H. Luque, B. S. Rubin, C. Sonnenschein, and A. M. Soto. 2005. Perinatal exposure to bisphenol-A alters peripubertal mammary gland development in mice. *Endocrinology* 146(9):4138–4147.

Murray, T. J., M. V. Maffini, A. A. Ucci, C. Sonnenschein, and A. M. Soto. 2007. Induction of mammary gland ductal hyperplasias and carcinoma in situ following fetal bisphenol A exposure. *Reprod Toxicol* 23(3):383–390.

Nagata, C., Y. Nagao, C. Shibuya, Y. Kashiki, and H. Shimizu. 2005. Urinary cadmium and serum levels of estrogens and androgens in postmenopausal Japanese women. *Cancer Epidemiol Biomarkers Prev* 14(3):705–708.

Nasca, P. C., M. S. Baptiste, N. A. Field, B. B. Metzger, M. Black, C. S. Kwon, and H. Jacobson. 1990. An epidemiological case–control study of breast cancer and alcohol consumption. *Int J Epidemiol* 19(3):532–538.

NCI (National Cancer Institute). 2008. *Antiperspirants/deodorants and breast cancer: Questions and answers.* http://www.cancer.gov/cancertopics/factsheet/Risk/Fs3_66.pdf (accessed October 24, 2011).

NCI. 2011. *Physician Data Query.* http://www.cancer.gov/cancertopics/pdq (accessed October 24, 2011).

Negri, E., C. Bosetti, E. Fattore, and C. La Vecchia. 2003. Environmental exposure to polychlorinated biphenyls (PCBs) and breast cancer: A systematic review of the epidemiological evidence. *Eur J Cancer Prev* 12(6):509–516.

Neuhouser, M. L., S. Wassertheil-Smoller, C. Thomson, A. Aragaki, G. L. Anderson, J. E. Manson, R. E. Patterson, T. E. Rohan, et al. 2009. Multivitamin use and risk of cancer and cardiovascular disease in the Women's Health Initiative cohorts. *Arch Intern Med* 169(3):294–304.

NIAAA (National Institute on Alcohol Abuse and Alcoholism). 2009. *Percent distribution of current drinking status, drinking levels, and heavy drinking days by sex for persons 18 years of age and older: United States, NHIS, 1997–2008.* http://www.niaaa.nih.gov/Resources/DatabaseResources/QuickFacts/AlcoholConsumption/Pages/dkpat25.aspx (accessed June 13, 2011).

Nie, J., J. Beyea, M. R. Bonner, D. Han, J. E. Vena, P. Rogerson, D. Vito, P. Muti, et al. 2007. Exposure to traffic emissions throughout life and risk of breast cancer: The Western New York Exposures and Breast Cancer (WEB) study. *Cancer Causes Control* 18(9):947–955.

NIEHS (National Institute of Environmental Health Sciences). 2009. *28 Oct 2009: NIEHS awards Recovery Act Funds to address bisphenol A research gaps.* http://www.niehs.nih.gov/news/newsroom/releases/2009/october28/index.cfm (accessed October 31, 2011).

NIH (National Institutes of Health), Office of Dietary Supplements. 2011. *Background information: Dietary supplements.* http://ods.od.nih.gov/factsheets/DietarySupplements/ (accessed October 24, 2011).

Nikaido, Y., K. Yoshizawa, R. J. Pei, T. Yuri, N. Danbara, T. Hatano, and A. Tsubura. 2003. Prepubertal zearalenone exposure suppresses N-methyl-N-nitrosourea-induced mammary tumorigenesis but causes severe endocrine disruption in female Sprague-Dawley rats. *Nutr Cancer* 47(2):164–170.

Nissen, S. B., A. Tjonneland, C. Stripp, A. Olsen, J. Christensen, K. Overvad, L. O. Dragsted, and B. Thomsen. 2003. Intake of vitamins A, C, and E from diet and supplements and breast cancer in postmenopausal women. *Cancer Causes Control* 14(8):695–704.

Norman, S. A., J. A. Berlin, K. A. Soper, B. F. Middendorf, and P. D. Stolley. 1995. Cancer incidence in a group of workers potentially exposed to ethylene oxide. *Int J Epidemiol* 24(2):276–284.

NRC (National Research Council). 2006. *Health risks from dioxin and related compounds: Evaluation of the EPA reassessment.* Washington, DC: The National Academies Press.

NRC. 2008. *Phthalates and cumulative risk assessment: The task ahead.* Washington, DC: The National Academies Press.

NTP (National Toxicology Program). 1978. *Bioassays of DDT, TDE, and p,p'-DDE for possible carcinogenicity.* TR-131. http://ntp.niehs.nih.gov/ntp/htdocs/LT_rpts/tr131.pdf (accessed November 2, 2011).

NTP. 1982a. *Carcinogenesis bioassay of bisphenol A (CAS No. 80-05-7) in F344 rats and B6C3FX mice (feed study).* TR-215. Research Triangle Park, NC: NIEHS. http://ntp.niehs.nih.gov/ntp/htdocs/LT_rpts/tr215.pdf (accessed December 22, 2011).

NTP. 1982b. *Carcinogenesis bioassay of zeralenone in F344/N rats and B6C3F₁ mice (feed study).* NTP TR-235. Research Triangle Park, NC: NIEHS. http://ntp.niehs.nih.gov/ntp/htdocs/LT_rpts/tr235.pdf (accessed December 22, 2011).

NTP. 1986. *Toxicology and carcinogenesis studies of decabromodiphenyl oxide (CAS No. 1163-19-5) in R344/N Rats and B6C3F1 mice (feed studies)*. TR-309. Research Triangle Park, NC: NIEHS.

NTP. 1987. *Toxicology and carcinogenesis studies of ethylene oxide (CAS No. 75-21-8) in B6C3F1 mice (inhalation studies)*. NTP Technical Report Series No. 326. NIH Pub. No. 88-2582. Research Triangle Park, NC: NIEHS.

NTP. 2004. *Toxicology and carcinogenesis studies of 2,3,7,8-tetrachlorodibenzo-p-dioxin (TCDD) (CAS No. 1746-01-6) in female Harlan Sprague-Dawley rats (gavage studies)*. TR-521. Research Triangle Park, NC: NIEHS.

NTP. 2006a. *NTP-CERHR monograph on the potential human reproductive and developmental effects of butyl benzyl phthalate* (BBP). http://ntp.niehs.nih.gov/ntp/ohat/phthalates/bb-phthalate/BBP_Monograph_Final.pdf (accessed November 2, 2011).

NTP. 2006b. *Toxicology and carcinogenesis studies of 2,3,7,8-tetrachlorodibenzo-p-dioxin (TCDD) (CAS No. 1746–01–6) in female Harlan Sprague-Dawley rats (gavage studies)*. TR-521. Research Triangle Park, NC: NIEHS. http://ntp.niehs.nih.gov/files/521_web.pdf (accessed December 22, 2011).

NTP. 2008. *NTP–CERHR monograph on the potential human reproductive and developmental effects of bisphenol A*. NIH Pub. No. 08-5994. http://ntp.niehs.nih.gov/ntp/ohat/bisphenol/bisphenol.pdf (accessed November 17, 2011).

NTP. 2011a. *12th report on carcinogens (RoC)*. http://ntp.niehs.nih.gov/index.cfm?objectid=03C9AF75-E1BF-FF40-DBA9EC0928DF8B15 (accessed November 1, 2011).

NTP. 2011b. *Testing status of agents at NTP: 2,2′,4,4′,5,5′-hexabromodiphenyl ether (PDBE 153)*. http://ntp.niehs.nih.gov/INDEX951F_2.HTM (accessed November 1, 2011).

Ogden, C. L., M. M. Lamb, M. D. Carroll, and K. M. Flegal. 2010. Obesity and socioeconomic status in children and adolescents: United States, 2005–2008. *NCHS Data Brief* (51):1–8.

Oh, M. J., S. Paul, and S. J. Kim. 2008. Comparative analysis of gene expression pattern after exposure to nonylphenol in human cell lines. *Biochip J* 2:261–268.

O'Leary, E. S., E. R. Schoenfeld, R. G. Stevens, G. C. Kabat, K. Henderson, R. Grimson, M. D. Gammon, and M. C. Leske. 2006. Shift work, light at night, and breast cancer on Long Island, New York. *Am J Epidemiol* 164(4):358–366.

Olsen, G. W., J. L. Butenhoff, and L. R. Zobel. 2009. Perfluoroalkyl chemicals and human fetal development: An epidemiologic review with clinical and toxicological perspectives. *Reprod Toxicol* 27(3–4):212–230.

Palmer, J. R., and L. Rosenberg. 1993. Cigarette smoking and the risk of breast cancer. *Epidemiol Rev* 15(1):145–156.

Palmer, J. R., L. L. Adams-Campbell, D. A. Boggs, L. A. Wise, and L. Rosenberg. 2007. A prospective study of body size and breast cancer in black women. *Cancer Epidemiol Biomarkers Prev* 16(9):1795–1802.

Parmar, D., S. P. Srivastava, and P. K. Seth. 1985. Hepatic mixed function oxidases and cytochrome P-450 contents in rat pups exposed to di-(2-ethylhexyl)phthalate through mother's milk. *Drug Metab Dispos* 13(3):368–370.

Pesatori, A. C., D. Consonni, M. Rubagotti, P. Grillo, and P. A. Bertazzi. 2009. Cancer incidence in the population exposed to dioxin after the "Seveso accident": Twenty years of follow-up. *Environ Health* 8:39.

Peters, C. A. 1972. Photochemistry of zearalenone and its derivatives. *J Med Chem* 15(8): 867–868.

Petralia, S. A., J. E. Vena, J. L. Freudenheim, M. Dosemeci, A. Michalek, M. S. Goldberg, J. Brasure, and S. Graham. 1999. Risk of premenopausal breast cancer in association with occupational exposure to polycyclic aromatic hydrocarbons and benzene. *Scand J Work Environ Health* 25(3):215–221.

Petreas, M., J. She, F. R. Brown, J. Winkler, G. Windham, E. Rogers, G. Zhao, R. Bhatia, et al. 2003. High body burdens of 2,2′,4,4′-tetrabromodiphenyl ether (BDE-47) in California women. *Environ Health Perspect* 111(9):1175–1179.

Petreas, M., D. Smith, S. Hurley, S. S. Jeffrey, D. Gilliss, and P. Reynolds. 2004. Distribution of persistent, lipid-soluble chemicals in breast and abdominal adipose tissues: Lessons learned from a breast cancer study. *Cancer Epidemiol Biomarkers Prev* 13(3):416–424.

Petreas, M., D. Nelson, F. R. Brown, D. Goldberg, S. Hurley, and P. Reynolds. 2011. High concentrations of polybrominated diphenylethers (PBDEs) in breast adipose tissue of California women. *Environ Int* 37(1):190–197.

Pezzotti, A., P. Kraft, S. E. Hankinson, D. J. Hunter, J. Buring, and D. G. Cox. 2009. The mitochondrial A10398G polymorphism, interaction with alcohol consumption, and breast cancer risk. *PLoS One* 4(4):e5356.

Platek, M. E., P. G. Shields, C. Marian, S. E. McCann, M. R. Bonner, J. Nie, C. B. Ambrosone, A. E. Millen, et al. 2009. Alcohol consumption and genetic variation in methylenetetrahydrofolate reductase and 5-methyltetrahydrofolate-homocysteine methyltransferase in relation to breast cancer risk. *Cancer Epidemiol Biomarkers Prev* 18(9):2453–2459.

Pliskova, M., J. Vondracek, R. F. Canton, J. Nera, A. Kocan, J. Petrik, T. Trnovec, T. Sanderson, et al. 2005. Impact of polychlorinated biphenyls contamination on estrogenic activity in human male serum. *Environ Health Perspect* 113(10):1277–1284.

Polanco, T. A., C. Crismale-Gann, K. R. Reuhl, D. K. Sarkar, and W. S. Cohick. 2010. Fetal alcohol exposure increases mammary tumor susceptibility and alters tumor phenotype in rats. *Alcohol Clin Exp Res* 34(11):1879–1887.

Pompa, G., C. Montesissa, F. M. Di Lauro, L. Fadini, and C. Capua. 1988. Zearanol metabolism by subcellular fractions from lamb liver. *J Vet Pharmacol Ther* 11(2):197–203.

Quach, T., K. D. Nguyen, P. A. Doan-Billings, L. Okahara, C. Fan, and P. Reynolds. 2008. A preliminary survey of Vietnamese nail salon workers in Alameda County, California. *J Community Health* 33(5):336–343.

Quach, T., P. A. Doan-Billing, M. Layefsky, D. Nelson, K. D. Nguyen, L. Okahara, A. N. Tran, J. Von Behren, et al. 2010. Cancer incidence in female cosmetologists and manicurists in California, 1988–2005. *Am J Epidemiol* 172(6):691–699.

Quach, T., R. Gunier, A. Tran, J. Von Behren, P. A. Doan-Billings, K. D. Nguyen, L. Okahara, B. Lui, et al. 2011. Characterizing workplace exposures in Vietnamese women working in California nail salons. *Am J Public Health* 101 (Suppl 1):S271–S276.

Rato, A. G., J. G. Pedrero, M. A. Martinez, B. del Rio, P. S. Lazo, and S. Ramos. 1999. Melatonin blocks the activation of estrogen receptor for DNA binding. *FASEB J* 13(8):857–868.

Rayner, J. L., R. R. Enoch, and S. E. Fenton. 2005. Adverse effects of prenatal exposure to atrazine during a critical period of mammary gland growth. *Toxicol Sci* 87(1):255–266.

Reynolds, P., S. E. Hurley, M. Petreas, D. E. Goldberg, D. Smith, D. Gilliss, M. E. Mahoney, and S. S. Jeffrey. 2005. Adipose levels of dioxins and risk of breast cancer. *Cancer Causes Control* 16(5):525–535.

Reynolds, P., D. Goldberg, S. Hurley, D. O. Nelson, J. Largent, K. D. Henderson, and L. Bernstein. 2009. Passive smoking and risk of breast cancer in the California Teachers Study. *Cancer Epidemiol Biomarkers Prev* 18(12):3389–3398.

Rhomberg, L. R., and T. A. Lewandoski. 2004. *Methods for identifying a default cross-species scaling factor*. Prepared for Risk Assessment Forum, U.S. Environmental Protection Agency. http://www.epa.gov/raf/publications/pdfs/RHOMBERGSPAPER.PDF (accessed December 22, 2011).

Rivara, M. I., M. Cebrian, G. Corey, M. Hernandez, and I. Romieu. 1997. Cancer risk in an arsenic-contaminated area of Chile. *Toxicol Ind Health* 13(2–3):321–338.

Robien, K., G. J. Cutler, and D. Lazovich. 2007. Vitamin D intake and breast cancer risk in postmenopausal women: The Iowa Women's Health Study. *Cancer Causes Control* 18(7):775–782.

Robinson, C. F., and J. T. Walker. 1999. Cancer mortality among women employed in fast-growing U.S. occupations. *Am J Ind Med* 36(1):186–192.

Romero-Corral, A., V. K. Somers, J. Sierra-Johnson, R. J. Thomas, M. L. Collazo-Clavell, J. Korinek, T. G. Allison, J. A. Batsis, et al. 2008. Accuracy of body mass index in diagnosing obesity in the adult general population. *Int J Obes (Lond)* 32(6):959–966.

Rose, M., D. H. Bennett, A. Bergman, B. Fangstrom, I. N. Pessah, and I. Hertz-Picciotto. 2010. PBDEs in 2–5 year-old children from California and associations with diet and indoor environment. *Environ Sci Technol* 44(7):2648–2653.

Rosenbaum, P. F., J. E. Vena, M. A. Zielezny, and A. M. Michalek. 1994. Occupational exposures associated with male breast cancer. *Am J Epidemiol* 139(1):30–36.

Routledge, E. J., R. White, M. G. Parker, and J. P. Sumpter. 2000. Differential effects of xenoestrogens on coactivator recruitment by estrogen receptor (ER) alpha and ER beta. *J Biol Chem* 275(46):35986–35993.

Rudel, R. A., J. M. Gray, C. L. Engel, T. W. Rawsthorne, R. E. Dodson, J. M. Ackerman, J. Rizzo, J. L. Nudelman, et al. 2011. Food packaging and bisphenol A and bis(2-ethyhexyl) phthalate exposure: Findings from a dietary intervention. *Environ Health Perspect* 119(7):914–920.

Safe, S. 2005. Clinical correlates of environmental endocrine disruptors. *Trends Endocrinol Metab* 16(4):139–144.

Safe, S., and M. Wormke. 2003. Inhibitory aryl hydrocarbon receptor-estrogen receptor alpha cross-talk and mechanisms of action. *Chem Res Toxicol* 16(7):807–816.

Sagiv, S. K., M. M. Gaudet, S. M. Eng, P. E. Abrahamson, S. Shantakumar, S. L. Teitelbaum, P. Bell, J. A. Thomas, et al. 2009. Polycyclic aromatic hydrocarbon-DNA adducts and survival among women with breast cancer. *Environ Res* 109(3):287–291.

Salehi, F., M. C. Turner, K. P. Phillips, D. T. Wigle, D. Krewski, and K. J. Aronson. 2008. Review of the etiology of breast cancer with special attention to organochlorines as potential endocrine disruptors. *J Toxicol Environ Health B Crit Rev* 11(3–4):276–300.

Santodonato, J. 1997. Review of the estrogenic and antiestrogenic activity of polycyclic aromatic hydrocarbons: Relationship to carcinogenicity. *Chemosphere* 34(4):835–848.

Sathiakumar, N., P. A. MacLennan, J. Mandel, and E. Delzell. 2011. A review of epidemiologic studies of triazine herbicides and cancer. *Crit Rev Toxicol* 41(Suppl 1):1–34.

Sathyanarayana, S., C. J. Karr, P. Lozano, E. Brown, A. M. Calafat, F. Liu, and S. H. Swan. 2008. Baby care products: Possible sources of infant phthalate exposure. *Pediatrics* 121(2):e260–e268.

Schecter, A., O. Papke, K. C. Tung, J. Joseph, T. R. Harris, and J. Dahlgren. 2005. Polybrominated diphenyl ether flame retardants in the U.S. population: Current levels, temporal trends, and comparison with dioxins, dibenzofurans, and polychlorinated biphenyls. *J Occup Environ Med* 47(3):199–211.

Schernhammer, E. S., and S. E. Hankinson. 2005. Urinary melatonin levels and breast cancer risk. *J Natl Cancer Inst* 97(14):1084–1087.

Schernhammer, E. S., F. Laden, F. E. Speizer, W. C. Willett, D. J. Hunter, I. Kawachi, and G. A. Colditz. 2001. Rotating night shifts and risk of breast cancer in women participating in the Nurses' Health Study. *J Natl Cancer Inst* 93(20):1563–1568.

Schoenfeld, E. R., E. S. O'Leary, K. Henderson, R. Grimson, G. C. Kabat, S. Ahnn, W. T. Kaune, M. D. Gammon, et al. 2003. Electromagnetic fields and breast cancer on Long Island: A case–control study. *Am J Epidemiol* 158(1):47–58.

Schoental, R. 1974. Letter: Role of podophyllotoxin in the bedding and dietary zearalenone on incidence of spontaneous tumors in laboratory animals. *Cancer Res* 34(9):2419–2420.

Schrader, T. J., I. Langlois, K. Soper, and W. Cherry. 2002. Mutagenicity of bisphenol A (4,4′-isopropylidenediphenol) in vitro: Effects of nitrosylation. *Teratog Carcinog Mutagen* 22(6):425–441.

Schwartzbaum, J., A. Ahlbom, and M. Feychting. 2007. Cohort study of cancer risk among male and female shift workers. *Scand J Work Environ Health* 33(5):336–343.

Schweikl, H., G. Schmalz, and K. Rackebrandt. 1998. The mutagenic activity of unpolymerized resin monomers in *Salmonella typhimurium* and V79 cells. *Mutat Res* 415(1–2):119–130.

Secretan, B., K. Straif, R. Baan, Y. Grosse, F. El Ghissassi, V. Bouvard, L. Benbrahim-Tallaa, N. Guha, et al. 2009. A review of human carcinogens—Part E: Tobacco, areca nut, alcohol, coal smoke, and salted fish. *Lancet Oncol* 10(11):1033–1034.

Seitz, H. K., and B. Maurer. 2007. The relationship between alcohol metabolism, estrogen levels, and breast cancer risk. *Alcohol Res Health* 30(1):42–43.

Seitz, H. K., and U. A. Simanowski. 1988. Alcohol and carcinogenesis. *Annu Rev Nutr* 8:99–119.

Seitz, H. K., and F. Stickel. 2010. Acetaldehyde as an underestimated risk factor for cancer development: Role of genetics in ethanol metabolism. *Genes Nutr* 5:121–128.

Shier, V. K., C. J. Hancey, and S. J. Benkovic. 2001. Identification of the active oligomeric state of an essential adenine DNA methyltransferase from *Caulobacter crescentus*. *J Biol Chem* 276(18):14744–14751.

Shipley, J. M., and D. J. Waxman. 2006. Aryl hydrocarbon receptor-independent activation of estrogen receptor-dependent transcription by 3-methylcholanthrene. *Toxicol Appl Pharmacol* 213(2):87–97.

Sibinski, L. J. 1987. *Final report of a two-year oral (diet) toxicity and carcinogenicity study of fluorochemical FC-143 (perfluorooctane ammonium carboxylate) in rats.* Study No. 0281CR0012; 8EHQ-1087-0394. Report prepared for 3M, St Paul, Minnesota by Riker Laboratories, Inc.

Silva, M. J., J. A. Reidy, A. R. Herbert, J. L. Preau, Jr., L. L. Needham, and A. M. Calafat. 2004. Detection of phthalate metabolites in human amniotic fluid. *Bull Environ Contam Toxicol* 72(6):1226–1231.

Silva, E., M. J. Lopez-Espinosa, J. M. Molina-Molina, M. Fernandez, N. Olea, and A. Kortenkamp. 2006. Lack of activity of cadmium in in vitro estrogenicity assays. *Toxicol Appl Pharmacol* 216(1):20–28.

Silver, S. R., E. A. Whelan, J. A. Deddens, N. K. Steenland, N. B. Hopf, M. A. Waters, A. M. Ruder, M. M. Prince, et al. 2009. Occupational exposure to polychlorinated biphenyls and risk of breast cancer. *Environ Health Perspect* 117(2):276–282.

Singleton, J. A., and J. J. Beaumont. 1989. *COMS II: California occupational mortality 1979–1981, adjusted for smoking, alcohol, and socioeconomic status.* Sacramento, CA: California Department of Health Services.

Sjodin, A., R. S. Jones, J. F. Focant, C. Lapeza, R. Y. Wang, E. E. McGahee, III, Y. Zhang, W. E. Turner, et al. 2004. Retrospective time-trend study of polybrominated diphenyl ether and polybrominated and polychlorinated biphenyl levels in human serum from the United States. *Environ Health Perspect* 112(6):654–658.

Sjodin, A., L. Y. Wong, R. S. Jones, A. Park, Y. Zhang, C. Hodge, E. Dipietro, C. McClure, et al. 2008. Serum concentrations of polybrominated diphenyl ethers (PBDEs) and polybrominated biphenyl (PBB) in the United States population: 2003–2004. *Environ Sci Technol* 42(4):1377–1384.

Smith-Warner, S. A., D. Spiegelman, S. S. Yaun, P. A. van den Brandt, A. R. Folsom, R. A. Goldbohm, S. Graham, L. Holmberg, et al. 1998. Alcohol and breast cancer in women: A pooled analysis of cohort studies. *JAMA* 279(7):535–540.

Snedeker, S. M. 2001. Pesticides and breast cancer risk: A review of DDT, DDE, and dieldrin. *Environ Health Perspect* 109(Suppl 1):35–47.

Soni, M. G., I. G. Carabin, and G. A. Burdock. 2005. Safety assessment of esters of p-hydroxybenzoic acid (parabens). *Food Chem Toxicol* 43(7):985–1015.

Soto, A. M., and C. Sonnenschein. 2010. Environmental causes of cancer: Endocrine disruptors as carcinogens. *Nat Rev Endocrinol* 6(7):363–370.

Stahlhut, R. W., W. V. Welshons, and S. H. Swan. 2009. Bisphenol A data in NHANES suggest longer than expected half-life, substantial nonfood exposure, or both. *Environ Health Perspect* 117(5):784–789.

Starek, A. 2003. Estrogens and organochlorine xenoestrogens and breast cancer risk. *Int J Occup Med Environ Health* 16(2):113–124.

Stark, A., D. Schultz, A. Kapke, P. Nadkarni, M. Burke, M. Linden, and U. Raju. 2009. Obesity and risk of the less commonly diagnosed subtypes of breast cancer. *Eur J Surg Oncol* 35(9):928–935.

Steck, S. E., M. M. Gaudet, S. M. Eng, J. A. Britton, S. L. Teitelbaum, A. I. Neugut, R. M. Santella, and M. D. Gammon. 2007. Cooked meat and risk of breast cancer—lifetime versus recent dietary intake. *Epidemiology* 18(3):373–382.

Steenland, K., E. Whelan, J. Deddens, L. Stayner, and E. Ward. 2003. Ethylene oxide and breast cancer incidence in a cohort study of 7576 women (United States). *Cancer Causes Control* 14(6):531–539.

Steenland, K., T. Fletcher, and D. A. Savitz. 2010. Epidemiologic evidence on the health effects of perfluorooctanoic acid (PFOA). *Environ Health Perspect* 118(8):1100–1108.

Steinetz, B. G., T. Gordon, S. Lasano, L. Horton, S. P. Ng, J. T. Zelikoff, A. Nadas, and M. C. Bosland. 2006. The parity-related protection against breast cancer is compromised by cigarette smoke during rat pregnancy: Observations on tumorigenesis and immunological defenses of the neonate. *Carcinogenesis* 27(6):1146–1152.

Stenlund, C., and B. Floderus. 1997. Occupational exposure to magnetic fields in relation to male breast cancer and testicular cancer: A Swedish case–control study. *Cancer Causes Control* 8(2):184–191.

Stolzenberg-Solomon, R. Z., S. C. Chang, M. F. Leitzmann, K. A. Johnson, C. Johnson, S. S. Buys, R. N. Hoover, and R. G. Ziegler. 2006. Folate intake, alcohol use, and postmenopausal breast cancer risk in the Prostate, Lung, Colorectal, and Ovarian Cancer Screening Trial. *Am J Clin Nutr* 83(4):895–904.

Straif, K., R. Baan, Y. Grosse, B. Secretan, F. El Ghissassi, V. Bouvard, A. Altieri, L. Benbrahim-Tallaa, et al. 2007. Carcinogenicity of shift-work, painting, and fire-fighting. *Lancet Oncol* 8(12):1065–1066.

Straif, K. L Benbrahim-Tallaa, R. Baan, Y. Grosse, B. Secretan, F El Ghissassi, B. Bouvard, N. Guha, C. Freeman, L. Galichet, et al. 2009. A review of human carcinogens—Part C: Metals, arsenic, dusts, and fibres. *Lancet Oncol* 10: 453–454.

Sun, H., X. L. Xu, J. H. Qu, X. Hong, Y. B. Wang, L. C. Xu, and X. R. Wang. 2008. 4-Alkylphenols and related chemicals show similar effect on the function of human and rat estrogen receptor alpha in reporter gene assay. *Chemosphere* 71(3):582–588.

Suvorov, A., and L. Takser. 2008. Facing the challenge of data transfer from animal models to humans: The case of persistent organohalogens. *Environ Health* 7:58.

Suzuki, R., W. Ye, T. Rylander-Rudqvist, S. Saji, G. A. Colditz, and A. Wolk. 2005. Alcohol and postmenopausal breast cancer risk defined by estrogen and progesterone receptor status: A prospective cohort study. *J Natl Cancer Inst* 97(21):1601–1608.

Suzuki, R., N. Orsini, L. Mignone, S. Saji, and A. Wolk. 2008. Alcohol intake and risk of breast cancer defined by estrogen and progesterone receptor status—a meta-analysis of epidemiological studies. *Int J Cancer* 122(8):1832–1841.

Suzuki, R., N. Orsini, S. Saji, T. J. Key, and A. Wolk. 2009. Body weight and incidence of breast cancer defined by estrogen and progesterone receptor status—a meta-analysis. *Int J Cancer* 124(3):698–712.

Takemura, H., J. Y. Shim, K. Sayama, A. Tsubura, B. T. Zhu, and K. Shimoi. 2007. Characterization of the estrogenic activities of zearalenone and zeranol in vivo and in vitro. *J Steroid Biochem Mol Biol* 103(2):170–177.

Takkouche, B., M. Etminan, and A. Montes-Martinez. 2005. Personal use of hair dyes and risk of cancer: A meta-analysis. *JAMA* 293(20):2516–2525.

Talsness, C. E., S. N. Kuriyama, A. Sterner-Kock, P. Schnitker, S. W. Grande, M. Shakibaei, A. Andrade, K. Grote, et al. 2008. In utero and lactational exposures to low doses of polybrominated diphenyl ether-47 alter the reproductive system and thyroid gland of female rat offspring. *Environ Health Perspect* 116(3):308–314.

Tannheimer, S. L., S. L. Barton, S. P. Ethier, and S. W. Burchiel. 1997. Carcinogenic polycyclic aromatic hydrocarbons increase intracellular Ca2+ and cell proliferation in primary human mammary epithelial cells. *Carcinogenesis* 18(6):1177–1182.

Tarone, R. E. 2008. DDT and breast cancer trends. *Environ Health Perspect* 116(9):A374.

Teeguarden, J. G., A. M. Calafat, X. Ye, D. R. Doerge, M. I. Churchwell, R. Gunawan, and M. K. Graham. 2011. Twenty-four hour human urine and serum profiles of bisphenol A during high-dietary exposure. *Toxicol Sci* 123(1):48–57.

Tempfer, C. B., G. Froese, G. Heinze, E. K. Bentz, L. A. Hefler, and J. C. Huber. 2009. Side effects of phytoestrogens: A meta-analysis of randomized trials. *Am J Med* 122(10): 939–946.

Tennant, R. W., B. H. Margolin, M. D. Shelby, E. Zeiger, J. K. Haseman, J. Spalding, W. Caspary, M. Resnick, et al. 1987. Prediction of chemical carcinogenicity in rodents from in vitro genetic toxicity assays. *Science* 236(4804):933–941.

Teras, L. R., M. Goodman, A. V. Patel, W. R. Diver, W. D. Flanders, and H. S. Feigelson. 2011. Weight loss and postmenopausal breast cancer in a prospective cohort of overweight and obese US women. *Cancer Causes Control* 22(4):573–579.

Terasaka, S., A. Inoue, M. Tanji, and R. Kiyama. 2006. Expression profiling of estrogen-responsive genes in breast cancer cells treated with alkylphenols, chlorinated phenols, parabens, or bis- and benzoylphenols for evaluation of estrogenic activity. *Toxicol Lett* 163(2):130–141.

Terry, P. D., and M. Goodman. 2006. Is the association between cigarette smoking and breast cancer modified by genotype? A review of epidemiologic studies and meta-analysis. *Cancer Epidemiol Biomarkers Prev* 15(4):602–611.

Terry, M. B., M. D. Gammon, F. F. Zhang, J. A. Knight, Q. Wang, J. A. Britton, S. L. Teitelbaum, A. I. Neugut, and R. M. Santella. 2006a. ADH3 genotype, alcohol intake and breast cancer risk. *Carcinogenesis* 27(4):840–847.

Terry, M. B., F. F. Zhang, G. Kabat, J. A. Britton, S. L. Teitelbaum, A. I. Neugut, and M. D. Gammon. 2006b. Lifetime alcohol intake and breast cancer risk. *Ann Epidemiol* 16(3):230–240.

Teta, M. J., J. Walrath, J. W. Meigs, and J. T. Flannery. 1984. Cancer incidence among cosmetologists. *J Natl Cancer Inst* 72(5):1051–1057.

Theriault, G., M. Goldberg, A. B. Miller, B. Armstrong, P. Guenel, J. Deadman, E. Imbernon, T. To, et al. 1994. Cancer risks associated with occupational exposure to magnetic fields among electric utility workers in Ontario and Quebec, Canada, and France: 1970–1989. *Am J Epidemiol* 139(6):550–572.

Thompson, P. D., and V. Lim. 2003. Physical activity in the prevention of atherosclerotic coronary heart disease. *Curr Treat Options Cardiovasc Med* 5(4):279–285.

Thomsen, C., E. Lundanes, and G. Becher. 2002. Brominated flame retardants in archived serum samples from Norway: A study on temporal trends and the role of age. *Environ Sci Technol* 36(7):1414–1418.

Tinfo, N. S., M. G. Hotchkiss, A. R. Buckalew, L. M. Zorrilla, R. L. Cooper, and S. C. Laws. 2011. Understanding the effects of atrazine on steroidogenesis in rat granulosa and H295R adrenal cortical carcinoma cells. *Reprod Toxicol* 31(2):184–193.

Tjonneland, A., J. Christensen, B. L. Thomsen, A. Olsen, C. Stripp, K. Overvad, and J. H. Olsen. 2004. Lifetime alcohol consumption and postmenopausal breast cancer rate in Denmark: A prospective cohort study. *J Nutr* 134(1):173–178.

Trivers, K. F., M. J. Lund, P. L. Porter, J. M. Liff, E. W. Flagg, R. J. Coates, and J. W. Eley. 2009. The epidemiology of triple-negative breast cancer, including race. *Cancer Causes Control* 20(7):1071–1082.

Trudel, D., L. Horowitz, M. Wormuth, M. Scheringer, I. T. Cousins, and K. Hungerbuhler. 2008. Estimating consumer exposure to PFOS and PFOA. *Risk Anal* 28(2):251–269.

Tsai, S. M., T. N. Wang, and Y. C. Ko. 1999. Mortality for certain diseases in areas with high levels of arsenic in drinking water. *Arch Environ Health* 54(3):186–193.

Tsutsui, T., Y. Tamura, E. Yagi, K. Hasegawa, M. Takahashi, N. Maizumi, F. Yamaguchi, and J. C. Barrett. 1998. Bisphenol-A induces cellular transformation, aneuploidy and DNA adduct formation in cultured Syrian hamster embryo cells. *Int J Cancer* 75(2):290–294.

Tworoger, S. S., S. A. Missmer, A. H. Eliassen, D. Spiegelman, E. Folkerd, M. Dowsett, R. L. Barbieri, and S. E. Hankinson. 2006. The association of plasma DHEA and DHEA sulfate with breast cancer risk in predominantly premenopausal women. *Cancer Epidemiol Biomarkers Prev* 15(5):967–971.

Tynes, T., M. Hannevik, A. Andersen, A. I. Vistnes, and T. Haldorsen. 1996. Incidence of breast cancer in Norwegian female radio and telegraph operators. *Cancer Causes Control* 7(2):197–204.

Ueno, Y., F. Tashiro, and T. Kobayashi. 1983. Species differences in zearalenone-reductase activity. *Food Chem Toxicol* 21(2):167–173.

Uppala, P. T., S. K. Roy, A. Tousson, S. Barnes, G. R. Uppala, and D. A. Eastmond. 2005. Induction of cell proliferation, micronuclei and hyperdiploidy/polyploidy in the mammary cells of DDT- and DMBA-treated pubertal rats. *Environ Mol Mutagen* 46(1):43–52.

van Noord, P. A. 2004. Breast cancer and the brain: A neurodevelopmental hypothesis to explain the opposing effects of caloric deprivation during the Dutch famine of 1944–1945 on breast cancer and its risk factors. *J Nutr* 134(12 Suppl):3399S–3406S.

Van Wijngaarden, E., L. A. Nylander-French, R. C. Millikan, D. A. Savitz, and D. Loomis. 2001. Population-based case–control study of occupational exposure to electromagnetic fields and breast cancer. *Annals of Epidemiology* 11(5):297–303.

Vandenberg, L. N., R. Hauser, M. Marcus, N. Olea, and W. V. Welshons. 2007. Human exposure to bisphenol A (BPA). *Reprod Toxicol* 24(2):139–177.

Vandenberg, L. N., I. Chahoud, J. J. Heindel, V. Padmanabhan, F. J. Paumgartten, and G. Schoenfelder. 2010. Urinary, circulating, and tissue biomonitoring studies indicate widespread exposure to bisphenol A. *Environ Health Perspect* 118(8):1055–1070.

van't Veer, P., F. J. Kok, R. J. Hermus, and F. Sturmans. 1989. Alcohol dose, frequency and age at first exposure in relation to the risk of breast cancer. *Int J Epidemiol* 18(3):511–517.

Vasiliu, O., J. Muttineni, and W. Karmaus. 2004. In utero exposure to organochlorines and age at menarche. *Hum Reprod* 19(7):1506–1512.

Vena, J. E., S. Graham, R. Hellmann, M. Swanson, and J. Brasure. 1991. Use of electric blankets and risk of postmenopausal breast cancer. *Am J Epidemiol* 134(2):180–185.

Vena, J. E., J. L. Freudenheim, J. R. Marshall, R. Laughlin, M. Swanson, and S. Graham. 1994. Risk of premenopausal breast cancer and use of electric blankets. *Am J Epidemiol* 140(11):974–979.

Verhoeven, D. T., N. Assen, R. A. Goldbohm, E. Dorant, P. van't Veer, F. Sturmans, R. J. Hermus, and P. A. van den Brandt. 1997. Vitamins C and E, retinol, beta-carotene and dietary fibre in relation to breast cancer risk: A prospective cohort study. *Br J Cancer* 75(1):149–155.

Verner, M. A., D. Bachelet, R. McDougall, M. Charbonneau, P. Guenel, and S. Haddad. 2011. A case study addressing the reliability of polychlorinated biphenyl levels measured at the time of breast cancer diagnosis in representing early-life exposure. *Cancer Epidemiol Biomarkers Prev* 20(2):281–286.

Villeneuve, S., D. Cyr, E. Lynge, L. Orsi, S. Sabroe, F. Merletti, G. Gorini, M. Morales-Suarez-Varela, et al. 2010. Occupation and occupational exposure to endocrine disrupting chemicals in male breast cancer: A case–control study in Europe. *Occup Environ Med* 67(12):837–844.

Visvanathan, K., R. M. Crum, P. T. Strickland, X. You, I. Ruczinski, S. I. Berndt, A. J. Alberg, S. C. Hoffman, et al. 2007. Alcohol dehydrogenase genetic polymorphisms, low-to-moderate alcohol consumption, and risk of breast cancer. *Alcohol Clin Exp Res* 31(3):467–476.

vom Saal, F. S., B. T. Akingbemi, S. M. Belcher, L. S. Birnbaum, D. A. Crain, M. Eriksen, F. Farabollini, L. J. Guillette, Jr., et al. 2007. Chapel Hill bisphenol A expert panel consensus statement: Integration of mechanisms, effects in animals and potential to impact human health at current levels of exposure. *Reprod Toxicol* 24(2):131–138.

Vorderstrasse, B. A., S. E. Fenton, A. A. Bohn, J. A. Cundiff, and B. P. Lawrence. 2004. A novel effect of dioxin: Exposure during pregnancy severely impairs mammary gland differentiation. *Toxicol Sci* 78(2):248–257.

Wang, T., H. M. Gavin, V. M. Arlt, B. P. Lawrence, S. E. Fenton, D. Medina, and B. A. Vorderstrasse. 2011. Aryl hydrocarbon receptor activation during pregnancy, and in adult nulliparous mice, delays the subsequent development of DMBA-induced mammary tumors. *Int J Cancer* 128(7):1509–1523.

Warburton, D. E., C. W. Nicol, and S. S. Bredin. 2006. Health benefits of physical activity: The evidence. *CMAJ* 174(6):801–809.

Warner, M., B. Eskenazi, P. Mocarelli, P. M. Gerthoux, S. Samuels, L. Needham, D. Patterson, and P. Brambilla. 2002. Serum dioxin concentrations and breast cancer risk in the Seveso Women's Health Study. *Environ Health Perspect* 110(7):625–628.

Warner, M., P. Mocarelli, S. Samuels, L. L. Needham, P. Brambilla, and B. Eskenazi. 2011. Dioxin exposure and cancer risk in the Seveso Women's Health Study. *Environ Health Perspect* 119(12):1700–1705.

Watson, C. S., N. N. Bulayeva, A. L. Wozniak, and R. A. Alyea. 2007. Xenoestrogens are potent activators of nongenomic estrogenic responses. *Steroids* 72(2):124–134.

WCRF/AICR (World Cancer Research Fund/American Institute for Cancer Research). 2007. *Food, nutrition, physical activity, and the prevention of cancer: A global perspective.* Washington, DC: AICR.

WCRF/AICR. 2008. The associations between food, nutrition and physical activity and the risk of breast cancer. *WCRF/AICR Systematic literature review, Continuous update report.* Washington, DC: AICR.

WCRF/AICR. 2010. The associations between food, nutrition and physical activity and the risk of breast cancer. *WCRF/AICR Systematic literature review, Continuous update report.* Washington, DC: AICR.

Wen, S., F. Yang, J. G. Li, Y. Gong, X. L. Zhang, Y. Hui, Y. N. Wu, Y. F. Zhao, et al. 2009. Polychlorinated dibenzo-p-dioxin and dibenzofurans (PCDD/Fs), polybrominated diphenyl ethers (PBDEs), and polychlorinated biphenyls (PCBs) monitored by tree bark in an E-waste recycling area. *Chemosphere* 74(7):981–987.

Weng, Y. I., P. Y. Hsu, S. Liyanarachchi, J. Liu, D. E. Deatherage, Y. W. Huang, T. Zuo, B. Rodriguez, et al. 2010. Epigenetic influences of low-dose bisphenol A in primary human breast epithelial cells. *Toxicol Appl Pharmacol* 248(2):111–121.

Wetherill, Y. B., B. T. Akingbemi, J. Kanno, J. A. McLachlan, A. Nadal, C. Sonnenschein, C. S. Watson, R. T. Zoeller, et al. 2007. In vitro molecular mechanisms of bisphenol A action. *Reprod Toxicol* 24(2):178–198.

White, S. S., A. M. Calafat, Z. Kuklenyik, L. Villanueva, R. D. Zehr, L. Helfant, M. J. Strynar, A. B. Lindstrom, et al. 2007. Gestational PFOA exposure of mice is associated with altered mammary gland development in dams and female offspring. *Toxicol Sci* 96(1):133–144.

White, S. S., S. E. Fenton, and E. P. Hines. 2011a. Endocrine disrupting properties of perfluorooctanoic acid. *J Steroid Biochem Mol Biol* 127(1–2):16–26.

White, S. S., J. P. Stanko, K. Kato, A. M. Calafat, E. P. Hines, and S. E. Fenton. 2011b. Gestational and chronic low-dose PFOA exposures and mammary gland growth and differentiation in three generations of CD-1 mice. *Environ Health Perspect* 119(8):1070–1076.

Williams, J. A., and D. H. Phillips. 2000. Mammary expression of xenobiotic metabolizing enzymes and their potential role in breast cancer. *Cancer Res* 60(17):4667–4677.

Windham, G. C., S. M. Pinney, A. Sjodin, R. Lum, R. S. Jones, L. L. Needham, F. M. Biro, R. A. Hiatt, et al. 2010. Body burdens of brominated flame retardants and other persistent organo-halogenated compounds and their descriptors in US girls. *Environ Res* 110(3):251–257.

Witorsch, R. J., and J. A. Thomas. 2010. Personal care products and endocrine disruption: A critical review of the literature. *Critical Reviews in Toxicology* 40(Suppl 3):1–30.

Wolff, M. S., J. A. Britton, L. Boguski, S. Hochman, N. Maloney, N. Serra, Z. Liu, G. Berkowitz, et al. 2008. Environmental exposures and puberty in inner-city girls. *Environ Res* 107(3):393–400.

Wolff, M. S., S. L. Teitelbaum, S. M. Pinney, G. Windham, L. Liao, F. Biro, L. H. Kushi, C. Erdmann, et al. 2010. Investigation of relationships between urinary biomarkers of phytoestrogens, phthalates, and phenols and pubertal stages in girls. *Environ Health Perspect* 118(7):1039–1046.

Woodruff, T., and R. Morello-Frosch. 2011. Communicating about chemical body burden, with Tracey Woodruff and Rachel Morello-Frosch. *Environ Health Perspect* 119(5). http://ehp03.niehs.nih.gov/article/fetchArticle.action?articleURI=info%3Adoi%2F10.1289%2Fehp.trp050111 (accessed December 22, 2011).

Writing Group for the Women's Health Initiative Investigators. 2002. Risks and benefits of estrogen plus progestin in healthy postmenopausal women: Principal results from the Women's Health Initiative randomized controlled trial. *JAMA* 288(3):321–333.

Xu, X., A. B. Dailey, E. O. Talbott, V. A. Ilacqua, G. Kearney, and N. R. Asal. 2010. Associations of serum concentrations of organochlorine pesticides with breast cancer and prostate cancer in U.S. adults. *Environ Health Perspect* 118(1):60–66.

Xue, F., W. C. Willett, B. A. Rosner, S. E. Hankinson, and K. B. Michels. 2011. Cigarette smoking and the incidence of breast cancer. *Arch Intern Med* 171(2):125–133.

Yang, X. R., J. Chang-Claude, E. L. Goode, F. J. Couch, H. Nevanlinna, R. L. Milne, M. Gaudet, M. K. Schmidt, et al. 2011. Associations of breast cancer risk factors with tumor subtypes: A pooled analysis from the Breast Cancer Association Consortium studies. *J Natl Cancer Inst* 103(3):250–263.

Yetley, E. A. 2007. Multivitamin and multimineral dietary supplements: Definitions, characterization, bioavailability, and drug interactions. *Am J Clin Nutr* 85(1):269S–276S.

Young, T. B. 1989. A case–control study of breast cancer and alcohol consumption habits. *Cancer* 64(2):552–558.

Yuan, J., L. Chen, D. Chen, H. Guo, X. Bi, Y. Ju, P. Jiang, J. Shi, et al. 2008. Elevated serum polybrominated diphenyl ethers and thyroid-stimulating hormone associated with lymphocytic micronuclei in Chinese workers from an e-waste dismantling site. *Environ Sci Technol* 42(6):2195–2200.

Zava, D. T., M. Blen, and G. Duwe. 1997. Estrogenic activity of natural and synthetic estrogens in human breast cancer cells in culture. *Environ Health Perspect* 105(Suppl 3):637–645.

Zhang, Y., J. P. Wise, T. R. Holford, H. Xie, P. Boyle, S. H. Zahm, J. Rusiecki, K. Zou, et al. 2004. Serum polychlorinated biphenyls, cytochrome P-450 1A1 polymorphisms, and risk of breast cancer in Connecticut women. *Am J Epidemiol* 160(12):1177–1183.

Zhang, S. M., N. R. Cook, C. M. Albert, J. M. Gaziano, J. E. Buring, and J. E. Manson. 2008. Effect of combined folic acid, vitamin B6, and vitamin B12 on cancer risk in women: A randomized trial. *JAMA* 300(17):2012–2021.

Zhang, S., P. Lei, X. Liu, X. Li, K. Walker, L. Kotha, C. Rowlands, and S. Safe. 2009. The aryl hydrocarbon receptor as a target for estrogen receptor-negative breast cancer chemotherapy. *Endocr Relat Cancer* 16(3):835–844.

Zhang, J., L. X. Qiu, Z. H. Wang, J. L. Wang, S. S. He, and X. C. Hu. 2010. NAT2 polymorphisms combining with smoking associated with breast cancer susceptibility: A meta-analysis. *Breast Cancer Res Treat* 123(3):877–883.

Zhang, L., Q. Lan, W. Guo, A. E. Hubbard, G. Li, S. M. Rappaport, C. M. McHale, M. Shen, et al. 2011. Chromosome-wide aneuploidy study (CWAS) in workers exposed to an established leukemogen, benzene. *Carcinogenesis* 32(4):605–612.

Zhao, G., Z. Wang, M. H. Dong, K. Rao, J. Luo, D. Wang, J. Zha, S. Huang, et al. 2008. PBBs, PBDEs, and PCBs levels in hair of residents around e-waste disassembly sites in Zhejiang Province, China, and their potential sources. *Sci Total Environ* 397(1–3):46–57.

Zhao, Y. X., X. F. Qin, Y. Li, P. Y. Liu, M. Tian, S. S. Yan, Z. F. Qin, X. B. Xu, et al. 2009. Diffusion of polybrominated diphenyl ether (PBDE) from an e-waste recycling area to the surrounding regions in Southeast China. *Chemosphere* 76(11):1470–1476.

Zhao, Y., Y. S. Tan, S. Z. Haslam, and C. Yang. 2010. Perfluorooctanoic acid effects on steroid hormone and growth factor levels mediate stimulation of peripubertal mammary gland development in C57BL/6 mice. *Toxicol Sci* 115(1):214–224.

Zheng, T., T. R. Holford, S. T. Mayne, P. H. Owens, B. Zhang, P. Boyle, D. Carter, B. Ward, et al. 2000. Exposure to electromagnetic fields from use of electric blankets and other in-home electrical appliances and breast cancer risk. *Am J Epidemiol* 151(11):1103–1111.

Zheng, T., T. R. Holford, S. T. Mayne, P. H. Owens, P. Boyle, B. Zhang, Y. W. Zhang, and S. H. Zahm. 2002. Use of hair colouring products and breast cancer risk: A case–control study in Connecticut. *European Journal of Cancer* 38(12):1647–1652.

Zinedine, A., J. M. Soriano, J. C. Molto, and J. Manes. 2007. Review on the toxicity, occurrence, metabolism, detoxification, regulations and intake of zearalenone: An oestrogenic mycotoxin. *Food Chem Toxicol* 45(1):1–18.

Zota, A. R., R. A. Rudel, R. A. Morello-Frosch, and J. G. Brody. 2008. Elevated house dust and serum concentrations of PBDEs in California: Unintended consequences of furniture flammability standards? *Environ Sci Technol* 42(21):8158–8164.

4

Challenges of Studying Environmental Risk Factors for Breast Cancer

The committee was asked to review the methodologic challenges involved in conducting research on breast cancer and the environment. New insights into carcinogenesis are giving researchers new opportunities to explore both the biology and the epidemiology of breast cancer in relation to environmental exposures. Although progress has been made in investigating the role (whether adverse or not) of environmental factors in breast cancer, the scope of the potential research questions is vast and the questions to be answered are complex. This chapter reviews challenges facing researchers on a variety of fronts, including the nature of the various forms of breast cancer; the diversity and complexity of environmental factors; identifying and measuring exposures at appropriate times; genetic complexity that is still unfolding; and the inherent limitations of the laboratory and epidemiologic tools available to evaluate associations between environmental exposures and disease.

COMPLEXITY OF BREAST CANCER

As noted in Chapter 2, breast cancer is a term that captures what is likely to be several diseases. Tumor types have been categorized based on several different characteristics, including age or menopausal status of the woman at the time of diagnosis; the state of the tumor as in situ or invasive; the extent of spread from the initial tumor site; cell type (lobular, ductal); and molecular features of the cells, such as the presence or absence of hormone or growth factor receptors (e.g., estrogen or progesterone receptors [ER or PR], human epidermal growth factor receptor 2 [HER2]). Within each of these broad categories is considerable variability in tumor characteristics and gene expression. A study examining the gene expression of 65 surgical samples of breast tumors from 42 individuals' cancers found each tumor to have a distinctive molecular portrait. The tumors showed great variation in their patterns of gene expression, and the variation was multidimensional: different sets of genes showed largely independent patterns of variation (Perou et al., 2000). Further study of the molecular pathology of breast cancer has shed additional light on the possible divergent evolutionary pathways of breast cancer progression, revealing still more complexity (Bombonati and Sgroi, 2011), as discussed in Chapter 5.

While characterizations of tumor and cancer types, such as those noted above, are proving increasingly useful as guides to clinical care and prognosis, their relevance to etiology is not clear. Some associations have been observed between certain tumor types and risk factors (e.g., obesity and ER-positive [ER+]) tumors, but for the most part, the mechanistic basis for these relationships remains to be clarified, as described further in Chapter 5.

Various schematics have been used to illustrate the complexity and interconnectedness of potential factors in breast cancer causation. Howell et al. (2005), for example, illustrate possible roles for genes, pathways, risk factors, modifiable variables, and life events (Figure 4-1). In this representation of some of the known modifiable and unmodifiable risk factors for breast cancer, alcohol serves as an example of a factor that might alter risk for breast cancer in multiple ways. Through induction of aromatase activity, it may foster conversion of androgens to estrogens that have a causal role in breast cancer (Etique et al., 2004). It has also been hypothesized to contribute to genomic instability (Benassi-Evans and Fenech, 2011). Furthermore, it may act indirectly in that its calories can contribute to obesity that itself is associated with breast cancer.

Another illustration (Figure 4-2) of the numerous interrelated factors important in the etiology of breast cancer comes from a complex systems model developed by Robert Hiatt and colleagues as part of a project sponsored by the California Breast Cancer Research Program.[1] The developers of this model used expert opinion to select causal factors from four large domains (Societal/Cultural, Physical/Chemical, Behavioral, and Biological) to illustrate the multiple levels of causation that must be considered along with how the factors are integrated across levels and over time. Even though multiple key factors are present, all possible etiologic factors were not included for relative simplicity in interpretation. The model focuses solely on postmenopausal breast cancer because of the different etiologic factors and pathways for premenopausal disease. It takes into account both

---

[1]Personal communication, R. A. Hiatt, University of California, San Francisco, May 21, 2011.

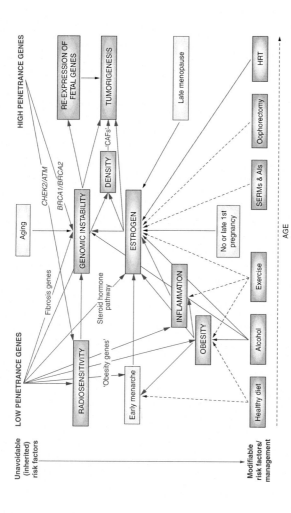

FIGURE 4-1 Overview of risk factors associated with breast cancer. "The diagram summarizes the unavoidable (inherited) and modifiable risk factors that can ultimately lead to tumorigenesis. Genes/pathways/risk factors are shown in red; inherited or unmodifiable factors are shown in green; modifiable variables are shown in blue; life events are represented by gray boxes; increased/positive effects are denoted by solid arrows; and reduced/negative effects are denoted by dashed arrows. AIs, aromatase inhibitors; *ATM*, ataxia telangiectasia mutated; *BRCA1* and *BRCA2* (genes in which deleterious germline mutations increase the risk of cancer); *CHEK2*, CHK2, checkpoint homolog; HRT, hormone replacement therapy; SERMs, selective estrogen receptor modulators."

SOURCE: Adapted from Howell et al. (2005, p. 638). Used with permission; Howell, A., A. H. Sims, K. R. Ong, M. N. Harvie, D. G. Evans, and R. B. Clarke. 2005. Mechanisms of disease: Prediction and prevention of breast cancer—cellular and molecular interactions. *Nat Clin Pract Oncol* 2(12):635–646.

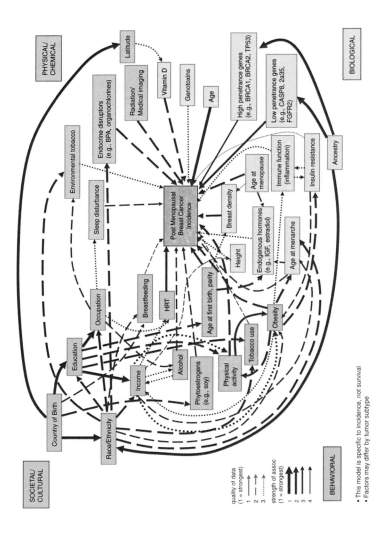

FIGURE 4-2 Illustration of an evidence-based complex-systems model of postmenopausal breast cancer causation. This model displays multiple factors associated with postmenopausal breast cancer causation in four broad domains and shows their interconnections across levels (genes to society) by arrows that indicate variations in the strength of the associations and the quality of the data.
SOURCE: Personal communication, R. A. Hiatt, University of California, San Francisco, May 21, 2011. Developed with support from the California Breast Cancer Research Program.

the strength of the associations as well as the quality of the data in the size and hatching of the interconnecting arrows.

Diagrams such as these, which attempt to depict the multiplicity of factors that seem to have a role in breast cancer, help underline the biological complexity of the pathways along which those factors may be acting, the difficulty of distinguishing truly causal effects from associations with intermediate factors, and the challenges of designing, conducting, and interpreting studies that try to evaluate risk factors for the various forms of this disease.

Although these challenges share similarities across the spectrum of risk factors evaluated in this report, they may be particularly acute for evaluating risk relationships from exposures to environmental chemicals. For studies in humans, these include the issues inherent to estimating and assessing exposures, the study design and analytic challenges of environmental epidemiology, and efforts to account for genetic differences in susceptibility to cancer and potential gene–environment interactions. The next portion of this chapter pays particular attention to the challenges in studying environmental chemicals. Studies in animals and in vitro systems pose their own technical obstacles and challenges of interpretation and extrapolation to humans, which are discussed in a subsequent portion of the chapter.

STUDYING ENVIRONMENTAL CHEMICAL AND PHYSICAL EXPOSURES THROUGH HUMAN STUDIES

As noted previously, the committee has adopted a broad approach to the definition of "environment." A subset of environmental exposures that are of potential concern in the etiology of breast cancer is that of specific chemical and physical agents that might influence breast cancer development. Although information on exposure and the toxicology of many chemicals may be incomplete, for many other chemicals, knowledge of some their properties indicates that they are unlikely to be mutagenic or carcinogenic.

Whether other agents in the environment are able to causally contribute to breast cancer is highly dependent upon both the duration and magnitude (dose) of exposure. One of the most difficult problems in conducting epidemiologic studies on environmental exposures and health effects is to obtain reasonably accurate measurements or estimates of exposures relevant to the disease process. These exposures may occur at low or varying levels or both, for which the relevant time period—the window when the exposure might influence the development of a tumor—is unknown, or they may have occurred years or decades previously. The sections that follow address some of the specific challenges associated with assessing exposures to environmental and physical agents and illustrate the need for additional or more refined tools to aid in disentangling the possible contributions of these environmental factors to breast cancer.

Assessing Exposures to Chemical and Physical Agents in the Environment

Both the nature of the exposures to chemical and physical agents and the limited means for measuring or assessing them pose challenges for observational research. Human exposures to substances in the environment take place throughout the life course, and in all settings. People are exposed to myriad substances in air, water, and food encountered in homes, schools, workplaces, and even before birth via in utero exposures. A person is exposed not only to individual chemicals, but to mixtures of many different substances, at varying doses simultaneously or at different times. Sometimes it is possible to identify individuals or groups, such as workers in particular occupations, whose typical exposures are considerably higher than those of the average person.

Epidemiologic studies assess whether groups with higher exposures are more likely to experience the outcome of interest, cancer for example, than groups with lower exposures. Determining who is exposed and the degree of their exposures are critical to accurately assessing the association with the health outcome. However, errors in classifying who is more and who is less exposed (exposure misclassification) could limit the ability of a study to show an association with the risk factor where there is one. Thus, accurate exposure assessment is a critical component of human studies to evaluate risk factors for breast cancer or any health outcome.

Historically, studies in occupational settings have been an important means for identifying most chemical carcinogens because in occupational settings, chemical use is often documented and exposure levels tend to be higher than elsewhere. Assessment of exposures in occupational studies are facilitated by extensive sources of data, such as job histories, understanding of production processes and chemicals used, and data from personal or area sampling to measure exposures, as required by the Occupational Safety and Health Administration (OSHA) and standard industrial hygiene practices. Exposure of certain workers to some chemicals may be thousands (or more) times greater than that experienced by the general public, while other workers with different job tasks might experience a wide range of exposures. This variability makes it easier to distinguish people who are exposed to very high levels from those with lesser exposure. The greater the contrast, the firmer the conclusions that can be drawn about differences in risk of disease. When exposure levels are low, contrasts are smaller and exposure misclassification is likely to be relatively greater. Determining exposures can be more difficult in environmental settings, particularly for chemicals that are not regularly monitored in air or food, or for chemicals for which exposure occurs indoors as a result of specific behaviors or products used. For these reasons, environmental epidemiologic studies are a less effective or efficient approach than occupational epidemiologic studies for demonstrating associations between chemicals and increased rates of disease.

Few of the chemicals identified by the International Agency for Research on Cancer (IARC) or the U.S. Environmental Protection Agency (EPA) as human carcinogens have been classified as such on the basis of studies showing breast cancer in humans. One cannot conclude, however, that these chemicals do not contribute to breast cancer. For virtually all carcinogens identified by IARC and EPA, the evidence base has primarily been from occupational epidemiologic studies for reasons described. For the vast majority of these chemicals, the cohorts were assembled and followed during the 1940s through the 1970s, periods when most industrial firms employed only men.

Historically, therefore, most epidemiologic studies of cancer in the workplace omitted women from the analysis because there were too few present to observe an effect. Because breast cancer is rare in men, such studies lacked the power to detect breast carcinogens. (Power is a function of the expected number of cases of disease in the studies, the level and variability of exposure, the validity of the exposure assessment, and the strength of the true underlying association.) Not only are studies of breast cancer in men underpowered, but also, extrapolation of cancer findings from men to women, which may be justified for other forms of cancer, might not be appropriate for breast cancer.

Beyond the Workplace: Environmental Chemical Exposures

Outside the workplace, exposures to chemicals arise in multiple locations (home, car, ambient air pollution); from multiple activities, including commuting, cleaning, gardening, and smoking; and through different routes of exposure (ingestion, inhalation, dermal absorption).

The home, where people typically spend most of their time[2] (Klepeis et al., 1995), provides opportunities for exposure to many chemicals, including naturally occurring chemicals in the diet as well as chemicals from food packaging, processing, or cooking; the release of volatile chemicals from carpets, furniture, clothing treatments, and cleaning products; home use of pesticides; use of cosmetics and personal care products; tobacco smoke; and infiltration of ambient air pollution. Typically, thousands of synthetic and naturally occurring chemicals are present in people's homes and diet, most at relatively low concentrations.

---

[2]Survey data indicate that on average people spend 69 percent of their time in a residence and 87 percent of their time in enclosed buildings (Klepeis et al., 1995).

The 20th century saw a substantial increase in the synthesis of new chemicals. Tens of thousands of chemicals are used in commerce, and more than 3,000 industrial chemicals (excluding polymers), mostly organic compounds, are produced or imported into the United States at rates exceeding 1 million pounds per year (EPA, 1998b). These are known as high production volume chemicals. A 1998 EPA report found that insufficient testing had been done to evaluate the health effects of all but a few of these chemicals. Of 2,800 chemicals investigated, 93 percent lacked one or more of the six basic toxicity tests,[3] and 43 percent of the chemicals had undergone none of these tests, which are considered necessary for a minimum understanding of a chemical's toxicity. The percentage of chemicals with complete or at least some toxicity information was considerably higher for chemicals with potential for greater exposure through industrial releases or for those in consumer or children's products. In addition, not all of these 3,000 chemicals are of high priority for testing, because they belong to chemical classes or structural groups for which there is less concern regarding mutagenicity, carcinogencity, or endocrine effects. The High Production Volume Chemicals Program (HPV Program) is an international program to assess the potential hazard of chemicals produced in high volumes. Production levels of specific chemicals can change over time as demand for them increases or declines.

Other chemicals of potential concern are by-products of industrial processes (e.g., dioxins), and the amounts produced cannot be measured as directly as those of deliberately produced chemicals. Opportunities for exposure may change in line with changes in production volumes, but they also may vary independently if industrial processes become more effective in reducing environmental release of a chemical during production. Among the substances reviewed in this report as potential risk factors for breast cancer, environmental releases from different sources have varied, and some have declined over recent years (e.g., dioxin, Figure 4-3 [EPA, 2006]; or perfluorooctanoic acid, Figure 4-4 [Paul et al., 2009]).

Hazard Versus Risk

In the assessment of the impact of environmental chemicals on humans, there is an important distinction between hazard and risk. A chemical may be identified as harmful or a *hazard*, but the risk it poses to people depends on both its toxic potency and the nature of the exposure, especially the amount to which people are exposed but also potentially the timing of the exposure. While thousands of chemicals are produced in or imported into

---

[3]The tests evaluate acute toxicity, chronic toxicity, developmental and reproductive toxicity, mutagenicity, ecotoxicity, and environmental fate.

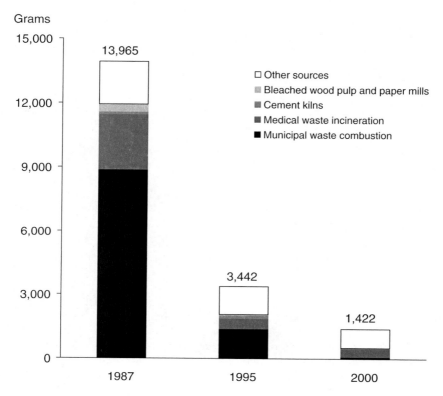

FIGURE 4-3 Sources and amounts (g/yr) of dioxin-like compounds released in the United States in 1987, 1995, and 2000.
SOURCE: EPA (2006).

the United States, not all of them pose risks to the general population. Some are used only in specific industrial processes, where potential exposure is limited to those in the workplace. Some chemicals have low potency, generally causing health effects only at very high exposures. Thus, a chemical known to be a hazard on the basis of toxicologic studies, but with low potency and to which people are exposed at low concentrations, may present little risk of cancer or other adverse health effects.

Route of Exposure

In occupational settings, inhalation and dermal contact are frequently the primary routes of exposure (Eaton and Klaason, 1996), although incidental ingestion pathways can occur. In the home, opportunities exist for

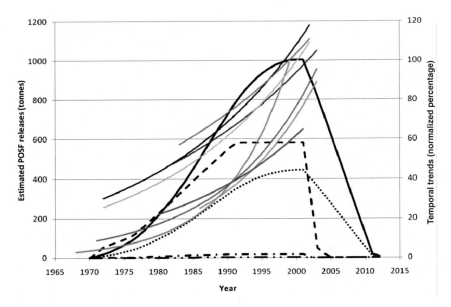

FIGURE 4-4 Estimated releases of perfluorooctane sulfonyl fluoride (POSF) from 1970 to 2012 and exponential temporal trends in biota. POSF breaks down into perfluorooctanesulfonic acid (PFOS). Note: 2012 is when aqueous fire-fighting foams (AFFFs) are scheduled to be restricted and treated carpets end their natural life. The projection to zero is based on 3M's production only, therefore some emissions will continue from remaining producers. Temporal trends in biota have been normalized to 100 percent for each species/dataset. Usage is depicted as follows: carpets (—), paper and packaging (- • -), apparel (- - -), performance chemicals (– • •), AFFFs (• • •). Biota trend lines are as follows: ringed seals from Arctic locations, Qeqertarsuaq (purple) and Ittoqqortoormiit (yellow); Baltic guillemot eggs (pooled: light green; and mean: dark green); polar bears from western (light blue) and eastern Canadian Arctic (dark blue); herring gulls from Norway (orange); and lake trout from Lake Ontario (red).
SOURCE: Paul et al. (2009, p. 390). Published in: Alexander G. Paul; Kevin C. Jones; Andrew J. Sweetman; *Environ Sci Technol* 2009, 43, 386–392. Copyright © 2008 American Chemical Society.

exposure via ingestion, inhalation, and dermal contact. Pesticide exposures, for example, can occur through consumption of food (from agricultural applications), inhalation (directly from exposure to sprays and foggers or subsequently from volatilization of residues of past use or resuspension of contaminated dust), and dermal absorption (from contact with residues on the surfaces of tables, countertops, or household objects). Various assessments have found that concentrations of some volatile and semivolatile chemicals are much higher in indoor spaces, such as homes and schools, than in outdoor areas around the home (Sax et al., 2006; Turpin et al., 2007; Ward et al., 2009; Rudel et al., 2010). Dermal exposure may be the predominant exposure pathway for chemicals in some cleaning or personal care products.

Each chemical must be examined for how it is used as well as its volatility and ability to pass through the skin. Sometimes potential routes of exposure can be overlooked—for example, in taking showers, people may experience both dermal and inhalational exposure to some volatile organic compounds (VOCs) in the water supply. Typically, however, this exposure to VOCs is primarily via inhalation and may equal the exposure from drinking water (Jo et al., 1990).

Measurement of Exposure

In occupational studies, job titles and records from industrial hygiene measurements (individual air monitoring, or air sampling from work areas) are frequently used to estimate exposures. For population studies, researchers may use location of residence or distance from a source of concern (transmission wires, freeways, factories); structured questionnaires relying on participants to report product use; measurements taken in air, water, soil, or other environmental media; and measurements in biological specimens (e.g., blood lead, urinary metabolites of pesticides, cotinine from the breakdown of nicotine to indicate tobacco smoke exposure). The utility of these chemical measurements in both environmental and biological samples depends on when the samples are taken relative to the disease in question; the half-life in the environment or human body, respectively; and the variability in actual exposures over time. In the 1990s, researchers began to develop biomarkers as a means not only to improve estimation of exposure, but also to document intermediate steps along the pathway between exposure and effect. For example, markers of oxidative stress, DNA adducts, and epigenetic marks such as methyl groups can provide evidence that tissues have been affected. Such markers may suggest a mechanism by which an exposure may increase or decrease the risk of breast cancer; however, it can be difficult to demonstrate a direct relationship between the exposure and the marker, and between the marker and subsequent disease.

Importance of Timing of Exposure

Understanding the link between chemical exposure and disease is especially challenging when studying chronic diseases that develop gradually over many years, such as cancer. Because the first steps in carcinogenesis may begin decades before the diagnosis of a cancer, relevant exposures for breast cancer may include those that occurred in childhood or perhaps before birth. Assessing past environmental exposures poses serious challenges.

Chemicals and other factors may act differently or have different exposure routes at different stages of a woman's life. The breast may be more vulnerable to carcinogenic exposures during in utero development, in the interval between menarche and a first full-term pregnancy, or during key windows of proliferation and maturation. Such periods of increased susceptibility would also imply that total lifetime exposure is not the appropriate metric, but, rather, that exposures need to be measured during critical life stages, some of which may be harder to capture than others. One of the more classic examples of the importance of timing of exposure comes from studies of atomic bomb survivors. Early reports suggested that increased breast cancer risk appeared to be limited primarily to women who were exposed during puberty. Although more recent analyses suggest elevated risks even among those exposed later in life, early exposure remains particularly important (Land et al., 2003; Preston et al., 2007). The potential importance of timing of exposure to breast carcinogenesis is discussed in greater detail in Chapter 5.

Factors for Consideration in Measures of Exposure

A key consideration in epidemiologic studies is persistence of the risk factor or exposure. Some substances are unstable in the environment. Others are retained in environmental media, but have short residence times in the human body. For example, pyrethroid pesticides can persist indoors away from sunlight for months or years, but they are rapidly excreted by humans (CDC, 2009). Some chemicals or their metabolites can be retained in the human body for decades. Levels of internal exposure to these stored chemicals can be influenced by changes in the body that are unrelated to current levels of external exposure. Lead, for example, is stored in bone for decades, but it is released during pregnancy or menopause. Some endocrine-disrupting compounds, such as certain polychlorinated biphenyls (PCBs) and polybrominated diphenyl ethers (PBDEs), are persistent in both the environment and in humans (Brown, 1994; Sudaryanto et al., 2008; EPA, 2010).

If the substance is retained in either environmental media or the human body, it may be possible to make measurements in one time period and infer aspects of exposure in other time periods. To do so requires knowing not only the rate at which a chemical and its metabolites are eliminated from the body (so-called half-lives), but also the variation in exposure levels over time and the determinants of variability in retention (both for the ecosystem and the human organism). Perhaps the most promising situation for inferring previous exposures for the persistent compounds is when they have been banned and no new exposures are occurring, so that half-lives are the prime determinant of change over time. However, even if a typical half-life has been established for a persistent chemical, measurements for individuals can be influenced by such factors as age-related or genetic influences on metabolism of the chemical or experiences that may affect remobilization of the stored chemical (e.g., lactation, substantial change in weight).

If the chemical is not retained, then any one-time measurement will likely be inadequate to capture the true exposure, unless the exposures are consistent. This could occur, for instance, when there is a product whose formulation has not changed and the contact is consistent over time (e.g., used daily in the same amounts and in the same way).

To effectively study exposures over long time periods, research protocols may require measurements of exposure at multiple time points. The number of measurements required will depend on the variation in exposure over time. Compounds that are rapidly excreted may require a large number of measurements, even to obtain estimates of short-term exposures. Because this can be prohibitively burdensome, alternative strategies that rely on external indicators of exposure may provide more accurate estimates of exposure. For instance, if 50 percent of the body burden of a chemical is from consumption of one particular food source, then questionnaires about such behavior patterns may be more reliable than measurements of urinary metabolites.

Thus, new methodologies for the measurement of suspected breast carcinogens in the environment can lead to higher quality epidemiologic studies, both retrospective and prospective. The modalities needed include improving measurements in the environment and assessing variation over time and space; determining routes of exposures and how they vary over time and over the life course; using emissions inventories along with environmental dispersion modeling; measuring compounds and their metabolites in biospecimens; understanding pharmacodynamics and pharmacokinetics and how they vary by age, body weight, nutrition, comorbidity, or other factors; developing biomarkers for early biologic effects (DNA adducts, methylation, tissue changes, gene expression, etc.); using human exposure biomonitoring programs (e.g., breast milk repositories) by geographic areas; and validating exposure assessment questionnaires through various strategies.

Although reductionist science has generally driven typical chemical risk assessments to examine risk "one chemical at a time," humans are not exposed to just one chemical at a time. Indeed, there is a need to establish new, innovative approaches that allow quantitative assessment of multiple concurrent exposures with various disease end points, such as breast cancer. An interesting conceptual model that addresses the reality of multiple concurrent exposures is the "exposome." As proposed by Rappaport (2011, p. 5), the exposome represents "the totality of exposures received by a person during life, encompasses all sources of toxicants and, therefore, offers scientists an agnostic approach for investigating the environmental causes of chronic diseases." Other examples of recent thinking on the complexities of multiple human exposures include the use of "environment-wide" association studies, such as was done recently for 266 different environmental factors and risk for developing type 2 diabetes (Patel et al., 2010). This study reported small, but statistically significant, associations between development of type 2 diabetes and hepatachlor epoxide, PCBs, and the vitamin gamma-tocopheral. A protective (inverse) effect was seen with beta-carotene. The application of such approaches to the environment and breast cancer is potentially feasible. However, even with such holistic approaches to exposure analysis, assessment of exposure during early life stages would be important, as discussed elsewhere in this report, making it particularly challenging for breast cancer and other diseases that often occur later in life.

Other novel approaches to unraveling the complex association between genes, multiple environmental exposures, and complex diseases such as breast cancer are needed. A better mechanistic understanding of the role of environmental factors in disease etiology, including especially "pathway analyses" and other tools that can be used to identify key regulatory pathways that integrate genetic and environmental modulators, are needed to help set priorities for future research (Gohlke et al., 2009).

HUMAN EPIDEMIOLOGIC STUDY DESIGN AND IMPLEMENTATION

As introduced in Chapter 2, various types of human epidemiologic studies are conducted: (1) double-blind randomized controlled trials, (2) observational designs such as cohort and case–control studies, and (3) the less reliable "ecologic" studies that do not use individual-level data. Each type of study design has different strengths and weaknesses. Several factors are discussed here that can interfere with the execution of studies and interpretation of their findings.

Bias

A major goal of epidemiologic studies is to estimate the effect of a suspected causal factor by measuring how strong its association is with the disease under study. Bias occurs when the estimate of effect systematically misses the mark and is artificially higher or lower than the true association in the population. Bias can arise in several ways. Selection bias occurs when study participants differ from the population of interest with regard to the joint distribution of exposure and disease. For example, different exclusion criteria applied to cases and controls skewed the prevalences of exposure differently in the two groups in a hospital-based study in Helsinki. This led to an apparent association between the use of reserpine, an antihypertensive agent, and breast cancer (Heinonen et al., 1974). A later study helped to illustrate the false findings resulting from differentially applied selection criteria (Horwitz and Feinstein, 1985; Gordis, 2000).

Information bias can occur when methods for gathering information about study participants are fallible, such that either exposure or disease outcome information is incorrect. Information bias can arise from either random or systematic errors, for example, in information abstracted from medical records or obtained by use of surrogates (e.g., spouses or other family members when the study subject is deceased or too ill to provide information). Recall or reporting bias is thought to be a common type of information bias in case–control studies, but it can occur in any study in which information is obtained after occurrence of disease. One factor that can contribute to recall bias is the tendency of people who develop a disease (cases) to think harder about and recall more potential exposures than those in the comparison group (controls), who have not developed a disease and may be less likely to recollect past events and activities. This type of recall bias would typically result in a bias toward an increased association between the exposures and disease, if those exposures are harmful. However, in general, bias can be either toward the null (the "null" refers to "no effect," such as an odds ratio [OR] or relative risk [RR] of 1.0) or away from the null. In other words, bias can either attenuate or exaggerate a measure of association.

Another type of bias is referred to as "confounding" and is often considered a category of its own. It occurs when another risk factor for the disease under study occurs more or less frequently in those who are exposed as compared with the unexposed. An association observed between the exposure under study and the disease outcome might be the result of the alternative risk factor that is associated with, but not the result of, the exposure being studied. For example, in the United States, the incidence of breast cancer is generally higher among women with higher incomes, but higher income itself is not a causal factor. Instead, higher income is associated with having fewer children, having children at later ages, and other factors that are more clearly associated with the biologic processes that contribute to breast cancer. For this reason, scientists are skeptical of results unless potential confounders have been taken into account. If confounders are known and measurable, it is often possible to limit the effect of confounding ("control" for it) through appropriate design, data collection, and statistical analysis. For example, matching of controls to cases on factors such as age is a method that can improve the ability to control confounding when the data are analyzed correctly. But such steps will not be sufficient if confounders are unrecognized or difficult to measure, or if their relationship to exposures is poorly understood.

Restriction of study participants can serve to improve control of confounding, but may result in lower generalizability if the restricted group is characterized by a different relationship between exposure and disease (e.g., a study of women ages 40–60 may yield results that would not apply to women ages 60–80 or 20–40). In matched or unmatched designs, appropriate statistical methods must be applied to effectively control confounding. These include stratification and statistical adjustment. Stratifying the data involves grouping by levels of the suspected confounder and deriving an adjusted measure of association across all levels. Adjustment can involve use of a statistical model that is assumed to reflect the relationship of multiple independent variables in relation to the outcome. Statistical models can provide simultaneous adjustment for multiple potential confounders. However, adjustment for confounders is not always sufficient or adequate, especially when information on all potential confounders has not been collected or the confounding factors are not even known to the investigator.

Moreover, confounders need to be understood as operating, not one-by-one, but rather in a complex network of causal relationships. Graphical tools, such as directed acyclic graphs (DAGs), are sometimes used to identify the appropriate confounders for control, and to identify which factors should not be controlled (Greenland et al., 1999; Hernán et al., 2002). This latter group consists of two categories of variables: (1) factors that are downstream of the exposure and (2) factors that block a pathway between exposure and disease (e.g., they have antecedents, one that is associated with exposure and the other with disease). Some factors that are downstream of exposure may be intermediates on a causal pathway, but whether they are or not, control for them can introduce bias, except in very specific circumstances (Petersen et al., 2006). In most instances, factors that block an exposure–disease pathway should also not be controlled, in order to obtain unbiased measures of the association of interest. But if they are, at least one of the antecedents will also require control. The importance of taking prior knowledge into account when selecting variables has been clearly demonstrated (Hernán et al., 2002). Stated differently, the use of empirical data alone for selection of confounders to control can be misleading and can be used to justify what will be a biased model.

Potential confounding by "unknown factors" is often cited as a precaution in ascribing causality to an exposure associated with disease and, in fact, was a central argument used in questioning whether smoking was causing lung cancer (e.g., Fisher, 1958a,b). However, hypothesized explanatory confounders are subject to some stringent constraints, such as that the association between the confounder and disease must be (much) larger than that of the exposure and disease, in order to "explain" the observed exposure–disease association (Cornfield et al., 1959; Langholz, 2001; Goodman et al., 2002). Consideration of whether such factors exist is warranted when this criticism is expressed. An example was when smoking was argued to explain the association of an occupational exposure with lung cancer. At first blush, this seemed plausible, since smoking has such high relative risks while the occupational exposure showed a lower association. However, smoking levels were found to be very similar when comparing exposed to unexposed workers, and hence smoking was not a strong confounder (Axelson and Sundell, 1978).

Yet another type of bias is statistical bias, which can occur as a result of unavoidable limitations in the methods of analysis. Statistical bias may also result from use of an inappropriate method of analysis, such as using unmatched methods for a matched design, or failing to employ survival analysis when follow-up in a cohort study is highly variable. In addition, a study may be uninformative because it has inadequate statistical power to detect differences in risk when the anticipated effect is small. This may happen because the sample size is too small or, for a rare outcome in a large cohort, the number of expected cases is small or an exposure is rare. Because the study of rare exposures requires such large sample sizes, studies are often conducted in populations that are more highly exposed than the general population, such as occupational groups exposed in the course of their work.

In general, when a study has inadequate statistical power, "random error" could lead to over- or underestimation of the true effect, even though with repeated sampling the average effect estimate would converge on the true value. This imprecision will be reflected in wide confidence intervals around the risk estimate. A wide confidence interval might have an upper limit that is, for instance, 8 or 10 times larger than the lower limit. Another cause of low statistical power is an underlying association that is so small that it is difficult to distinguish from the null effect. For instance, if the exposures encountered by the population truly increase risk by, say, 5 to 10 percent, even a very large study with a few thousand cases will generally not be of sufficient size to reliably generate a statistically significant estimate of increased risk.

Interpretation of Attributable Risk and Population Attributable Risk

Chapter 2 introduced some of the measures that are used to estimate the disease risk associated with factors of interest, including attributable risk (AR) and population attributable risk (PAR). In its simplest form, the AR is a measure (percentage) of the cases that occur in the exposed group that are in excess of those in the comparison group and that are considered to have occurred because of a given factor. This measure can be interpreted as the maximum potential for risk reduction among those currently exposed to the factor of interest if the exposure could be eliminated and if the association is truly causal. The PAR is a population-based measure of the percentage of excess cases associated with the exposure of interest that also takes into account the distribution of the exposure within the population, again, assuming that the relationship between the exposure and the disease outcome is causal. If an exposure is rare, it may contribute only a small proportion of a population's disease risk, even if disease incidence is much higher among those who are exposed. The AR is a statement about disease among people who have an exposure, not about exposure among people who have a disease. The PAR is a statement about disease risk ostensibly due to the exposure in the entire population, not about exposure among people with disease.

Several methods have been used to estimate the PAR. A comprehensive review of methodological developments up to 2000 is given in a series of papers, including Benichou (2001), Eide and Heuch (2001), and Uter and Pfahlberg (2001), while more recent developments are described in Steenland and Armstrong (2006) and Eide (2008). With respect to breast cancer, an insightful commentary, "Use and Misuse of Population Attributable Fraction" by Rockhill et al. (1998), succinctly summarizes the definitional and estimation issues and discusses how attributable risk has been misinterpreted in the breast cancer epidemiologic literature.

Although PAR numbers appear fairly frequently in scientific literature, they are prone to misinterpretation by health professionals as well as lay people. It is common to interpret the PAR as the change in disease risk if exposure were reduced in the exposed individuals of the study population. However, exposure may be correlated with other factors that are also risk factors for disease. A change in the exposure under consideration may or may not result in making the "exposure altered" population similar to the unexposed population in all respects relevant to breast cancer. For instance, nulliparous women (those who have never had children) are more likely to be unmarried than parous women. If some of the factors related to being married but not related to childbearing are associated with breast cancer risk, then single nulliparous women who "become parous" to reduce breast cancer risk may not have the same risk as the general parous population, which has more married people in it.

ARs or PARs can be calculated separately for several risk factors related to a disease such as breast cancer, but these separate estimates cannot be added together. It is possible, however, to calculate an AR or PAR for a group of risk factors together. When an AR or PAR is calculated for multiple risk factors combined, the result is likely to be smaller than the sum of the ARs or PARs for the individual factors if the correlations among those factors are positive. Risk factors are often related to each other or to common disease pathways so their contributions to disease risk are not independent. For example, people who smoke may also consume more alcohol. It is also possible to calculate an AR or PAR for a single factor holding others fixed. The results of some of these studies that have calculated PARs are discussed in Chapter 6.

It is also crucial that ARs and PARs be seen as a function of the characteristics of the source population and mix of exposures from which they are estimated. Because they capture as-yet undetermined factors that contribute to risk, they may not apply to other populations that differ in their mix of risk factors.

Rockhill et al. (1998) point out several common errors in interpretation and communication of the PAR and discuss them in relation to estimated PAR for breast cancer (Seidman et al., 1982; Madigan et al., 1995). The first problem is confusing the attributable risk with the proportion of cases who have any of the risk factors included in the PAR. In the examples they cite, a PAR is reported (say, 25 percent) and then it is erroneously concluded that the PAR proportion (25 percent) of cases have one or more of the risk factors while the remaining cases have no risk factors. As noted, the PAR is a statement about disease risk considered to be due to exposure in the entire population, not about exposure among those with the disease. Rockhill and colleagues noted that in each of these studies, the proportion of controls or of the underlying population who were exposed to at least one of the risk factors was 90 percent or more.

The PAR also does not mean that the causes(s) of breast cancer can be identified for the percentage equivalent to the PAR of those with the disease. The PAR "does not address probability of causation for a specific case of disease, nor does its estimation enable epidemiologists to discriminate between those cases caused by, and not caused by, the risk factors under consideration" (Rockhill et al., 1998, p. 17). A further problem is the lack of distinction between factors that are likely to be causally related to disease risk and those that capture a whole set of lifestyle and exposure characteristics. For instance, they note that denying women a college education (a risk factor for breast cancer) is not going to reduce breast cancer risk if the "more causally proximate exposures and behaviors remain the same" (Rockhill et al., 1998, p. 18).

The third issue has to do with the definition of the exposed group. If the message is that the exposed cases could be prevented, then defining the unexposed group to have characteristics that are unattainable in the exposed population is not useful. In a related note, Rockhill et al. (1998) cite the point by Rose (1985) that susceptibility to chronic disease is rarely confined to a high-risk minority within the population; this would certainly seem to hold for breast cancer.

The last issue is the common practice of "equating the AR with the proportion of disease cases that are 'explained' by the risk factors" (Rockhill et al., 1998, p. 18). Their concern is that the term "explained" is equated with, and misinterpreted as, "cause." A better interpretation might be that the PAR represents the proportion of cases that "are associated with" the risk factors in question. Rockhill and colleagues (1998) note that breast cancer risk factors are poor predictors of breast cancer occurrence; the vast majority of women with these risk factors do not develop breast cancer. As an example to make their point, they consider "age greater than 15" as the exposure variable. Nearly 100 percent of cases are exposed, but so are the vast majority of noncases; therefore, considering age greater than 15 as a risk factor is of little value. This issue can be described in statistical terms as having a defined exposure that has high sensitivity (i.e., a large proportion of breast cancer patients are exposed), but very low specificity (i.e., a large proportion of women without breast cancer are also exposed).

The committee believes that many of the problems in interpretation of the PAR arise when there is either an expressed or implied causal relationship between the exposure and disease. A definition that is more reflective of what may be estimated from observational data is that the PAR is the relative difference in the risk of disease between the whole population and the unexposed portion of the population. With this interpretation, it should be better understood that the lower breast cancer risk in married women is not necessarily due to "marriage" per se, but to some constellation of characteristics of the population of married women.

Experimental Studies in Humans

Many of the various sources of bias and confounding that can affect observational studies are eliminated or reduced in experimental studies or clinical trials that are randomized. In humans, the gold standard for a study to examine a potential causal relationship between an exposure and disease is the randomized controlled trial, in which study participants are randomly assigned to groups receiving (or not) an intervention or exposure. Randomized controlled trials may also be "blinded" when either study subjects or investigators carrying out the study, or both, are unaware of the intervention group assignments. When participants are randomized to receive (or not) a treatment, the likelihood of confounding is reduced, but unless the trial is large, analyses still need to control for the possibility that some imbalance in risk factors (confounding) occurred despite randomization. The randomized trial design is most often used for studies of treatment efficacy, but it is rarely used for etiologic studies.

If large enough, such randomized trials could in theory resolve outstanding questions regarding causal relationships for breast cancer. However, such studies are not practical or appropriate for most of the environmental exposures of greatest concern. Randomized trials are not ethical if it would be necessary to subject participants to an exposure anticipated by the investigators or the scientific community to be harmful. In many cases, randomly assigning participants to an "unexposed" group would also be infeasible because many substances of interest are widely present in the environment. Nevertheless, randomized trials of intervention strategies to mitigate exposure (e.g., to increase smoking cessation or to reduce worker exposures to a suspected carcinogen) could produce very strong causal evidence if a difference in health outcomes were found.

Another challenge for intervention trials for disease prevention is that assignment to a given intervention may coincide with other changes occurring in the population under study that are not part of the intervention being tested. The Multiple Risk Factor Intervention Trial (the MRFIT study), for example, was designed to test the impact of several interventions on mortality from coronary heart disease (MRFIT Research Group, 1982). After the 7-year study period, investigators found no statistically significant difference in mortality between those who had and had not been part of the intervention group. This unanticipated result was attributed, in part, to those who were assigned to receive "usual care" (the group that did not receive the interventions) experiencing risk-reducing changes (e.g., smoking cessation) that were independent of the study. These studies may also be limited if people who are willing to participate, and to be randomized to the condition of interest, are not particularly representative of the general population at risk. Thus, generalizability of results is often lacking.

Of critical importance for the study of breast cancer are studies that compare women who were "exposed" and "unexposed" during the early life stages for which there is growing concern about higher sensitivities or vulnerability. If such studies were to rely on following women from the time of these exposures, they would have to be carried out over decades to discern differences in rates of breast cancer. A strategy to circumvent this need is for epidemiologists to examine whether the exposure influences an intermediate marker of breast cancer risk, such as age of pubertal onset, that is measureable long before the usual onset of breast cancer.

Interpretation of Group Differences

The interpretation of trends in cancer incidence and mortality in epidemiology requires consideration of multiple simultaneously changing cancer determinants, confounding factors, and even unrelated coincidental trends. Some studies examining statistical associations between health outcomes and exposures or other characteristics make assessments using data at the population or group level rather than the individual level.

Studies that examine population-level associations of disease rates with potential causal factors are termed ecologic studies. They do not look at individuals, but instead look at grouped data for both disease and exposure, such as county rates of cancer (an outcome) and percentages of the county population who have a characteristic (an exposure). Ecologic studies are prone to "ecological fallacy" (Lilienfeld and Lilienfeld, 1980) because sometimes the association seen in the group does not apply to the individuals. For example, counties with high breast cancer rates might also be counties with more women in the workforce, even though within counties, breast cancer might actually occur more often in women who are not employed, or might occur equally in those who are and are not employed. Because of this problem, ecologic studies are considered to be one of the weakest study designs in epidemiology. These designs are best viewed as "hypothesis generating" (a kind of "brainstorming") rather than "hypothesis testing."

Impact of Disease Screening

Cancer screening detects asymptomatic cancers. Uptake of screening or dissemination of a more sensitive screening test increases the detection of silent tumors as quickly as the enthusiasm for the new test builds or insurers agree to pay for the test, as occurred in the case of the dissemination of prostate-specific antigen (PSA) screening for prostate cancer. Very few factors other than screening or a sudden shift in diagnostic criteria for cancer can account for rapid changes in cancer incidence. Thus, the implementation of a new screening program or method can account for rapid increases in cancer incidence (Kramer and Croswell, 2009). If the screening process is detecting tumors sooner than they would otherwise have been found, incidence rates are likely to return to previous levels, assuming other factors are not contributing to changes in incidence. New or more extensive screening may also result in a sustained increase in incidence if it detects a reservoir of tumors that routinely exist and would never otherwise have become evident.

Another distinguishing characteristic of screening-mediated increases in cancer incidence compared to appearance of a new carcinogen is the spectrum of tumor stages found at diagnosis. In the absence of screening, the introduction of a new carcinogen would be associated with an increase in the incidence of cancer diagnoses at both localized and advanced stages of cancer. However, screening tests tend to have a disproportionate impact on the incidence of localized stages versus advanced stages because they are finding tumors that cannot otherwise be readily discovered (Kramer and Croswell, 2010).

Screening can also lead to increased detection of indolent cancers that are not life-threatening, a phenomenon known as "overdiagnosis"

(Morrison, 1985). There are two prerequisites for cancer overdiagnosis, and both have been met in the case of breast cancer: (1) the existence of a silent reservoir of tumors that would ordinarily not come to clinical attention during the life span of a given person, and (2) surveillance or screening activities that lead to detection of the reservoir (Esserman et al., 2009; Welch and Black, 2010). Estimates of breast cancer overdiagnosis vary widely (ranging from 7 to 50 percent), depending in part on whether ductal carcinoma in situ (DCIS) is included in the estimate and whether the denominator of the estimate is all cancers or only screen-detected cancers (Gøtzsche and Nielsen, 2006; Zackrisson et al., 2006; Duffy et al., 2008; Gøtzsche et al., 2009). As noted in Chapter 2, at present there is no way to know which instances of DCIS might progress to invasive cancers (Allred, 2010), so most women with in situ tumors receive treatment that is similar to the treatment for early-stage invasive tumors.

Overdiagnosis may also affect the interpretation of study results or surveillance data. Identifying modifiable risk factors that are disproportionately associated with indolent tumors might make it possible to reduce the nominal incidence of breast cancer and spare some women what is essentially unnecessary treatment, but it would have limited benefit for women with more aggressive tumors. Also, because overdiagnosis associated with cancer screening leads to an increase in incidence without necessarily changing the risk of dying of the cancer, it can artifactually inflate survival rates and cure rates of cancer, independent of any actual benefits of screening or improvements in therapy over time (Welch et al., 2000).

Long Latency and Intermediate Markers in Breast Cancer

The process of carcinogenesis usually takes place over many years or even decades. Even the most potent cancer-causing exposure, tobacco smoke, provides an example of long latency in its action. For lung cancer, tobacco smoke is a "complete" carcinogen, meaning that no other exposures are needed beyond smoking to cause cancer. Nevertheless, historically there was a delay of about two decades between widespread uptake of cigarette smoking and the subsequent epidemic of lung cancer, reflecting the latency of the disease.

Because the process of carcinogenesis usually spans years, studying early life exposures that might contribute to or cause cancer is particularly challenging. Of great use would be intermediate outcomes known to be in the causal pathway to cancer, so that studies could use these as endpoints for studying early causes for breast cancer or interventions that could ultimately lower the risk for breast cancer. Currently, candidate intermediate outcomes include early menarche, anovulatory menstrual cycles, greater maximum attained height, late age at first pregnancy, and a small number of pregnancies. While all are associated with increased risk of breast cancer, the mechanisms or explanations for these associations are not yet established; consequently, it is unknown whether altering the intermediate outcome will also alter the risk for breast cancer later in life. This question may be at the crux of the search for future intervention and prevention strategies.

Moving the Research Agenda Forward

Given the challenges of exposure assessment, timing, and intermediate outcomes outlined above, what are the options for approaching these important questions in human populations using the discipline of epidemiology?

Perhaps an ideal study design would be prospective/longitudinal follow-up of girls from intrauterine life to maturity and past menopause when breast cancer incidence, especially ER+ breast cancer, becomes most common. At frequent intervals over the life course and especially during critical windows of susceptibility, exposure would be assessed by various methods, including self-reports from parents and the girls and women themselves; environmental sampling and measurement; biospecimen assays; and other indicators of impact on physiologic, cellular, or molecular processes. Likewise, assessments of intermediate outcomes that suggest increased risk of breast cancer would also be recorded using the most accurate methods possible.

The obvious problems with this approach are the length of time and the expense needed to capture the data for decades until breast cancers are prevalent in the population. In addition, the type of exposures assessed or targeted along the way may no longer be of interest or relevant a half century later, and the collection, processing, or storage of appropriate specimens may not have been possible at the critical time window. Short of this ideal, what then are the useful, realistic study design alternatives?

No easy answers are apparent. Potentially viable options that could offer useful information are well-conceived monitoring systems and studies addressing intermediate outcomes. For example, large-scale monitoring systems with individual- or group-level information (or a combination of the two) could be leveraged for both prospective and retrospective studies. They could focus on environmental exposures, medical information from electronic health system data, or other sources of relevant exposures, covariates, and intermediate outcomes. Studies could also be designed to systematically focus on (1) relationships between exposures at early phases of development and biomarkers or intermediate outcomes, and linked to other studies of (2) these biomarkers or intermediate outcomes and later risk of breast cancer. Again, leveraging existing databases might lead more rapidly to results.

STUDYING THE ROLE OF GENETICS IN BREAST CANCER

As introduced in Chapter 2, statistical modeling of the potential cumulative effect of the inheritance of multiple risk variants, each of small effect, suggests that low-penetrance gene variants, that is, variants that do not give rise to a strong burden of breast cancer in families, could be associated with a substantial fraction of breast cancer risk. Epidemiologic studies using "candidate gene" approaches have been widely used to assess polymorphic variants in genes that plausibly influence breast cancer risk. More recently, genome-wide association studies (GWAS) have provided a more comprehensive search for associations across the genome, independent of hypotheses about specific genes. GWAS use key variants in single DNA components, called "tag single nucleotide polymorphisms" (tagSNPs), to efficiently evaluate common SNP variations in the human genome (Manolio, 2010). In studies of cases and controls of European ancestry, genotyping of 500,000–600,000 tagSNPs in each study subject permits genome-wide studies of susceptibility to breast cancer. Larger sample sizes are needed for studies of the more variant genomes of persons of African ancestry. As noted in Chapter 2, extreme levels of statistical significance are needed to identify true positive results because of the very large number of statistical tests being performed. Therefore, large sample sizes of thousands, even tens of thousands, of cases and controls are needed for these studies (Hunter et al., 2008).

Thus far, approximately 20 risk variants have been robustly associated with breast cancer risk in GWAS (Easton et al., 2007; Hunter et al., 2007; Stacey et al., 2007; Zheng et al., 2009; Turnbull et al., 2010). A number of these variants are not in regions of genes that code for gene products, and most of the others are not in genes that were previously strong candidates to be associated with breast cancer. Thus, the GWAS approach identifies variation in intergenic regions as potentially important, and it discloses new genes not previously associated with breast cancer, potentially providing new insights into mechanisms of breast cancer causation. Although it is possible that stronger associations may exist for rarer genetic variants (e.g., those with minor allele frequencies of <5 percent) that have not been tested with the technologies available to date, it is unlikely that stronger associations with common variants exist.

GENE–ENVIRONMENT INTERACTIONS

Much of breast cancer causation is assumed to be due to the interplay between inherited susceptibility to the disease and exposure to environmental risk factors or lifestyle choices. This interplay is often summarized loosely in the term "gene–environment interaction." Unfortunately, the term has several different meanings and mathematical formulations. The generally accepted meaning (Rothman et al., 2008) is that the strength of the association with a given outcome for those with both the high-risk gene polymorphism and the harmful exposure is greater than the sum of the associations for each factor alone. This type of interaction is referred to as "synergistic." Another use of the term interaction is "statistical," and this is model dependent. For studies of breast cancer (or any binary outcome measure such as yes/no), the typically used statistical models are all multiplicative, and interaction occurs when the associations for those with both the high-risk gene and the harmful exposure is not multiplicative (e.g., higher or lower). Under this approach, the two factors "interact statistically" if women exposed to both are at much higher risk than would be expected based on multiplying the individual relative risks together.[4] Generally speaking, this approach requires a much stronger combined effect than would be necessary to conclude that a synergistic relationship exists. As a result, it can be difficult to replicate findings of statistical interaction.

Because this statistical approach has dominated the breast cancer field, the examples given here test "multiplicative" interaction. The committee notes, however, that biologic interaction can occur through a variety of mechanisms (see Chapter 5), and the synergistic "additive" definition is consistent with factors acting through many of these biologic mechanisms.

Investigating Gene–Environment Interactions

Complex diseases are often the result of both genetic and environmental factors. Few researchers, however, have seriously undertaken the examination of their combined effects. Partly this is because very large studies are needed to identify interactions—for a binary exposure and a gene with two functional forms, the sample sizes need to be at least four times larger than those needed to assess any two-level factor alone with the same statistical power. However, the complexity takes on more dimensions. Even when a study is not investigating the interactions of genes and environmental factors, the ability of those studies to identify environmental risk factors may be compromised by those relationships. Moreover, when exposures become pervasive, all the variability will tend to appear to be due to genetic factors.

For some exposures, investigations of genetic interactions are drawing attention to specific genetic features. The evidence on smoking in conjunction with variants in the N-acetyltransferase 2 (*NAT2*) gene is discussed

---

[4]This corresponds to the *P* value for the "multiplicative" interaction term in a logistic regression model. A statistically significant *P* value with a positive regression coefficient indicates that the joint exposure is associated with higher risk than expected simply by multiplying the relative risks.

in Chapter 3. Some, but not all, studies have suggested that the "slow acetylator" form of *NAT2* appears to increase the risk of breast cancer for heavy smokers. Genetic characteristics investigated for interactions with exposure to PCBs and ionizing radiation are discussed here as examples of gene–environment interactions that are being studied.

Polychlorinated Biphenyls

An example of the difficulty of investigating gene–environment interactions is offered by data on polymorphisms in the *CYP1A1* gene and exposure to higher blood levels of PCBs. *CYP1A1* polymorphisms do not, by themselves, appear to be associated with alterations in risk for breast cancer (Laden et al., 2002; Masson et al., 2005). Similarly, a meta-analysis based on 1,400 case patients with breast cancer and 1,642 control subjects suggests no relation of higher blood levels of PCBs with risk of breast cancer (Laden et al., 2001). Although blood levels of PCBs found in the reviewed studies reflect many years of exposure, the blood samples were mostly collected at the time of breast cancer diagnosis, or less than 10 years before diagnosis, and thus do not exclude an influence of exposures in early life or adolescence.

Other studies have included data on *CYP1A1* polymorphisms in the analysis. Moysich et al. (1999) observed in a case–control study with data on 154 postmenopausal cases that women who carried at least one Val allele at codon 462 in the *CYP1A1* gene and whose blood levels of total PCB concentration were above the study median had an increased risk of breast cancer (OR = 2.9, 95% CI, 1.2–7.5) compared with women carrying two copies of the Ile allele and below the median for total PCBs (the test for statistical interaction was not statistically significant: P = .13). In a subsequent study based on 293 cases, Laden et al. (2002) reported that postmenopausal women who carried at least one Val allele at codon 462 and were in the highest third of total plasma PCB concentrations had a relative risk for breast cancer of 2.8 (95% CI, 1.0–7.8), compared with women carrying two copies of the Ile allele and in the bottom third of plasma total PCBs (the test for statistical interaction was marginally significant: P = .05). When premenopausal women were included (a combined total of 367 cases), no suggestion of increased risk was evident. In a third study of this association, Li et al. (2005) also observed an increase in risk among women who were above the median value for total plasma PCBs if they were carriers of the codon 462 Val allele (P, interaction .02), although the association was limited to premenopausal cases, not postmenopausal cases.

Thus, these three published studies of this association have reported an elevation in the risk of breast cancer among women jointly exposed to higher plasma PCB levels and the *CYP1A1* codon 462 polymorphism, although the statistical interactions were not all statistically significant, and there was some inconsistency according to menopausal status. Given the possibility that the relevant exposures may have occurred many years before the subjects' PCB levels were measured, that the particular PCB(s) of concern may not correlate well with the "total PCB" measurements used, and that the small size of the studies limited their power to detect multiplicative interaction, the findings from three different populations are, at a minimum, intriguing and worthy of further investigation.

Ionizing Radiation

Another relation that has been explored is that between mutations in a gene important for DNA repair and exposure to ionizing radiation, which can induce DNA damage. The *ATM* gene is critical in signaling the occurrence of double-strand breaks and directing repair of the damaged DNA. Furthermore, ataxia–telangiectasia (A-T), an autosomal recessive disorder characterized by extreme sensitivity to radiation, is the result of truncation mutations at the *ATM* gene. There is strong biological plausibility to the hypothesis that women with *ATM* mutations, who are less able to respond to DNA damage, will be at higher risk of breast cancer generally and that the breast cancer risk from a given dose of radiation will be greater in women who carry the *ATM* mutation than in those who do not. In addition, mothers of A-T patients are obligate (heterozygous) carriers of the A-T mutation, and studies indicate that these women are at higher risk of breast cancer than women who do not carry the mutation.

One well-designed study has been done to address the issue of radiation sensitivity. The Women's Environment, Cancer, and Radiation Epidemiology (WECARE) case–control study draws on women with breast cancer from a consortium of five cancer registries who were followed for a second primary breast cancer (in the contralateral breast). The details about the study design, patient population, and the study results have been reported (Bernstein et al., 2004, 2010; Concannon et al., 2008; Stovall et al., 2008; Langholz et al., 2009). Briefly, blood samples were taken and *ATM* genotyping performed to locate *ATM* SNP variants, as well as splicing and truncation mutations, the types of mutations most commonly associated with A-T (Concannon et al., 2008). About 40 percent of women with breast cancer received radiation therapy in the treatment of their disease. The ionizing radiation exposure to the healthy (contralateral) breast could be estimated with a fair degree of accuracy, based on standard practice and treatment records (Stovall et al., 2006). Overall, there was little evidence of variation in risk for a second breast cancer due to radiation across types of *ATM* variants. However, radiation susceptibility was found in one subset of women whose *ATM* gene had at least one "rare" variant at SNPs that both (1) resulted in a change in protein coding (i.e., a "missense" variant) and (2) were highly conserved over species (Bernstein et al., 2010). About 10 percent of breast cancer patients are in this subset, and, in the WECARE Study, the effect of radiation was found to be about twice that observed in women with the same radiation exposure who were not in this subset. Rare variants were defined as those occurring in less than 1 percent of controls, and variation in radiation effect was not seen in more common variants. It is notable that very few cases (15 out of 708) or controls (23 out of 1,397) had the truncation and splicing mutations associated with A-T (Bernstein et al., 2010).

If replicable, these findings suggest that evolutionarily recent changes in the DNA code may impair DNA repair mechanisms, resulting in increased risk of breast cancer due to radiation, but DNA repair is not affected by changes that have "propagated" into the population and become more common. GWAS studies, the current focus of genetic-epidemiologic research, are poorly suited to detect rare SNP changes as they use known marker sites of common genetic variation, and rely on correlation of genetic code locally to detect genetic (or gene–environment interaction) effects. Disease associations with isolated rare SNP changes are not readily detectable by this technique. The WECARE Study results exemplify the difficulty of establishing gene–environment interactions even in the context of an established breast carcinogen and a gene known to be involved in biological mediation of the effect of the carcinogen.

SNP Variants with Robust Associations with Breast Cancer Risk

Another approach to the study of gene–environment interaction is to assess whether the SNP variants robustly associated with breast cancer risk in the GWAS are modified by established environmental or lifestyle risk factors. This approach is not motivated by knowledge of the biological function of specific genes or their relevance to specific exposures. In the largest published report to date, an analysis of 12 such polymorphisms and 10 established risk factors among 7,610 breast cancer cases found no statistically significant interactions after accounting for the 120 interaction comparisons that were made (Travis et al., 2010).

Thus, evidence is limited for robust, replicable "synergistic" interactions between inherited genetic variants of unknown function and established environmental and lifestyle risk factors in breast cancer causation. This by no means makes it irrelevant to quantify these individual associations. For the purposes of risk prediction, all these risk factors appear largely to multiply together—the more genetic or environmental risk factors a women has, the higher her risk. The current evidence merely suggests that the known risk factors do not synergize with the genes that rise to attention from GWAS approaches, for which the functions are unknown, in a manner that amplifies risk beyond the expectation of multiplying the relative risks for individual factors. Notably, the relative risks associated with the 12 polymorphisms identified by Travis et al. (2010) were small (the maximum was RR = 1.22).

Studies that more systematically address gene–environment interaction are warranted, but the difficulties are considerable. Most published studies have assessed established lifestyle factors, and have limited or no information on hypothesized environmental factors that may be risk factors only in a genetically susceptible subset of women. Very large studies such as the National Institute of Environmental Health Sciences' Sister Study (www.sisterstudy.org), which has enrolled 50,000 women and collected blood, urine, toenail, and household dust specimens, may provide more information on gene–environment interactions in the future.

Implications of Genetic Variability for Understanding Risk for Breast Cancer

Women who carry a bona fide mutation in *BRCA1* or *BRCA2* are clearly at such substantially elevated risk of breast cancer that their medical care is altered (Narod and Offit, 2005). Women who carry a higher number of low-penetrance risk alleles are at higher risk than women who carry a lower number (about a 10 percent increase in lifetime risk), but the incremental increase in lifetime risks is far smaller than that of *BRCA1/2* carriers compared to noncarriers (estimated to be a 50 percent or more increase in lifetime risk). In an analysis of 5,590 cases and 5,998 controls from the Cancer Genetic Markers of Susceptibility (CGEMS) collaboration, in which the first 10 GWAS-associated SNPs were genotyped, Wacholder et al. (2010a,b) observed that women who carried 13 or more of the 20 risk-conferring variant markers (seen in about 4 percent of the population) had a nearly three-fold increase in risk (OR = 2.90, 95% CI, 2.37–3.55) compared with women carrying six or fewer markers (12 percent of the population). In terms of risk prediction using receiver operating characteristic (ROC) analysis, discrimination between cases and controls on the basis of the number of risk SNPs was relatively poor, but it was equivalent to the discriminatory ability of the clinical standard, the Gail model (Gail et al., 1989), which uses established breast cancer risk factors. Predictions derived from a model that included interactions among the SNPs and the factors used in the Gail model were no better than those from the simpler models. However, women who were classified as at high risk by both the Gail model and the genetic model were at modestly higher risk than women who were at high risk on only one.

In an analysis of 10,306 women with breast cancer and 10,393 controls, in which 7 of the GWAS-associated SNPs were independently confirmed (Reeves et al., 2010), a relative risk of approximately two-fold was observed between the highest quintile of genetic risk and the lowest quintile. The cumulative risk to age 70 for women in the highest of the five genetic risk groups was approximately 7.8 percent, much lower than the 50 to 85 percent cumulative risk associated with *BRCA1* or *BRCA2* mutations. These authors have previously reported that none of these SNPs were involved in statistical interactions with any established risk factors for breast cancer (Travis et al., 2010).

Thus, when a genetics-based tool for risk estimation becomes available, estimates based on currently established genetic associations of gene variants with breast cancer will contribute to the identification of women at higher risk of breast cancer. However, the relative risks observed are similar to those obtained from risk estimates derived from clinical history and established nongenetic breast cancer risk factors, without the need for genotyping. At present, the combination of genetic and nongenetic risk factors is not deemed to provide sufficient information to enable enough risk stratification to alter the medical care women receive (Reeves et al., 2010; Wacholder et al., 2010a), but this situation may change as more genetic risk loci are discovered.

Another hope for the use of genetic variability in understanding risk is that environmental risk factors that have not been convincingly associated with risk of breast cancer among all women will be convincingly associated with risk among a genetically defined "susceptible" stratum. According to this argument, if only some women are genetically predisposed to breast cancer after exposure to an environmental agent, the environmental signal will not be detected in studies that assess the effect of the environmental factor among all women, whereas if the susceptible women can be identified, the environmental signal will be convincing. At this point, a possible example appears to be that of slow acetylators based on variants in the *NAT2* gene and exposure to cigarette smoking. In breast cancer etiology research, most of the assessments have been based on random pairings rather than biologically plausible interactions and have not yielded potentially fruitful leads.

Studies of gene–environment interactions among women with the highest inherited risk, that is, women with pathogenic *BRCA1* or *BRCA2* mutations, are limited by small sample sizes, and by the fact that women with these mutations identified in the course of clinical care or investigation of familial risk are not systematically enrolled in research studies that include assessment of environmental exposures. Furthermore, prospective studies are limited by many of these women choosing to reduce their future risk of cancer by having bilateral prophylactic mastectomies, removal of their ovaries, or both, once they are identified as *BRCA1/2* carriers.

Identifying Important Genetic Factors

As summarized earlier, about 20 genetic loci associated with breast cancer have been discovered since 2007 using the new genome-wide association study approach. Given the sample sizes tested so far, many more loci are likely to be discovered in the near future. Based on the relative risks for SNP variants discovered so far, for a cancer like breast cancer, approximately 50 risk variants with similar characteristics remain to be discovered in larger sample sizes (Park et al., 2010). However, even once these are discovered, much less than half of the inherited variability in breast cancer risk will have been explained (Park et al., 2010). The unexplained variability is sometimes called the "dark matter" of the genome (Manolio et al., 2009). There is a wide variety of opinion on what type of genetic variation may underlie this "missing heritability," or whether the fraction of breast cancer due to inherited causes may have been overestimated (i.e., more of the true fraction due to inherited susceptibility has been explained than is currently estimated). It remains possible that common and rare variants operating in combination with additional "hits" from environmental factors hold the key to the remaining genetic contributions.

One lesson is clear from the GWAS studies: very large sample sizes of tens of thousands of cases and controls are needed to detect the small to modest relative risks that have typically been associated with common (>5 percent minor allele frequency) variants. This has necessitated the formation of large-scale consortia, such as the Breast Cancer Association Consortium (BCAC), the National Cancer Institute Breast and Prostate Cancer Cohort Consortium (BPC3), and the CGEMS study. Continued enrollment of cases of breast cancer, and appropriate controls, along with collection of DNA, blood, and environmental and lifestyle information, will be necessary to facilitate discovery of more genetic variants, particularly those that confer small additional risk of the disease or act in concert with exogenous (or endogenous) exposures.

Nearly all of the minor allele frequencies of the low-penetrance GWAS-discovered variants are greater than 10 percent. The frequency of disease-associated variants in high-penetrance genes such as *BRCA1*, *BRCA2*, *p53*, and *PTEN* are much less than 1 percent. The spectrum of genetic variation between 1 and 10 percent is still largely unexplored, and is likely to contribute to an unknown fraction of the "missing heritability" (Figure 4-5). New technologies, notably higher density SNP arrays that allow up to 5 million genotypes to be determined on a single DNA sample, and the advent of sequencing of the whole genome at $1,000 or less, will permit exploration of this genomic territory, but will require large sample sizes to reach statistically robust conclusions.

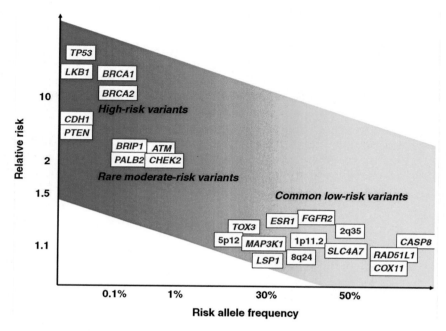

FIGURE 4-5 Genetic variants associated with breast cancer arrayed on the basis of their frequency and their impact on breast cancer risk.
SOURCE: Varghese and Easton (2010). Used with permission: Varghese, J. S., and D. F. Easton. 2010. Genome-wide association studies in common cancers—what have we learnt? *Curr Opin Genet Dev* 20(3):201–209.

GWAS-style studies of gene–environment interaction require even larger sample sizes, and are limited by the lack of environmental data in many of the case series that have been used for GWAS discovery. A major limitation is the difficulty of establishing exposure to many hypothesized environmental factors, such as exposures that may have occurred in utero, childhood, or early adulthood, or exposures that require sophisticated and potentially expensive biological measurements, especially of biological samples such as adipose or breast tissue. Despite the need for new and improved methods to estimate these exposures, most of the established breast cancer risk factors can be readily ascertained at interview or by questionnaire. Greater emphasis is needed on systematic collection of data on these established risk factors in studies that will be used to generate information on genotypes in order to maximize the sample sizes available for assessment of gene–environment interaction.

STUDYING ENVIRONMENTAL RISK FACTORS THROUGH WHOLE ANIMAL AND IN VITRO EXPERIMENTS

Laboratory animals and human and nonhuman cells and tissue cultures have provided powerful experimental systems for investigating mammary cancer development and mechanisms. As described in Chapter 2, these test systems are also used to identify chemicals that may pose human cancer risks, and provide critical information on how the magnitude and timing of exposure may affect the process of mammary carcinogenesis. This section discusses some challenges in using these model systems, and some emerging approaches that may ultimately provide for improvements in identifying and understanding environmental factors that contribute to breast cancer risk in humans.

In Vivo (Live Animal) Bioassays to Identify Potential Chemical Carcinogens

As briefly outlined in Chapter 2, rodents have long been used to screen chemicals for carcinogenic potential. Standard protocols have been developed that use well-characterized strains of rats and mice and carefully defined approaches for dose selection, length of study, and age windows of exposure. In rodents as in humans, mammary tumors may arise through several modes of action, such as from endocrine-related effects on tissue development and growth or through genotoxic effects on breast cells. Studies in rodent models have also shown the importance of exposure during critical windows of development on mammary cancer risk later in life, in terms of either direct carcinogenic effects or alterations in susceptibility to subsequent exposure to carcinogens. However, the standard protocols for rodent carcinogenicity testing may not be adequate to identify the potential for early-life chemical exposures to induce mammary tumors later in life. Moreover, concerns have been raised about their adequacy for predicting human breast cancer occurrence because of potential differences between rodents and humans in biochemical, cellular, or developmental characteristics (Rudel et al., 2007, 2011; Thayer and Foster, 2007). The following section discusses issues related to these concerns, such as timing of exposure in the bioassay at susceptible age windows, possible confounding by high-dose testing, and differences in species and strain susceptibility.

Timing of Dosing and Windows of Susceptibility

Standard experimental carcinogenicity studies typically begin with young adult animals and thus do not include early life (in utero, perinatal, prepubertal) windows of exposure, which are potentially critical periods of exposure for some chemicals. The studies often conclude after ongoing exposures for 2 years, or 18 months for some mouse protocols. The average life span of most rodent species and strains used in testing is approximately 2 to 3.5 years. Studies that end well before the completion of the animals' natural life may miss potential impacts of chemicals that might occur later in life, a period when human breast cancer incidence is high and observed to be modulated, at least by hormone replacement therapy. Rodent bioassays are typically not extended beyond 2 years because of increases in mortality at the maximum tolerated dose and rising numbers of background tumors with age in the controls, both of which compromise statistical comparisons.

Given the existing protocol of beginning chemical dosing after weaning, the most susceptible time for rats to be exposed to mammary carcinogens is for young virgin female rats between the ages of 35 and 60 days (Russo and Russo, 1996; Ren et al., 2008). Rats treated with potent genotoxic carcinogens at 21 days or at over 150 days in age have a much lower frequency of hormone-dependent tumors than rats exposed at 50 days (Medina, 2007). A full-term pregnancy (or administration of pregnancy-related hormones) in rats either shortly before or after exposure to a chemical carcinogen confers considerable protection against development of mammary tumors (Blakely et al., 2006). Parity induces changes in gene expression that are highly conserved among rat strains, thereby conferring increased resistance to development of mammary tumors, even in susceptible rat strains (Blakely et al., 2006). Parity before chemical exposure in mice, likewise, increases resistance to mammary tumor formation (Medina, 2007). In this way, these rodent models are consistent with the observation in humans that a first pregnancy at a younger age and multiple full-term pregnancies are associated with a reduction in breast cancer risk.

Like humans, rats have reproductive cycles that decline with age. Young adult female rats experience estrous every 4 to 6 days. When female rats reach middle age, their estrous cycles become irregular and eventually stop (Morrison et al., 2006). This transitional process is conceptually and functionally similar to menopause in women. However, rodent "estropause" and human menopause are not identical. Some strains of acyclic rats, like the Sprague Dawley strain, actually have chronically high estradiol concentrations and thus are in a state of persistent estrus and lose capacity for a luteinizing hormone surge, whereas Fischer 344 rats and humans develop reduced estrogen secretion as they enter reproductive senescence (Chapin et al., 1996). The time point of initiation of reproductive senescence also differs by rat strain. Similar to humans, it begins at 60 to 70 percent of Fischer 344 rat life span, but starts much earlier in the Sprague Dawley rat, at 30 to 40 percent of its life span (Chapin et al., 1996). Keeping such considerations in mind is important when using rats as models for breast carcinogenesis.

To study "postmenopausal" effects, scientists often use the ovariectomized (OVX) rat model, although this model is not used in standard testing protocols for pesticides and industrial chemicals. Using age-appropriate rats at young, middle, and old ages, the ovaries are surgically removed, and a month later the OVX rats are given estradiol, with or without progestins, in physiologically relevant dosages (Yin and Gore, 2006). There are also other important differences between humans and rodents in postmenopausal hormonal changes, further complicating the use of rodents to study the effects of chemicals on postmenopausal aging, including breast cancer (Morrison et al., 2006).

Various other animal protocols are used to study how exposures during different age windows may affect susceptibility and to examine those chemicals that may have effects in a given age window. Puberty is an important exposure period for later susceptibility for human breast carcinogenesis because of the rapid rate of tissue growth and development. For example, a meta-analysis by Henderson et al. (2010) found that studies were very consistent in showing that exposure to X-irradiation during younger ages (<15) was associated with higher relative risks for breast cancer than was exposure of older age groups, and that the risk declined with increasing age at exposure. Similarly, in rats, the period between the ages of 5 and 8 weeks is one of rapid mammary ductal growth and branching. Fifty-six percent of the rats given a single large dose of the potent genotoxic carcinogen dimethylbenz[a]anthracene (DMBA) during this period developed mammary carcinoma, compared to 8 percent given the same dose at ages less than 2 weeks (Meranze et al., 1969). At this particular point in mammary gland development in the rat, there is a high mitotic index in the terminal ductal structures (Jenkins et al., 2009; Betancourt et al., 2010; La Merrill et al., 2010).

Indeed, administration of DMBA, N-methyl-N-nitrosourea (MNU), or other potent mammary carcinogens on postnatal day 50 is the basis for a widely used experimental rat model of mammary carcinogenicity (Russo and Russo, 1996). This model has also been used to explore the potential for an exposure to an agent or factors much earlier in life to protect against or contribute to mammary tumorigenesis (e.g., olive oil [Pereira et al., 2009]; increased birth weight [de Assis et al., 2006]; and ethyl alcohol [Hilakivi-Clarke et al., 2004]; also see Rudel et al., 2011). Specifically, a number of studies have used models that give a high single dose of a potent genotoxic carcinogen such as DMBA or MNU on or about postnatal day 50 to investigate the effect on mammary carcinogenesis of in utero or perinatal exposures of rats and mice to endocrine-active chemicals (e.g., dioxin plus high-fat diet or BPA) (Jenkins et al., 2009; La Merrill et al., 2010).

Several studies have also used this model to evaluate whether in utero or perinatal exposure to various components of the diet can alter (enhance or inhibit) mammary carcinogenesis. For example, early dietary exposure to genistein, a soy phytoestrogen (Fritz et al., 1998; Hilakivi-Clarke et al., 1999b), and zearalenone, an estrogenic mycotoxin that frequently contaminates cereal grains (Hilakivi-Clarke et al., 1999b), were observed to be protective; whereas exposure to flaxseed enhanced carcinogenesis (Khan et al., 2007). However, in another study with different exposure characteristics (Hilakivi-Clarke et al., 1999a), genistein enhanced DMBA-induced mammary carcinogenesis. The authors concluded that the direction of effects of genistein on mammary carcinogenesis in the DMBA model was apparently dependent on the dose, length of exposure, and timing.

Ultimately this study protocol may be helpful in identifying some candidate risk factors that may not directly cause tumors later in life themselves, but that may, when exposures occur early in life, modulate risk of mammary tumors associated with subsequent exposure to potent carcinogens. It also raises questions about the predictive value of the standard protocols used in pesticide, pharmaceutical, and industrial chemical testing, as the standard 2-year chronic bioassay protocols (e.g., FDA, 1997; EPA, 1998a; Makris, 2011) generally do not include exposures during in utero and prepubertal periods. Furthermore, standard approaches use exposure to only a single test agent at a time, and thus are not able to identify possible interactive effects with other chemicals, including potential promotional influences on chemically initiated mammary tumors. The animal model of in utero administration of a chemical followed by administration of a potent genotoxic carcinogen at day 50 potentially reflects the ability of early-life exposures to influence the carcinogenicity of other chemicals. However, young children are unlikely to experience a single high dose of a genotoxic substance except in a therapeutic circumstance, such as treatment of cancer with genotoxic chemotherapeutic agents or ionizing radiation. Nevertheless, compared to standard protocols, this model provides information on the potential for cocarcinogenesis in breast cancer—having one exposure that does not appear to alter cancer incidence itself increase the rate of cancer associated with a subsequent exposure.

The National Toxicology Program (NTP), with its emphasis on using 2-year rodent bioassays to identify chemicals that may increase cancer risk in humans, has recently proposed to address the "age at exposure" issue by extending the exposure period for rats to include both in utero and prepubertal exposures (Bucher, 2010). An NTP workgroup on hormonally induced reproductive tumors recommended changes to the standard protocol to include various chemical exposure periods that may be relevant for breast cancer (Thayer and Foster, 2007). NTP indicated in 2010 that it has begun testing some chemicals using its "perinatal protocol": exposure of the dams begins on gestation day 6 and continues through to postnatal day 21, when the pups are weaned and begin to receive the test compound directly (Bucher, 2010; NTP, 2010c; Foster, 2011).

Species Concordance of Sites of Carcinogen-Induced Tumors

When an excess of tumors is identified in an animal study, questions arise regarding the extent to which such tumors predict human cancer risk from the test agent. While species concordance in overall positive versus negative evidence for carcinogenicity (regardless of organ) is relatively high in standard cancer bioassays (Huff et al., 1991), differences in the specific target site affected are common. In NTP testing reports reviewed by the committee, 90 percent of 30 chemicals that were tested in both mice and rats and that showed mammary carcinogenesis in either species had clear evidence of carcinogenicity (but not necessarily in the mammary gland) in both species (based on cancer occurrence at one or more specific sites). Mammary tumors occurred in both rats and mice for only 17 percent of these 30 tested chemicals, although tumors were induced at other sites. In a committee compilation of information from IARC and NTP reports, 89 percent of agents that IARC has found to have sufficient or limited evidence of human breast cancer also showed evidence of mammary tumors in rats or mice.

Strain and Species Similarities and Differences in Mammary Tumor Susceptibility

Another consideration in the selection of animals for testing is whether the strain used is sufficiently sensitive to detect the carcinogenic activity of an agent that poses a risk to humans. Large differences in species and strain susceptibility for mammary tumors are apparent among rats and mice. Some rat strains are sensitive, while others are resistant to mammary tumors formed spontaneously or induced by hormonal or other agents (Kacew et al., 1995; Ullrich et al., 1996; Thayer and Foster, 2007). For example, in the 1950s diethylstilbestrol (DES) was found to induce mammary tumors in F344 and ACI rats, but the Copenhagen strain was resistant, as was the Sprague Dawley strain in initial studies (Kacew and Festing, 1996). After the link between DES and human breast cancer became clear (IARC, 1987), further testing revealed additional strain differences in sensitivity. Table 4-1 shows some rat and mouse strains that have exhibited sensitivity or resistance to mammary carcinogenesis in research or carcinogenesis screening when exposed to chemical carcinogens (e.g., DMBA, MNU, radiation), hormonal influences, or for mice, the mouse mammary tumor virus (MMTV). The assignment of the relative sensitivity of a strain is somewhat dependent on the agent, its mechanism of action, and tumor type. Rodent strain differences in mammary tumor susceptibility can arise from differences in sexual development, endocrine function, tissue metabolism, or other factors (Kacew et al., 1995; Kacew and Festing, 1996; Bennett and Davis, 2002; Ren et al., 2008).

TABLE 4-1 Examples of Rat and Mouse Strains of Differing Sensitivity to Mammary Tumor Formation in Response to Carcinogenic Agents

Animal Model	More Sensitive	Less Sensitive
Rat	Sprague Dawley	Copenhagen
	Lewis	Wistar-Kyoto
	F344/N (fibroadenoma)	Long Evans
	Wistar-Firth	F344/N (carcinoma)
Mouse	BALB/c	B6CF$_1$[a]
	B6C3F1[b]	C57BL/6
	C3H	O20
	RIII	

NOTE: The carcinogenic agents may include chemicals (e.g., dimethylbenz[a]anthracene [DMBA]), radiation, or hormonal influences. For mice, the agent can be the mouse mammary turmor virus (MMTV).
[a]Cross between C57BL/6J and BALB/c.
[b]Cross between C57BL/6J and C3Hf.
SOURCES: Gillette (1976); Kacew et al. (1995); Kacew and Festing (1996); Russo and Russo (1996); Ullrich et al. (1996); Bennett and Davis (2002); Blakely et al. (2006); Boorman and Everitt (2006); Medina (2007); Thayer and Foster (2007); Ren et al. (2008).

Some rat strains show consistency in being highly susceptible (e.g., Sprague Dawley) or resistant (e.g., Copenhagen, Wistar-Kyoto) to most chemical carcinogens, radiation, and hormonal agents, as well as having a corresponding high or low rate of spontaneous mammary tumors (Kacew et al., 1995; Russo and Russo, 1996; Ullrich et al., 1996; Ren et al., 2008). Other strains demonstrate more complex susceptibilities. Fischer 344 rats are susceptible to radiation, some chemical carcinogens such as MNU (although less than Sprague Dawley), and hormonal agents, and they have a high incidence of spontaneous fibroadenomas that increases with age. However, they are also resistant to other chemicals such as N-hydroxy-acetyl-aminofluorene and atrazine (Kacew et al., 1995; Russo and Russo, 1996; Blakely et al., 2006). Long Evans rats are resistant to most chemical carcinogens, but are susceptible (although less so than Sprague Dawley and Lewis) to radiation-induced mammary tumors (Russo and Russo, 1996). Among mice, the B6C3F1 strain is susceptible to the MMTV, radiation, and hormonal agents (Ullrich et al.,1996), although it tends to develop liver rather than mammary tumors from chemical exposure, which may make it less sensitive than the BALB/c strain for testing potential mammary carcinogens (Bennett and Davis, 2002). The BALB/c strain is more susceptible to mammary tumors from chemical and radiation exposure, but it is less sensitive to MMTV (Ullrich et al., 1996; Bennett and Davis, 2002).

Background Tumor Rates and Types

The incidence of specific tumor types in the unexposed animal (the background rate) is another consideration in judging sensitivity of the test animal. Rodent strains used in testing differ in background rates of tumors. High background rates, particularly those that are variable across experiments, may reflect an inherent susceptibility of the animal to exogenous agents and therefore increase the sensitivity of testing for carcinogenicity. Although use of genetically sensitive strains increases the possibility of "false positives" when extrapolating the results to other species, including humans. On the other hand, from a statistical point of view, high background rates can render the study less sensitive if fewer animals are at risk from the test chemical, and more need to develop tumors for the result to achieve statistical significance. Highly variable background rates could also result in the observation of a chemical-related effect by chance, or, conversely, missing a positive response because of an abnormally high background rate in the concurrent control animals. Such considerations make the interpretation of both "positive" and "negative" results challenging in terms of using rodent models to predict human cancer risk.

Table 4-2 shows the background incidences for different types of mammary tumors in strains used in the National Toxicology Program testing program (NTP, 2008, 2010a,b), and compares them to human rates. Overall, the spontaneous lifetime incidence of malignant mammary tumors in female mice and rats was at or below the 12.1 percent lifetime probability of a woman in the United States being diagnosed with breast cancer (NCI, 2010). The low rates in male F344/N rats of 0.16 percent and B6C3F1 mice of 0.09 percent were comparable to the lifetime probability of U.S. men being diagnosed with breast cancer, 0.14 percent (NCI, 2010). Female mice and rats have higher spontaneous lifetime incidence values than males, as occurs in humans.

Benign adenomas occur with a much lower spontaneous incidence in male animals, with rates falling below those for carcinomas for both rat strains and the mouse strain (Table 4-2). Adenomas are tumors of epithelial origin (Donegan, 2002). In women, tubular adenomas have been reported to account for 0.3 to 1.7 percent of benign lesions (Bellocq and Magro, 2003). During pregnancy, tubular adenomas may show secretory changes and are designated as "lactating adenomas." Lactating and tubular adenomas can be observed in invasive cancer and cancer in situ, although this is uncommon (Bellocq and Magro, 2003). In rodents, tubular adenomas are part of a progression that can be influenced by carcinogen exposure and give rise to adenocarcinoma (Russo and Russo, 1996).

Fibroadenoma is the predominant mammary lesion in certain rat strains, including some used in carcinogenicity screening (Table 4-2). The

TABLE 4-2 Lifetime Incidence of Breast Tumors in U.S. Men and Women and Overall Percentage of Control Animals That Developed Spontaneous Mammary Tumors in Recent Reports on National Toxicology Program Carcinogenesis Studies

| | Incidence by Tumor Type (%) | | | |
| | Malignant | Benign | | |
	Carcinoma	Adenoma	Fibroadenoma	Fibroma, adenoma, or fibroadenoma
Female				
Human	12.15	Uncommon	9–28	9–28 or more
B6C3F1 mouse (n = 1,298)	1.77	0.08	0.08	0.15
F344/N rat (n = 1,250)	5.20	2.08	52.40	53.60
Sprague Dawley rat (n = 473)	10.15	2.54	67.44	68.29
Male				
Human[a]	0.1[a]	Very rare	Very rare	Very rare
B6C3F1 mouse (n = 1,250)	0.08	0	0	0
F344/N rat (n = 1,298)	0.39	0.31	3.16	3.62
Sprague Dawley rat (n = 50)	0	0	0	0

[a]The majority of male breast cancers are ductal carcinoma; of these, 80 percent of cases are invasive and 2–17 percent are in situ.
SOURCES: Komenaka et al. (2006); NTP (2008, 2010a,b); NCI (2010).

importance of these lesions for prediction of human cancer risk is controversial. Certain rat strains used in testing have a very high background rate for them, whereas this tumor type is less common in women. In addition, rat fibroadenoma infrequently progresses to malignancy (Boorman and Everitt, 2006). In the untreated female rats commonly used in NTP studies, background fibroadenoma rates now fall above 50 percent (Table 4-2).

Increased obesity in test animals has contributed to a rise in background incidence of these tumors (Haseman et al., 1998). Harlan Sprague Dawley rats have a significantly higher incidence of mammary gland fibroadenoma (67.4 percent versus 48.4 percent) than the previously used Fisher 344/N strain (Dinse et al., 2010).

Fibroadenomas in women are a benign, biphasic tumor with an epithelial component and a predominant stromal component (Bellocq and Magro, 2003). They are the most common breast tumors in women under age 40, with highest incidence between ages 15 and 35 (Guray and Sahin, 2006), and low incidence in postmenopausal women (Goehring and Morabia, 1997; Kuiper, 2005). In autopsy studies, fibroadenomas were found in from 9 to 28 percent of women (Dixon, 1991; Goehring and Morabia, 1997). Other estimates are that they may be present in 25 percent of asymptomatic women (El-Wakeel and Umpleby, 2003). These estimates suggest that the rates in women are similar to those in earlier NTP rat studies (28 percent). In women, medically important multiple and large fibroadenomas in the breast are seen following treatment with the immunosuppressive drug cyclosporine following organ transplantation (Perera et al., 1986). Other modifiable risk factors are not apparent, and obesity was identified in one review as protective in women (Goehring and Morabia, 1997). Fibroadenomas also occur in men, but they are rare (Rosen, 2001), and they often coexist with gynecomastia (Shin and Rosen, 2007). Cases of fibroadenoma have been reported to follow hormonal treatment for prostate cancer (Shin and Rosen, 2007; Noolkar et al., 2010) and in people who undergo male-to-female sex conversion (Kanhai et al., 1999).

In rats, malignancies will occasionally arise in a fibroadenoma (Boorman and Everitt, 2006). In one series, carcinomas were reported as arising from fibroadenomas at varying low rates in different rat strains (0.5 percent in BN/BiRij, 2.1 percent in Sprague Dawley, 2.3 percent in WAG/Rij) (Huff et al., 1989). In humans, in situ lobular and ductal carcinomas occasionally develop within fibroadenomas (Bellocq and Magro, 2003). There is not agreement as to whether human fibroadenomas can progress to malignancy, and research continues to explore the extent of their relationship to phyllodes tumors, some of which are malignant, and various carcinomas (Markopoulos et al., 2004; Kuiper, 2005; Kabat et al., 2010). Some consider the finding of carcinoma within fibroadenoma as a chance occurrence, and as commonly arising out of similar cells elsewhere in the breast (Dixon, 1991), and that ultimately a fibroadenoma is not a premalignant lesion (Thayer and Foster, 2007). Carcinomas are rarely found within a juvenile fibroadenoma (Rosen, 2009); they are more common in the complex fibroadenoma subtype. Complex fibroadenomas are treated by others as a low probability but a possible source of malignancy (Kuiper, 2005).

High-Dose Testing

The high doses used in carcinogenicity bioassays may complicate interpretation of findings of chemically induced mammary gland tumors. Debates around animal testing often note the potential for tumor induction related to cell proliferation secondary to high-dose cytotoxicity, and that this may be exacerbated by high-dose saturation of metabolic detoxification pathways (Rudel et al., 2007). However, this potential mechanism may be less likely for mammary tumor formation, given the limited fraction of chemicals positive for mammary carcinogenesis in tests conducted with maximum tolerated dosing. Also, unlike the lungs, forestomach, or skin, the mammary gland is not subject to point-of-entry exposure to a chemical. Unlike the liver, the mammary gland also does not receive the first pass of absorbed chemicals from the gastrointestinal tract.

Receptor-mediated responses (in which a chemical must first bind with a cellular receptor to produce the initial biochemical effect), especially those involving multiple receptors and feedback, can first increase and then decrease as dose increases or have other complex, non-monotonic dose–response relationships, as occurs for a variety of endpoints with estrogen and xenoestrogens (Sergeev et al., 2001; Welshons et al., 2003; Watson et al., 2007; Kochukov et al., 2009). This leads to the possibility that the compound has a proportionally different activity in the high-dose experiment than it would at much lower doses. For chemicals that may cause cell proliferation by hormonal mechanisms, such activity may be diminished or eliminated at high doses but present at lower doses. Nevertheless, many of the animal studies demonstrating the hormonal action of environmental chemicals and thereby their influence on carcinogenic risk are conducted at relatively high doses compared to those experienced by humans.

For some chemicals, maximum "tolerated" dosing has the potential to have cytotoxic effects that could either exacerbate mammary carcinogenesis through enhanced cell replication as part of an adaptive response to tissue loss, or if sufficiently high, could potentially inhibit carcinogenesis by reducing the population of rapidly dividing cells through cytotoxicity. Either way, high-dose testing that results in cytotoxicity in mammary tissue can leave results of a bioassay for carcinogenesis difficult to interpret or to extrapolate to humans. Add to this the possibility that the carcinogenicity occurs only during a narrow window of development, and the interpretation of any shaped dose–response becomes even more complicated.

Toxicity related to maximum tolerated dosing may also reduce mammary tumor incidence because of reduced weight gain or weight loss (Haseman et al., 1998). Even decreased palatability of the diet due to high concentrations of a test chemical (in the absence of toxicity) can result in decreased weight gain, with attendant decrease in the rate of "spontaneous" mammary tumors. Over time, the body weights of rats have steadily increased, and spontaneous mammary tumor incidence correspondingly increased (Haseman et al., 1998). In studies with observed dose-dependent body weight decrements, due to caloric restriction protocols or other factors, there are corresponding reductions in mammary tumor incidence. Thus dose-related reductions in body weight may mask chemical-related increases in mammary tumors. Early mortality due to competing causes of death, including other cancers or toxicity, particularly at the highest dose tested, also decreases study power to observe mammary as well as other tumor types (Rudel et al., 2007).

High doses or more direct administration (e.g., injection) may also change the pharmacokinetics and metabolic pathways of chemicals such that there is a greater (or in some cases lower) proportionate amount of proximate carcinogen formed. High doses administered during pregnancy may result in less maternal sequestering of fat-soluble chemicals such as dioxin and higher proportionate delivery to the fetus than would occur at lower doses. Bolus dosing (rather than more continuous dosing in the diet or drinking water) can also affect the distribution and pharmacokinetics of a chemical. For example, a bolus dose administered directly by stomach tube (gavage) each day is not equivalent to the same amount of chemical ingested continuously over a day. Bolus dosing can result in a larger amount of the chemical or metabolites delivered to cellular targets at one time. Toxicity may be greatly increased if metabolic detoxification pathways become saturated, resulting in greater delivery of the chemical to tissues or greater production of more toxic metabolites through alternative pathways. If, on the other hand, metabolism to more toxic metabolites becomes saturated with bolus dosing, a less than proportional increase in toxicity with dose may result. High bolus dosing may be analogous to an industrial accident or some periodic occupational or medical exposures, but it is unlike much lower episodic or continuous environmental exposures, leaving challenges for interpretation for low-dose risk assessment.

Thus high-dose testing can lead to false positives or false negatives. This poses a dilemma for testing because studies performed using low doses comparable to environmental levels and a standard protocol for numbers of animals per dose group would not be sensitive. A study with 50 animals per dose group can at best detect a statistically significant difference in mammary tumor rate between treated and control animals of 10 to 15 percentage points; any smaller difference would not be detectable because of lack of statistical significance. Thus a hypothetical agent in the background that could be a predominant factor in half the mammary cancers and pose, for example, a difference in incidence of say 6 percentage points, would go undetected in such an animal study if tested at environmental levels. Table 4-3 illustrates this point. It shows the outcome of a hypothetical

TABLE 4-3 Theoretical Experimental Outcome from an Exposure That Induces an Increase in Absolute Risk of 6 Percentage Points

Species and Strain	Number (%) of Animals Expected to Develop Tumors per 50-Animal Dose Group[a]		Statistical Significance of Outcome[b]
	Control	Exposed	
B6C3F1 mouse	1/50 (1.8%)	4/50 (7.8%)	.18
Fischer 344 rat	3/50 (5.2%)	6/50 (11.2%)	.24
Sprague Dawley rat	5/50 (10.2%)	8/50 (16.8%)	.27

[a]Theoretical underlying probability of cancer.
[b]p value, Fisher's exact test, showing the probability that the outcome of the experiment is due to chance alone. A p value of .05 or less typically indicates statistical significance.

experiment in which the exposed group of animals receives an environmentally relevant dose of an agent that results in an increase in the incidence of tumors (6 percentage points) equivalent to half the average lifetime risk of breast cancer in the United States. The control groups reflect background rates of mammary carcinomas observed in NTP studies (see Table 4-2). The tumor rates in the exposed groups reflect the expected response in the controls, elevated by 6 percentage points—the effect expected from the chemical exposure. For no case would the results in the exposed group be found to be statistically significant, given a sample size of 50 animals and the background tumor rates in the controls.

Species Similarities and Differences in Mammary Gland Biology and Carcinogenesis

Rodents used in carcinogenicity experiments are generally similar to humans in mammary gland development, although some aspects of the timing may differ. For example, the epithelial bud and ductal outgrowth occurs late in gestation for rodents, but the evidence suggests ductal development occurs early in gestation for humans (Fenton, 2006; Table 4-4).

As with humans, female mice and rats have greater occurrence of spontaneous mammary tumors than males, reflecting endocrine-related influences (see Table 4-2). Estrogen plays an important role in breast cancer in rodents and humans. Certain other hormones—such as prolactin, progesterone alone, and androgens—appear to be involved in inducing mammary tumors in rodents, but have a less clear role in the induction of breast cancer in humans (Thayer and Foster, 2007). However, the recent evidence for the role of prolactin and progesterone in human cancer has become stronger (Fernandez et al., 2010; Jacobson et al., 2011).

TABLE 4-4 Timing of Events in Mammary Development in Humans and Rodents

Developmental Event	Human	Rodent
Milk streak evidence	EW 4–6	GD 10–11 (mice)
Mammary epithelial bud forms	EW 10–13	GD 12–14 (mice) GD 14–16 (rats)
Female nipple and areola form	EW 12–16	GD 18 (mice) GD 20 (rats)
Branching and canalization of epithelium	EW 20–32	GD 16 to birth (mice) GD 18 to birth (rats)
Secretion is possible	EW 32–40 (ability lost postnatally)	At birth with hormonal stimuli
Isometric development of ducts	Birth to puberty	Birth to puberty
Terminal end buds present (peri-pubertal)	8- to 13-year-old girls	23- to 60-day-old rodents
Formation of lobular units	EW 32–40, or within 1–2 years of first menstrual cycle	Puberty and into adulthood

NOTES: EW, embryonic week; GD, gestational day.
SOURCE: Fenton (2006, p. S19). Used with permission: Fenton, S. E. 2006. Endocrine-disrupting compounds and mammary gland development: Early exposure and later life consequences. *Endocrinology* 147(6 Suppl):S18–S24. *Copyright 2006, The Endocrine Society.*

As also noted above, tumor types in rats and humans are similar (Russo and Russo, 1996; Fenton, 2006), although the background rates of tumors may differ (see Table 4-2). Premalignant changes can also be observed in both rats and humans (Thayer and Foster, 2007). Rats, however, may be poor animal models for studying metastasis in humans because their mammary gland carcinomas rarely metastasize (Thayer and Foster, 2007). By contrast, mouse mammary tumors often metastasize to the lungs (Cardiff and Kenney, 2011). Although spontaneous and virally induced mouse mammary tumors differ in histology and morphology from breast tumors in humans, tumors in genetically engineered mice resemble those in humans (Cardiff and Kenney, 2011).

Certain modes of action hypothesized for the rat have been questioned regarding their relevance to humans, such as mammary tumors that result from atrazine exposure, presumably through lengthening of the estrous cycle and elevation of endogenous levels of prolactin and estradiol in Sprague Dawley but not F344 rats (Kacew et al., 1995). Induction of mammary tumors in mice by viral origin does not appear to be relevant for humans, but it remains a useful model in studies of expression of oncogenes (Medina, 2010).

Similarities and Differences Between Rodents and Humans in Metabolism of Carcinogens

One of the most common reasons why one species may differ from another in terms of tumor development following exposure to a chemical carcinogen is because of differences in the way the carcinogen is biotransformed (metabolized). More than 95 percent of known chemical carcinogens require biotransformation to reactive intermediates to exert their carcinogenic effects, typically through an enzyme-mediated oxidation reaction. These reactive intermediates are often quickly eliminated through other biotransformation processes, such as conjugation or hydrolysis. Because biotransformation of xenobiotics is in part a function of "adaptive response" to one's environment, evolutionary influences have resulted in relatively large genetic divergence in xenobiotic biotransformation pathways, giving rise to potentially important species differences in susceptibility to carcinogens. For example, mice express a particular form of glutathione S-transferase (GST) with remarkably high catalytic efficiency toward detoxification of the potent liver carcinogen, aflatoxin B1. Neither rats nor humans express this particular form of GST in the liver, and consequently both species are highly sensitive to the hepatocarcinogenic effects of aflatoxins (Eaton and Gallagher, 1994). Thus, understanding species- and tissue-specific pathways involved in carcinogen activation and detoxification is an important element of "predictive toxicology" that relies on animal models.

Biotransformation of xenobiotics can occur in virtually any tissue, although the majority of "clearance" of a chemical from the body through biotransformation reactions usually occurs in the liver. However, for some carcinogens that act at sites distant from the point of exposure, biotransformation in the target tissue may be critically important. Even if it does not contribute substantially to the overall elimination of the substance from the body, tissue-specific biotransformation could be significant for activating a compound to a carcinogenic form in a particular tissue. Thus, for potential mammary carcinogens, it is important to understand if there are tissue-specific differences in biotransformation of xenobiotics in breast tissue in humans versus experimental animals. In addition, it is important to understand potential differences in the formation and systemic transport of reactive intermediates from the liver (Ioannides, 2002).

For most lipophilic genotoxic carcinogens, activation to reactive intermediates occurs via the cytochrome P-450 (CYP) superfamily of enzymes (Nebert and Dalton, 2006). CYP expression in mammary tissue is potentially important in both activation and detoxification of polycyclic aromatic hydrocarbons (PAHs), aromatic amines, and other genotoxic "procarcinogens." Furthermore, CYPs may also play a role in the elimination of estrogenic compounds that partition into mammary lipids, thus potentially playing a protective role for estrogen-active compounds at the site of action. They also play a role in steroid hormone biosynthesis, and induction or inhibition of some CYPs can affect the levels of estrogens and other hormones.

There are 57 different genes, in 18 gene families, in the human genome that code for CYP enzymes (Nelson et al., 2004). A subset of these genes and their related enzymes are of particular relevance to breast cancer because they are involved in xenobiotic biotransformation to active or inactive agents or can be modulated by other xenobiotics to affect steroid hormone levels or steroid genotoxicity. These include the *CYP1* family (which has three members: *CYP1A1*, *CYP1A2*, and *CYP1B1*), as well as *CYP2E1*, *CYP3A4*, and *CYP19*. For example, the CYP1A1 enzyme is involved in the activation of PAHs to reactive diol-epoxides, and thus is very important in the carcinogenicity of PAHs. PAHs cause mammary tumors in rats (Cavalieri et al., 1988), and they are biotransformed by human mammary epithelial cells to metabolites that form PAH–DNA adducts (Calaf and Russo, 1993).

Whether activation of PAHs to mutagenic metabolites via CYP1A1 occurs directly in breast tissue is not completely clear, although it is likely because CYP1A1 mRNA (Huang et al., 1996) and CYP1A1 are identified in human breast tissue (Hellmold et al., 1998). PAH–DNA adducts have also been identified in normal breast tissue, and were reported to be higher in women with breast cancer than in healthy controls (Li et al., 1999). Thus, it appears that, at least in some individuals, human breast tissue has the capacity to activate PAHs and potentially other procarcinogens to DNA-reactive molecules. CYP1A1 is inducible in lung and oral mucosal tissue by exposure to cigarette smoke, although no published reports demonstrating that smoking induces expression in human breast tissue were found. It is also inducible in mouse and human mammary tissue by a number of other xenobiotics, including 2,3,7,8-tetrachlorodibenzo-p-dioxin (TCDD). The degree to which species differences in tissue-specific expression of xenobiotic biotransformation enzymes might contribute to species differences in response to breast carcinogens evaluated in chronic rodent bioassays is an issue for consideration in selection of animals for carcinogenicity testing.

Nonstandard Whole Animal Carcinogenicity and Related Studies

Mouse mammary tumors and the associated MMTV were among the first tumor types investigated in studies of animal carcinogenicity, including research on neoplasia, oncogenic viruses, host responses, role of endocrinology and stem cells, and progression. More recently, genetically engineered mice have been used to investigate the role of agents such as the MMTV in promoting *Myc* oncogene expression in the mammary gland. Genetically engineered mice have allowed considerable research into the molecular biology of mammary tumor formation and progression.

Since the development of the first model of breast cancer in a genetically modified mouse, more than 100 models have been developed to study breast cancer (Cardiff and Kenney, 2011). At least three basic types have been developed based on transgenes, combinations of transgenes, and targeted mutations. Gene targets have included growth factors and their receptors, cell signaling pathways, cell cycle regulators, and differentiation mediators (Bucher, 2010). The advantage of these models is that they have a defined genetic background, enabling the study of particular pathways, without variations attributable to differences in genotype. The mice develop disease after a predictable time period, and the stage-specific alterations directly translate to humans (Bucher, 2010). Finally, they correspond well to humans in that mammary tumors in mice are caused by the genes that are overexpressed or mutated in human breast cancer (Bucher, 2010).

In contrast to most breast cancers in humans, most mouse tumors (including those in many genetically modified mice)—whether of spontaneous, viral, or chemical origin—do not express hormone receptors (Lanari et al., 2009). To aid in the investigation of human breast cancers, however, specific mouse models have been developed in which mammary tumors do express hormone receptors (e.g., estrogen and progesterone receptors) and demonstrate other molecular features found in human tumors (e.g., Lanari et al., 2009; Herschkowitz et al., 2011; Nguyen et al., 2011). For one of these models, medroxyprogesterone acetate (MPA) is administered to BALB/c female mice, resulting in hormone-dependent, metastatic mammary ductal carcinomas that express both estrogen and progesterone receptors, as in humans (Lanari et al., 2009). Another model relies on transplantation of mammary cells without p53 tumor suppressor function into BALB/c mice (Herschkowitz et al., 2011). Models such as these allow investigation into factors affecting hormone dependence of carcinomas, as well as those that promote tumor progression or regression. Further research is needed on the use of genetically modified mouse models for assessing the effect of environmental exposures in inducing tumors.

Some studies are designed to assess the impact of a test compound on mammary development by periodic evaluation of whole mounts of mammary tissue from animals during the postnatal period (Fenton et al., 2002; Birnbaum and Fenton, 2003; Rudel et al., 2011). The test compound is typically given in utero or neonatally, and mammary structures are evaluated against those of control animals to determine treatment-related differ-

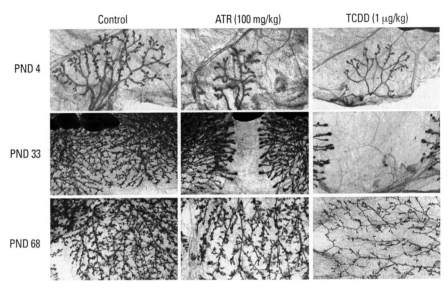

FIGURE 4-6 Development of the mammary gland in rats following in utero exposure to atrazine (ATR) and 2,3,7,8-tetrachlorodibenzo-p-dioxin (TCDD).
SOURCE: Birnbaum and Fenton (2003). Reproduced with permission from *Environmental Health Perspectives.*

ences in development. In particular, branching density, bud formation off the major ducts, and terminal end bud formation are evaluated. Figure 4-6 shows mammary gland whole mounts from Long Evans rats exposed during late gestation to atrazine or TCDD (Birnbaum and Fenton, 2003). The tissue preparations show the differences in key developmental parameters on different postnatal days (PNDs). Mammary development was observed to be disrupted as early as PND 4. For example, in treated animals, there was a lack of branching, fewer primary ducts from the nipple, and fewer terminal structures. These are examples of how such a test system could be used to evaluate the potential of a chemical to modulate mammary development. The degree to which changes in mammary structure that result from early exposures signal increased sensitivity to mammary tumor development is an area for study to increase the usefulness of these assays for detecting potential breast carcinogens (e.g., Rudel et al., 2011).

In Vitro Studies

In vitro tests, discussed briefly in Chapter 2, have been used as initial, putative predictors of carcinogenicity potential. Genotoxicity has long been treated as an indicator of possible carcinogenicity. Testing requirements for pesticides and pharmaceuticals include in vivo studies in rodents to test for chromosomal aberrations and micronuclei, and in vitro tests for mutations in mammalian cells and bacteria. Tests based on mechanisms other than genotoxicity are also used. The goal is to target an agent's ability to modulate pathways that underlie the basic mechanisms of toxicity. While regulatory agencies typically do not label chemicals that have tested positive in genotoxicity tests as possible carcinogens in the absence of supporting human or animal data, product development programs for pesticides and pharmaceuticals often avoid chemicals with strong signals of genotoxic potential, mostly out of concerns for potential carcinogenicity and mutagenicity.

Because the current standard whole-animal testing approach for carcinogenicity (as well as other endpoints) is time- and resource-consuming, initiatives to move toward reliance on in vitro and structure–activity relationships have been advocated and are under way (NRC, 2007; Krewski et al., 2009; EPA, 2011). The National Research Council (NRC) has envisioned a new toxicity testing system, focusing on upstream events: chemical perturbations of cellular response networks (i.e., complex biochemical interactions that maintain normal cellular function) (NRC, 2007). For example, testing might identify perturbation of estrogen signaling and the subsequent events that potentially result in cancer. The NRC vision was made possible because of the emerging scientific understanding of cellular response networks, and high-throughput technology that enables the exploration of the structure of these networks and rapid conduct of in vitro tests. NRC (2007) proposed the development of suites of predictive, high- and medium-throughput assays, emphasizing those based on cells of human origin, to evaluate perturbations. These would be complemented by assays of more integrated cellular responses and in vivo assays to cover uncertainties in the testing regimen, to test prototypic compounds, and to address metabolism. Other components of the framework include the use of physiologically based pharmacokinetic studies, human biomonitoring data, and epidemiologic data to evaluate and fine-tune the predictive ability of the tests. The NRC vision was accompanied by a long-term strategy for its realization, involving a substantial multidisciplinary research program.

Subsequently, in 2008, various federal institutions entered into a Memorandum of Understanding to "research, develop, validate and translate innovative chemical testing methods that characterize toxicity pathways" (EPA, 2011). The current "Tox21 collaboration," renewed for 5 years in 2010, includes EPA, the NTP, the National Human Genome Research Institute and the Chemical Genomics Center of the National Institutes of Health, and the FDA. The main work will be to explore high-throughput screening assays and tests using phylogenetically lower animal species (e.g., fish, worms), and high-throughput whole-genome analytical methods to evaluate mechanisms of toxicity. The ultimate end is to generate data with the new tools for use in the protection of human health and the environment.

An important consideration in the development of tests that have adequate coverage will be the degree to which they cover the pathways involved in the general mechanisms underlying breast cancer—mutagenesis, estrogen receptor signaling, epigenetic programming, growth promotion via mitogenic cell signaling, microenvironmental change, and modulation of immune functioning. This will require attention in selection of cell types and environments relevant to breast cancer.

SUMMARY

Better understanding of the contribution of environmental factors to breast cancer entails understanding the multiple challenges in carrying out and interpreting studies in humans, animals, and in vitro systems. For studies in humans, these include the issues inherent in estimating and assessing exposures, the study design and analytic challenges of environmental epidemiology, and efforts to account for genetic differences in susceptibility to cancer and potential gene–environment interactions. Studies in animals and in vitro systems bring with them their own technical obstacles and challenges of interpretation and extrapolation to humans. An understanding of these challenges informs understanding of the existing data and their implications for next steps for action and research.

REFERENCES

Allred, D. C. 2010. Ductal carcinoma in situ: Terminology, classification, and natural history. *J Natl Cancer Inst Monogr* 41:134–138.

Axelson, O., and L. Sundell. 1978. Mining, lung cancer and smoking. *Scand J Work Environ Health* 4(1):46–52.

Bellocq, J. P., and G. Magro. 2003. Fibroepithelial tumours. In *Pathlogy and genetics of tumours of the breast and female genital organs*. Edited by F. A. Tavassoli and P. Devilee. Lyon, France: IARC Press.

Benassi-Evans, B., and M. Fenech. 2011. Chronic alcohol exposure induces genome damage measured using the cytokinesis-block micronucleus cytome assay and aneuploidy in human B lymphoblastoid cell lines. *Mutagenesis* 26(3):421–429.

Benichou, J. 2001. A review of adjusted estimators of attributable risk. *Stat Methods Med Res* 10(3):195–216.

Bennett, L. M., and B. J. Davis. 2002. Identification of mammary carcinogens in rodent bioassays. *Environ Mol Mutagen* 39(2–3):150–157.

Bernstein, J. L., B. Langholz, R. W. Haile, L. Bernstein, D. C. Thomas, M. Stovall, K. E. Malone, C. F. Lynch, et al. 2004. Study design: Evaluating gene–environment interactions in the etiology of breast cancer—the WECARE study. *Breast Cancer Res* 6(3):R199–R214.

Bernstein, J. L., R. W. Haile, M. Stovall, J. D. Boice, Jr., R. E. Shore, B. Langholz, D. C. Thomas, L. Bernstein, et al. 2010. Radiation exposure, the *ATM* Gene, and contralateral breast cancer in the Women's Environmental Cancer and Radiation Epidemiology Study. *J Natl Cancer Inst* 102(7):475–483.

Betancourt, A. M., I. A. Eltoum, R. A. Desmond, J. Russo, and C. A. Lamartiniere. 2010. In utero exposure to bisphenol A shifts the window of susceptibility for mammary carcinogenesis in the rat. *Environ Health Perspect* 118(11):1614–1619.

Birnbaum, L. S., and S. E. Fenton. 2003. Cancer and developmental exposure to endocrine disruptors. *Environ Health Perspect* 111(4):389–394.

Blakely, C. M., A. J. Stoddard, G. K. Belka, K. D. Dugan, K. L. Notarfrancesco, S. E. Moody, C. M. D'Cruz, and L. A. Chodosh. 2006. Hormone-induced protection against mammary tumorigenesis is conserved in multiple rat strains and identifies a core gene expression signature induced by pregnancy. *Cancer Res* 66(12):6421–6431.

Bombonati, A., and D. C. Sgroi. 2011. The molecular pathology of breast cancer progression. *J Pathol* 223(2):307–317.

Boorman, G., and J. I. Everitt. 2006. Neoplastic disease. In *The laboratory rat*, 2nd ed. Edited by S. M. Suckow, S. H. Weisbroth, and C. L. Franklin. American College of Laboratory Animal Medicine Series. Boston, MA: Elsevier Academic Press.

Brown, J. F. 1994. Determination of PCB metabolic, excretion, and accumulation rates for use as indicators of biological response and relative risk. *Environ Sci Technol* 28(13):2295–2305.

Bucher, J. 2010. *Weighing evidence from National Toxicology Program cancer bioassays.* Presentation at the Institute of Medicine Breast Cancer and the Environment meeting, July 6–7, San Francisco, CA.

Calaf, G., and J. Russo. 1993. Transformation of human breast epithelial cells by chemical carcinogens. *Carcinogenesis* 14(3):483–492.

Cardiff, R. D., and N. Kenney. 2011. A compendium of the mouse mammary tumor biologist: From the initial observations in the house mouse to the development of genetically engineered mice. *Cold Spring Harb Perspect Biol* 3(6).

Cavalieri, E., E. Rogan, and D. Sinha. 1988. Carcinogenicity of aromatic hydrocarbons directly applied to rat mammary gland. *J Cancer Res Clin Oncol* 114(1):3–9.

CDC (Centers for Disease Control and Prevention). 2009. *Fourth national report on human exposure to environmental chemicals, 2009.* Atlanta, GA: CDC. http://www.cdc.gov/exposurereport/index.html (accessed October 16, 2011).

Chapin, R. E., J. T. Stevens, C. L. Hughes, W. R. Kelce, R. A. Hess, and G. P. Daston. 1996. Endocrine modulation of reproduction. *Fundam Appl Toxicol* 29(1):1–17.

Concannon, P., R. W. Haile, A. L. Borresen-Dale, B. S. Rosenstein, R. A. Gatti, S. N. Teraoka, T. A. Diep, L. Jansen, et al. 2008. Variants in the *ATM* gene associated with a reduced risk of contralateral breast cancer. *Cancer Res* 68(16):6486–6491.

Cornfield, J., W. Haenszel, E. C. Hammond, A. M. Lilienfeld, M. B. Shimkin, and E. L. Wynder. 1959. Smoking and lung cancer: Recent evidence and a discussion of some questions. *J Natl Cancer Inst* 22(1):173–203.

de Assis, S., G. Khan, and L. Hilakivi-Clarke. 2006. High birth weight increases mammary tumorigenesis in rats. *Int J Cancer* 119(7):1537–1546.

Dinse, G. E., S. D. Peddada, S. F. Harris, and S. A. Elmore. 2010. Comparison of NTP historical control tumor incidence rates in female Harlan Sprague Dawley and Fischer 344/N Rats. *Toxicol Pathol* 38(5):765–775.

Dixon, J. M. 1991. Cystic disease and fibroadenoma of the breast: Natural history and relation to breast cancer risk. *Br Med Bull* 47(2):258–271.

Donegan, W. L. 2002. Common benign conditions of the breast. In *Cancer of the breast*, 5th ed. Edited by W. L. Donegan and J. S. Spratt. Philadelphia, PA: Saunders.

Duffy, S. W., E. Lynge, H. Jonsson, S. Ayyaz, and A. H. Olsen. 2008. Complexities in the estimation of overdiagnosis in breast cancer screening. *Br J Cancer* 99(7):1176–1178.

Easton, D. F., K. A. Pooley, A. M. Dunning, P. D. Pharoah, D. Thompson, D. G. Ballinger, J. P. Struewing, J. Morrison, et al. 2007. Genome-wide association study identifies novel breast cancer susceptibility loci. *Nature* 447(7148):1087–1093.

Eaton, D. L., and E. P. Gallagher. 1994. Mechanisms of aflatoxin carcinogenesis. *Annu Rev Pharmacol Toxicol* 34:135–172.

Eaton, D. L., and C. D. Klaason. 1996. Principles of toxicology. In *Casarett & Doull's toxicology: The basic science of poisons*, 5th ed. Edited by M. J. Wonsiewicz and L. A. Sheinis. New York: McGraw Hill.

Eide, G. E. 2008. Attributable fractions for partitioning risk and evaluating disease prevention: A practical guide. *Clin Respir J* 2(Suppl 1):92–103.

Eide, G. E., and I. Heuch. 2001. Attributable fractions: Fundamental concepts and their visualization. *Stat Methods Med Res* 10(3):159–193.

El-Wakeel, H., and H. C. Umpleby. 2003. Systematic review of fibroadenoma as a risk factor for breast cancer. *Breast* 12(5):302–307.

EPA (Environmental Protection Agency). 1998a. *Health effects test guidelines. OPPTS 870.4200: Carcinogenicity*. EPA 712–C–98–211. Washington, DC: Government Printing Office. http://hero.epa.gov/index.cfm?action=search.view&reference_ID=6378 (accessed November 16, 2011).

EPA. 1998b. *High production volume (HPV) chemicals and SIDS testing*. http://www.epa.gov/HPV/pubs/general/hazchem.htm (accessed June 15, 2011).

EPA. 2006. *An inventory of sources and environmental releases of dioxin-like compounds in the U.S. for the years 1987, 1995, and 2000 (Final, Nov 2006)*. http://cfpub.epa.gov/ncea/cfm/recordisplay.cfm?deid=159286 (accessed November 2, 2011).

EPA. 2010. *Polychlorinated biphenyls (PCBs)*. http://www.epa.gov/epawaste/hazard/tsd/pcbs/pubs/about.htm (accessed November 17, 2011).

EPA. 2011. *Tox21*. http://www.epa.gov/ncct/Tox21/ (accessed November 17, 2011).

Esserman, L., Y. Shieh, and I. Thompson. 2009. Rethinking screening for breast cancer and prostate cancer. *JAMA* 302(15):1685–1692.

Etique, N., D. Chardard, A. Chesnel, J. L. Merlin, S. Flament, and I. Grillier-Vuissoz. 2004. Ethanol stimulates proliferation, ERα and aromatase expression in MCF-7 human breast cancer cells. *Int J Mol Med* 13(1):149–155.

FDA (Food and Drug Administration). 1997. *Guidance for industry: S1B testing for carcinogenicity of pharmaceuticals*. Rockville, MD: FDA. http://www.fda.gov/downloads/Drugs/GuidanceComplianceRegulatoryInformation/Guidances/ucm074916.pdf (accessed November 17, 2011).

Fenton, S. E. 2006. Endocrine-disrupting compounds and mammary gland development: Early exposure and later life consequences. *Endocrinology* 147(6 Suppl):S18–S24.

Fenton, S. E., J. T. Hamm, L. S. Birnbaum, and G. L. Youngblood. 2002. Persistent abnormalities in the rat mammary gland following gestational and lactational exposure to 2,3,7,8-tetrachlorodibenzo-p-dioxin (TCDD). *Toxicol Sci* 67(1):63–74.

Fernandez, I., P. Touraine, and V. Goffin. 2010. Prolactin and human tumourogenesis. *J Neuroendocrinol* 22(7):771–777.

Fisher, R. A. 1958a. Cancer and smoking. *Nature* 182(4635):596.

Fisher, R. A. 1958b. Lung cancer and cigarettes. *Nature* 182(4628):108.

Foster, P. 2011. NTP's modified one-generation reproduction study Board of Scientific Counselor's meeting. http://ntp.niehs.nih.gov/ntp/About_NTP/BSC/2011/April/MOGDesign.pdf (accessed November 13, 2011).

Fritz, W. A., L. Coward, J. Wang, and C. A. Lamartiniere. 1998. Dietary genistein: Perinatal mammary cancer prevention, bioavailability and toxicity testing in the rat. *Carcinogenesis* 19(12):2151–2158.

Gail, M. H., L. A. Brinton, D. P. Byar, D. K. Corle, S. B. Green, C. Schairer, and J. J. Mulvihill. 1989. Projecting individualized probabilities of developing breast cancer for white females who are being examined annually. *J Natl Cancer Inst* 81(24):1879–1886.

Gillette, R. W. 1976. Mouse mammary tumor virus as a model for viral carcinogenesis. *J Toxicol Environ Health* 1(4):545–550.

Goehring, C., and A. Morabia. 1997. Epidemiology of benign breast disease, with special attention to histologic types. *Epidemiol Rev* 19(2):310–327.

Gohlke, J. M., R. Thomas, Y. Zhang, M. C. Rosenstein, A. P. Davis, C. Murphy, K. G. Becker, C. J. Mattingly, and C. J. Portier. 2009. Genetic and environmental pathways to complex diseases. *BMC Syst Biol* 3:46.

Goodman, M., M. Kelsh, K. Ebi, J. Iannuzzi, and B. Langholz. 2002. Evaluation of potential confounders in planning a study of occupational magnetic field exposure and female breast cancer. *Epidemiology* 13(1):50–58.

Gordis, L. 2000. *Epidemiology*, 2nd ed. Philadelphia, PA: W. B. Saunders Company.

Gøtzsche, P. C., and M. Nielsen. 2006. Screening for breast cancer with mammography. *Cochrane Database Syst Rev* (4):CD001877.

Gøtzsche, P. C., K. J. Jørgensen, J. Maehlen, and P. H. Zahl. 2009. Estimation of lead time and overdiagnosis in breast cancer screening. *Br J Cancer* 100(1):219.

Greenland, S., J. Pearl, and J. M. Robins. 1999. Causal diagrams for epidemiologic research. *Epidemiology* 10(1):37–48.

Guray, M., and A. A. Sahin. 2006. Benign breast diseases: Classification, diagnosis, and management. *Oncologist* 11(5):435–449.

Haseman, J. K., J. R. Hailey, and R. W. Morris. 1998. Spontaneous neoplasm incidences in Fischer 344 rats and B6C3F1 mice in two-year carcinogenicity studies: A National Toxicology Program update. *Toxicol Pathol* 26(3):428–441.

Heinonen, O. P., S. Shapiro, L. Tuominen, and M. I. Turunen. 1974. Reserpine use in relation to breast cancer. *Lancet* 2(7882):675–677.

Hellmold, H., T. Rylander, M. Magnusson, E. Reihner, M. Warner, and J. A. Gustafsson. 1998. Characterization of cytochrome P450 enzymes in human breast tissue from reduction mammaplasties. *J Clin Endocrinol Metab* 83(3):886–895.

Henderson, T. O., A. Amsterdam, S. Bhatia, M. M. Hudson, A. T. Meadows, J. P. Neglia, L. R. Diller, L. S. Constine, et al. 2010. Systematic review: Surveillance for breast cancer in women treated with chest radiation for childhood, adolescent, or young adult cancer. *Ann Intern Med* 152(7):444–455; W144–W154.

Hernán, M. A., S. Hernandez-Diaz, M. M. Werler, and A. A. Mitchell. 2002. Causal knowledge as a prerequisite for confounding evaluation: An application to birth defects epidemiology. *Am J Epidemiol* 155(2):176-184.

Herschkowitz, J. I., W. Zhao, M. Zhang, J. Usary, G. Murrow, D. Edwards, J. Knezevic, S. B. Greene, et al. 2011. Comparative oncogenomics identifies breast tumors enriched in functional tumor-initiating cells. *Proc Natl Acad Sci U S A*. June 1. [Epub ahead of print].

Hilakivi-Clarke, L., E. Cho, I. Onojafe, M. Raygada, and R. Clarke. 1999a. Maternal exposure to genistein during pregnancy increases carcinogen-induced mammary tumorigenesis in female rat offspring. *Oncol Rep* 6(5):1089–1095.

Hilakivi-Clarke, L., I. Onojafe, M. Raygada, E. Cho, T. Skaar, I. Russo, and R. Clarke. 1999b. Prepubertal exposure to zearalenone or genistein reduces mammary tumorigenesis. *Br J Cancer* 80(11):1682–1688.

Hilakivi-Clarke, L., A. Cabanes, S. de Assis, M. Wang, G. Khan, W. J. Shoemaker, and R. G. Stevens. 2004. In utero alcohol exposure increases mammary tumorigenesis in rats. *Br J Cancer* 90(11):2225–2231.

Horwitz, R. I., and A. R. Feinstein. 1985. Exclusion bias and the false relationship of reserpine and breast cancer. *Arch Intern Med* 145(10):1873–1875.

Howell, A., A. H. Sims, K. R. Ong, M. N. Harvie, D. G. Evans, and R. B. Clarke. 2005. Mechanisms of disease: Prediction and prevention of breast cancer—cellular and molecular interactions. *Nat Clin Pract Oncol* 2(12):635–646.

Huang, Z., M. J. Fasco, H. L. Figge, K. Keyomarsi, and L. S. Kaminsky. 1996. Expression of cytochromes P450 in human breast tissue and tumors. *Drug Metab Dispos* 24(8):899–905.

Huff, J. E., S. L. Eustis, and J. K. Haseman. 1989. Occurrence and relevance of chemically induced benign neoplasms in long-term carcinogenicity studies. *Cancer Metastasis Rev* 8(1):1–22.

Huff, J., J. Cirvello, J. Haseman, and J. Bucher. 1991. Chemicals associated with site-specific neoplasia in 1394 long-term carcinogenesis experiments in laboratory rodents. *Environ Health Perspect* 93:247–270.

Hunter, D. J., P. Kraft, K. B. Jacobs, D. G. Cox, M. Yeager, S. E. Hankinson, S. Wacholder, Z. Wang, et al. 2007. A genome-wide association study identifies alleles in *FGFR2* associated with risk of sporadic postmenopausal breast cancer. *Nat Genet* 39(7):870–874.

Hunter, D. J., D. Altshuler, and D. J. Rader. 2008. From Darwin's finches to canaries in the coal mine—mining the genome for new biology. *N Engl J Med* 358(26):2760–2763.

IARC (International Agency for Research on Cancer). 1987. Diethylstilbestrol. In *Overall evaluations of carcinogenicity: An updating of IARC Monographs volumes 1 to 42.* Lyon, France: IARC.

Ioannides, C. 2002. *Enzyme systems that metabolise drugs and other xenobiotics.* New York: John Wiley and Sons.

Jacobson, E. M., E. R. Hugo, D. C. Borcherding, and N. Ben-Jonathan. 2011. Prolactin in breast and prostate cancer: Molecular and genetic perspectives. *Discov Med* 11(59):315–324.

Jenkins, S., N. Raghuraman, I. Eltoum, M. Carpenter, J. Russo, and C. A. Lamartiniere. 2009. Oral exposure to bisphenol A increases dimethylbenzanthracene-induced mammary cancer in rats. *Environ Health Perspect* 117(6):910–915.

Jo, W. K., C. P. Weisel, and P. J. Lioy. 1990. Chloroform exposure and the health risk associated with multiple uses of chlorinated tap water. *Risk Anal* 10(4):581–585.

Kabat, G. C., J. G. Jones, N. Olson, A. Negassa, C. Duggan, M. Ginsberg, R. A. Kandel, A. G. Glass, and T. E. Rohan. 2010. A multi-center prospective cohort study of benign breast disease and risk of subsequent breast cancer. *Cancer Causes Control* 21(6):821–828.

Kacew, S., and M. F. Festing. 1996. Role of rat strain in the differential sensitivity to pharmaceutical agents and naturally occurring substances. *J Toxicol Environ Health* 47(1):1–30.

Kacew, S., Z. Ruben, and R. F. McConnell. 1995. Strain as a determinant factor in the differential responsiveness of rats to chemicals. *Toxicol Pathol* 23(6):701–714; discussion 714–715.

Kanhai, R. C., J. J. Hage, E. Bloemena, P. J. van Diest, and R. B. Karim. 1999. Mammary fibroadenoma in a male-to-female transsexual. *Histopathology* 35(2):183–185.

Khan, G., P. Penttinen, A. Cabanes, A. Foxworth, A. Chezek, K. Mastropole, B. Yu, A. Smeds, et al. 2007. Maternal flaxseed diet during pregnancy or lactation increases female rat offspring's susceptibility to carcinogen-induced mammary tumorigenesis. *Reprod Toxicol* 23(3):397–406.

Klepeis, N. E., A. M. Tsang, and J. V. Behar. 1995. *Analysis of the national human activity pattern survey (NHAPS) respondents from a standpoint of exposure assessment.* National Exposure Research Laboratory Office of Research and Development. Las Vegas, NV: EPA. http://www.exposurescience.org/pub/reports/NHAPS_Report1.pdf (accessed November 3, 2011).

Kochukov, M. Y., Y. J. Jeng, and C. S. Watson. 2009. Alkylphenol xenoestrogens with varying carbon chain lengths differentially and potently activate signaling and functional responses in GH3/B6/F10 somatomammotropes. *Environ Health Perspect* 117(5):723–730.

Komenaka, I. K., K. D. Miller, and G. W. Sledge. 2006. Male breast cancer. In *Textbook of uncommon cancer*, 3rd ed. Edited by D. Raghavan, M. L. Brecher, D. H. Johnson, N. J. Meropol, P. L. Moots, P. G. Rose, and I. A. Mayer. Hoboken, NJ: Wiley.

Kramer, B. S., and J. M. Croswell. 2009. Cancer screening: The clash of science and intuition. *Annu Rev Med* 60:125–137.

Kramer, B. S., and J. Croswell. 2010. Limitations and pitfalls in the early detection of cancer: History, current concerns, and the future. *AUA Update Series* 29(Lesson 7):65–76.

Krewski, D., M. E. Andersen, E. Mantus, and L. Zeise. 2009. Toxicity testing in the 21st century: Implications for human health risk assessment. *Risk Anal* 29(4):474–479.

Kuiper, A. 2005. Pathogenesis and progression of fibroepithelial breast tumors. Ph.D. diss., University of Utrecht, Netherlands.

La Merrill, M., R. Harper, L. S. Birnbaum, R. D. Cardiff, and D. W. Threadgill. 2010. Maternal dioxin exposure combined with a diet high in fat increases mammary cancer incidence in mice. *Environ Health Perspect* 118(5):596–601.

Laden, F., G. Collman, K. Iwamoto, A. J. Alberg, G. S. Berkowitz, J. L. Freudenheim, S. E. Hankinson, K. J. Helzlsouer, et al. 2001. 1,1-dichloro-2,2-bis(p-chlorophenyl)ethylene and polychlorinated biphenyls and breast cancer: Combined analysis of five U.S. studies. *J Natl Cancer Inst* 93(10):768–776.

Laden, F., N. Ishibe, S. E. Hankinson, M. S. Wolff, D. M. Gertig, D. J. Hunter, and K. T. Kelsey. 2002. Polychlorinated biphenyls, cytochrome P450 1A1, and breast cancer risk in the Nurses' Health Study. *Cancer Epidemiol Biomarkers Prev* 11(12):1560–1565.

Lanari, C., C. A. Lamb, V. T. Fabris, L. A. Helguero, R. Soldati, M. C. Bottino, S. Giulianelli, J. P. Cerliani, et al. 2009. The MPA mouse breast cancer model: Evidence for a role of progesterone receptors in breast cancer. *Endocr Relat Cancer* 16(2):333–350.

Land, C. E., M. Tokunaga, K. Koyama, M. Soda, D. L. Preston, I. Nishimori, and S. Tokuoka. 2003. Incidence of female breast cancer among atomic bomb survivors, Hiroshima and Nagasaki, 1950–1990. *Radiat Res* 160(6):707–717.

Langholz, B. 2001. Factors that explain the power line configuration wiring code–childhood leukemia association: What would they look like? *Bioelectromagnetics* (Suppl 5):S19–S31.

Langholz, B., D. C. Thomas, M. Stovall, S. A. Smith, J. D. Boice, Jr., R. E. Shore, L. Bernstein, C. F. Lynch, et al. 2009. Statistical methods for analysis of radiation effects with tumor and dose location-specific information with application to the WECARE study of asynchronous contralateral breast cancer. *Biometrics* 65(2):599–608.

Li, D., W. Zhang, A. A. Sahin, and W. N. Hittelman. 1999. DNA adducts in normal tissue adjacent to breast cancer: A review. *Cancer Detect Prev* 23(6):454–462.

Li, Y., R. C. Millikan, D. A. Bell, L. Cui, C. K. Tse, B. Newman, and K. Conway. 2005. Polychlorinated biphenyls, cytochrome P450 1A1 (CYP1A1) polymorphisms, and breast cancer risk among African American women and white women in North Carolina: A population-based case–control study. *Breast Cancer Res* 7(1):R12–R18.

Lilienfeld, A. M., and D. E. Lilienfeld. 1980. Laying the foundations: The epidemiologic approach to disease. In *Foundations of Epidemiology*. New York: Oxford University Press. Pp. 3–22.

Madigan, M. P., R. G. Ziegler, J. Benichou, C. Byrne, and R. N. Hoover. 1995. Proportion of breast cancer cases in the United States explained by well-established risk factors. *J Natl Cancer Inst* 87(22):1681–1685.

Makris, S. L. 2011. Current assessment of the effects of environmental chemicals on the mammary gland in guideline rodent studies by the U.S. Environmental Protection Agency (U.S. EPA), Organisation For Economic Co-Operation And Development (OECD), and National Toxicology Program (NTP). *Environ Health Perspect* 119(8):1047–1052.

Manolio, T. A. 2010. Genomewide association studies and assessment of the risk of disease. *N Engl J Med* 363(2):166–176.

Manolio, T. A., F. S. Collins, N. J. Cox, D. B. Goldstein, L. A. Hindorff, D. J. Hunter, M. I. McCarthy, E. M. Ramos, et al. 2009. Finding the missing heritability of complex diseases. *Nature* 461(7265):747–753.

Markopoulos, C., E. Kouskos, D. Mantas, K. Kontzoglou, K. Antonopoulou, Z. Revenas, and V. Kyriakou. 2004. Fibroadenomas of the breast: Is there any association with breast cancer? *Eur J Gynaecol Oncol* 25(4):495–497.

Masson, L. F., L. Sharp, S. C. Cotton, and J. Little. 2005. Cytochrome P-450 1A1 gene polymorphisms and risk of breast cancer: A HuGE review. *Am J Epidemiol* 161(10):901–915.

Medina, D. 2007. Chemical carcinogenesis of rat and mouse mammary glands. *Breast Dis* 28:63–68.

Medina, D. 2010. Of mice and women: A short history of mouse mammary cancer research with an emphasis on the pardigms inspired by the transplantation method. *Cold Spring Harb Perspect Biol* 2(10):a004523.

Meranze, D. R., M. Gruenstein, and M. B. Shimkin. 1969. Effect of age and sex on the development of neoplasms in Wistar rats receiving a single intragastric instillation of 7,13-dimethylbenz(a)anthracene. *Int J Cancer* 4(4):480–486.

Morrison, A. 1985. *Screening in chronic disease*. New York: Oxford University Press.

Morrison, J. H., R. D. Brinton, P. J. Schmidt, and A. C. Gore. 2006. Estrogen, menopause, and the aging brain: How basic neuroscience can inform hormone therapy in women. *J Neurosci* 26(41):10332–10348.

Moysich, K. B., P. G. Shields, J. L. Freudenheim, E. F. Schisterman, J. E. Vena, P. Kostyniak, H. Greizerstein, J. R. Marshall, et al. 1999. Polychlorinated biphenyls, cytochrome P4501A1 polymorphism, and postmenopausal breast cancer risk. *Cancer Epidemiol Biomarkers Prev* 8(1):41–44.

MRFIT Research Group. 1982. Multiple risk factor intervention trial. Risk factor changes and mortality results. Multiple Risk Factor Intervention Trial Research Group. *JAMA* 248(12):1465–1477.

Narod, S. A., and K. Offit. 2005. Prevention and management of hereditary breast cancer. *J Clin Oncol* 23(8):1656–1663.

NCI (National Cancer Institute). 2010. *SEER cancer statistics review, 1975–2007*. Edited by S. F. Altekruse, C. L. Kosary, M. Krapcho, N. Neyman, R. Aminou, W. Waldron, J. Ruhl, N. Howlader, Z. Tatalovich, A. Cho, A. Mariotto, M. P. Eisner, D. R. Lewis, K. Cronin, H. S. Chen, E. J. Feuer, D. G. Stinchcomb, and B. K. Edwards. Bethesda, MD: NCI. http://seer.cancer.gov/csr/1975_2007/ (accessed January 6, 2011).

Nebert, D. W., and T. P. Dalton. 2006. The role of cytochrome P450 enzymes in endogenous signalling pathways and environmental carcinogenesis. *Nat Rev Cancer* 6(12):947–960.

Nelson, D. R., D. C. Zeldin, S. M. Hoffman, L. J. Maltais, H. M. Wain, and D. W. Nebert. 2004. Comparison of cytochrome P450 (CYP) genes from the mouse and human genomes, including nomenclature recommendations for genes, pseudogenes and alternative-splice variants. *Pharmacogenetics* 14(1):1–18.

Nguyen, D. H., H. A. Oketch-Rabah, I. Illa-Bochaca, F. C. Geyer, J. S. Reis-Filho, J. H. Mao, S. A. Ravani, J. Zavadil, et al. 2011. Radiation acts on the microenvironment to affect breast carcinogenesis by distinct mechanisms that decrease cancer latency and affect tumor type. *Cancer Cell* 19(5):640–651.

Noolkar, A., D. G. Mote, and S. D. Deshpande. 2010. Case report: A case of primary bilateral fibroadenoma of breasts in a male 20 years of age. In *Proceedings of the World Medical Conference, Malta, September 15–17, 2010.* Edited by V. V. Sumbayev, A. Pica, S.-P. Luh, H. K. Lim, K. Natori, and S. Meshitsuka. WSEAS Press. http://www.wseas.us/e-library/conferences/2010/Malta/MEDICAL/MEDICAL-29.pdf (accessed November 11, 2011).

NRC (National Research Council). 2007. *Toxicity testing in the 21st century: A vision and a strategy.* Washington, DC: The National Academies Press.

NTP (National Toxicology Program). 2008. *NTP historical controls report, all routes and vehicles: Harlan Sprague-Dawley rats.* http://ntp.niehs.nih.gov/ntp/Historical_Controls/NTP2000_2009/2008_HC_Report_SD_Rats_ByRoute.pdf (accessed December 14, 2011).

NTP. 2010a. *NTP historical controls report, all routes and vehicles: Rats.* http://ntp.niehs.nih.gov/ntp/Historical_Controls/NTP2000_2010/2010-03-22-HIST-RatsAllRoutes.pdf (accessed December 22, 2011).

NTP. 2010b. *NTP historical controls report, all routes and vehicles: Mice.* http://ntp.niehs.nih.gov/ntp/Historical_Controls/NTP2000_2010/2010-03-22-HIST-MiceAllRoutes.pdf (accessed December 22, 2011).

NTP. 2010c. *Toxicology/carcinogenicity.* http://ntp.niehs.nih.gov/?objectid=72015DAF-BDB7-CEBA-F9A7F9CAA57DD7F5 (accessed November 13, 2011).

Park, J. H., S. Wacholder, M. H. Gail, U. Peters, K. B. Jacobs, S. J. Chanock, and N. Chatterjee. 2010. Estimation of effect size distribution from genome-wide association studies and implications for future discoveries. *Nat Genet* 42(7):570–575.

Patel, C. J., J. Bhattacharya, and A. J. Butte. 2010. An environment-wide association study (EWAS) on type 2 diabetes mellitus. *PLoS One* 5(5):e10746.

Paul, A. G., K. C. Jones, and A. J. Sweetman. 2009. A first global production, emission, and environmental inventory for perfluorooctane sulfonate. *Environ Sci Technol* 43(2):386–392.

Pereira, P. C., A. F. Vicente, A. S. Cabrita, and M. F. Mesquita. 2009. The influence of olive oil on Sprague Dawley rats DMBA-induced mammary tumors. *International Journal of Cancer Research* 5(4):144–152.

Perera, M. I., H. W. Kunz, T. J. Gill, 3rd, and H. Shinozuka. 1986. Enhancement of induction of intestinal adenocarcinomas by cyclosporine in rats given a single dose of N-methyl N-nitrosourea. *Transplantation*, 42:297–302.

Perou, C. M., T. Sorlie, M. B. Eisen, M. van de Rijn, S. S. Jeffrey, C. A. Rees, J. R. Pollack, D. T. Ross, et al. 2000. Molecular portraits of human breast tumours. *Nature* 406(6797):747–752.

Petersen, M. L., S. E. Sinisi, and M. J. van der Laan. 2006. Estimation of direct causal effects. *Epidemiology* 17(3):276–284.

Preston, D. L., E. Ron, S. Tokuoka, S. Funamoto, N. Nishi, M. Soda, K. Mabuchi, and K. Kodama. 2007. Solid cancer incidence in atomic bomb survivors: 1958–1998. *Radiat Res* 168(1):1–64.

Rappaport, S. M. 2011. Implications of the exposome for exposure science. *J Expo Sci Environ Epidemiol* 21(1):5–9.

Reeves, G. K., R. C. Travis, J. Green, D. Bull, S. Tipper, K. Baker, V. Beral, R. Peto, et al. 2010. Incidence of breast cancer and its subtypes in relation to individual and multiple low-penetrance genetic susceptibility loci. *JAMA* 304(4):426–434.

Ren, X., X. Zhang, A. S. Kim, A. M. Mikheev, M. Fang, R. C. Sullivan, R. E. Bumgarner, and H. Zarbl. 2008. Comparative genomics of susceptibility to mammary carcinogenesis among inbred rat strains: Role of reduced prolactin signaling in resistance of the Copenhagen strain. *Carcinogenesis* 29(1):177–185.

Rockhill, B., B. Newman, and C. Weinberg. 1998. Use and misuse of population attributable fractions. *Am J Public Health* 88(1):15–19.

Rose, G. 1985. Sick individuals and sick populations. *Int J Epidemiol* 14(1):32–38.

Rosen, P. R. 2001. Benign proliferative lesions of the male breast. In *Rosen's breast pathology,* 2nd ed. Philadelphia, PA: Lippincott Williams & Wilkins.

Rosen, P. R. 2009. Fibroepithelial neoplasms. In *Rosen's breast pathology,* 3rd ed. Philadelphia, PA: Wolters Kluwer Health/Lippincott Williams & Wilkins.

Rothman, K. J., S. Greenland, and T. L. Lash. 2008. *Modern epidemiology,* 3rd ed. Philadelphia, PA: Lippincott Williams & Wilkins.

Rudel, R. A., K. R. Attfield, J. N. Schifano, and J. G. Brody. 2007. Chemicals causing mammary gland tumors in animals signal new directions for epidemiology, chemicals testing, and risk assessment for breast cancer prevention. *Cancer* 109(12 Suppl):2635–2666.

Rudel, R. A., R. E. Dodson, L. J. Perovich, R. Morello-Frosch, D. E. Camann, M. M. Zuniga, A. Y. Yau, A. C. Just, et al. 2010. Semivolatile endocrine-disrupting compounds in paired indoor and outdoor air in two northern California communities. *Environ Sci Technol* 44(17):6583–6590.

Rudel, R. A., S. E. Fenton, J. M. Ackerman, S. Y. Euling, and S. L. Makris. 2011. Environmental exposures and mammary gland development: State of the science, public health implications, and research recommendations. *Environ Health Perspect* 119(8):1053–1061.

Russo, I. H., and J. Russo. 1996. Mammary gland neoplasia in long-term rodent studies. *Environ Health Perspect* 104(9):938–967.

Sax, S. N., D. H. Bennett, S. N. Chillrud, J. Ross, P. L. Kinney, and J. D. Spengler. 2006. A cancer risk assessment of inner-city teenagers living in New York City and Los Angeles. *Environ Health Perspect* 114(10):1558–1566.

Seidman, H., S. D. Stellman, and M. H. Mushinski. 1982. A different perspective on breast cancer risk factors: Some implications of the nonattributable risk. *CA Cancer J Clin* 32(5):301–313.

Sergeev, P. V., T. V. Ukhina, K. G. Gurevich, and N. L. Shimanovskii. 2001. On the mechanisms of various dose–effect relationships for estrogens. *Pharm Chem J* 35(9):468–470.

Shin, S. J., and P. P. Rosen. 2007. Bilateral presentation of fibroadenoma with digital fibroma-like inclusions in the male breast. *Arch Pathol Lab Med* 131(7):1126–1129.

Stacey, S. N., A. Manolescu, P. Sulem, T. Rafnar, J. Gudmundsson, S. A. Gudjonsson, G. Masson, M. Jakobsdottir, et al. 2007. Common variants on chromosomes 2q35 and 16q12 confer susceptibility to estrogen receptor-positive breast cancer. *Nat Genet* 39(7): 865–869.

Steenland, K., and B. Armstrong. 2006. An overview of methods for calculating the burden of disease due to specific risk factors. *Epidemiology* 17(5):512–519.

Stovall, M., R. Weathers, C. Kasper, S. A. Smith, L. Travis, E. Ron, and R. Kleinerman. 2006. Dose reconstruction for therapeutic and diagnostic radiation exposures: Use in epidemiological studies. *Radiat Res* 166(1 Pt 2):141–157.

Stovall, M., S. A. Smith, B. M. Langholz, J. D. Boice, Jr., R. E. Shore, M. Andersson, T. A. Buchholz, M. Capanu, et al. 2008. Dose to the contralateral breast from radiotherapy and risk of second primary breast cancer in the WECARE study. *Int J Radiat Oncol Biol Phys* 72(4):1021–1030.

Sudaryanto, A., N. Kajiwara, S. Takahashi, S. Muawanah, and S. Tanabe. 2008. Geographical distribution and accumulation features of PBDEs in human breast milk from Indonesia. *Environ Pollut* 151(1):130–138.

Thayer, K. A., and P. M. Foster. 2007. Workgroup report: National Toxicology Program workshop on hormonally induced reproductive tumors—relevance of rodent bioassays. *Environ Health Perspect* 115(9):1351–1356.

Travis, R. C., G. K. Reeves, J. Green, D. Bull, S. J. Tipper, K. Baker, V. Beral, R. Peto, et al. 2010. Gene–environment interactions in 7610 women with breast cancer: Prospective evidence from the Million Women Study. *Lancet* 375(9732):2143–2151.

Turnbull, C., S. Ahmed, J. Morrison, D. Pernet, A. Renwick, M. Maranian, S. Seal, M. Ghoussaini, et al. 2010. Genome-wide association study identifies five new breast cancer susceptibility loci. *Nat Genet* 42(6):504–507.

Turpin, B. J., C. P. Weisel, M. Morandi, S. Colome, T. Stock, S. Eisenreich, and B. Buckley. 2007. *Relationships of indoor, outdoor, and personal air (RIOPA): Part II. Analyses of concentrations of particulate matter species.* HEI Research Report 130; NUATRC Research Report 10. Boston MA: Health Effects Institute, and Houston TX: Mickey Leland National Urban Air Toxics Research Center. http://pubs.healtheffects.org/getfile. php?u=379 (accessed November 8, 2011).

Ullrich, R. L., N. D. Bowles, L. C. Satterfield, and C. M. Davis. 1996. Strain-dependent susceptibility to radiation-induced mammary cancer is a result of differences in epithelial cell sensitivity to transformation. *Radiat Res* 146(3):353–355.

Uter, W., and A. Pfahlberg. 2001. The application of methods to quantify attributable risk in medical practice. *Stat Methods Med Res* 10(3):231–237.

Varghese, J. S., and D. F. Easton. 2010. Genome-wide association studies in common cancers—what have we learnt? *Curr Opin Genet Dev* 20(3):201–209.

Wacholder, S., P. Hartge, R. Prentice, M. Garcia-Closas, H. S. Feigelson, W. R. Diver, M. J. Thun, D. G. Cox, et al. 2010a. Performance of common genetic variants in breast-cancer risk models. *N Engl J Med* 362(11):986–993.

Wacholder, S., P. Hartge, R. Prentice, M. Garcia-Closas, H. S. Feigelson, W. R. Diver, M. J. Thun, D. G. Cox, et al. 2010b. Supplement to: Performance of common genetic variants in breast-cancer risk models. *N Engl J Med* 362(11):986–993.

Ward, T. J., H. Underberg, D. Jones, R. F. Hamilton, Jr., and E. Adams. 2009. Indoor/ambient residential air toxics results in rural western Montana. *Environ Monit Assess* 153(1–4):119–126.

Watson, C. S., N. N. Bulayeva, A. L. Wozniak, and R. A. Alyea. 2007. Xenoestrogens are potent activators of nongenomic estrogenic responses. *Steroids* 72(2):124–134.

Welch, H. G., and W. C. Black. 2010. Overdiagnosis in cancer. *J Natl Cancer Inst* 102(9): 605–613.

Welch, H. G., L. M. Schwartz, and S. Woloshin. 2000. Are increasing 5-year survival rates evidence of success against cancer? *JAMA* 283(22):2975–2978.

Welshons, W. V., K. A. Thayer, B. M. Judy, J. A. Taylor, E. M. Curran, and F. S. vom Saal. 2003. Large effects from small exposures. Mechanisms for endocrine-disrupting chemicals with estrogenic activity. *Environ Health Perspect* 111(8):994–1006.

Yin, W., and A. C. Gore. 2006. Neuroendocrine control of reproductive aging: Roles of GnRH neurons. *Reproduction* 131(3):403–414.

Zackrisson, S., I. Andersson, L. Janzon, J. Manjer, and J. P. Garne. 2006. Rate of over-diagnosis of breast cancer 15 years after end of Malmo mammographic screening trial: Follow-up study. *BMJ* 332(7543):689–692.

Zheng, W., J. Long, Y. T. Gao, C. Li, Y. Zheng, Y. B. Xiang, W. Wen, S. Levy, et al. 2009. Genome-wide association study identifies a new breast cancer susceptibility locus at 6q25.1. *Nat Genet* 41(3):324–328.

5

Examining Mechanisms of Breast Cancer Over the Life Course: Implications for Risk

The preceding chapters have summarized the available evidence for the relationship between environmental exposures and breast cancer, as well as the many challenges inherent in studying this issue. Although there is strong evidence of a modest role for a handful of modifiable environmental exposures as risk factors for breast cancer, many unanswered questions remain. These unanswered questions require new research approaches, which are discussed in Chapter 7.

Meanwhile, remarkable progress has been made in understanding the fundamentals of carcinogenesis, manifested in mechanisms at the genetic, epigenetic, cellular, and tissue levels. Scientific advances are revealing complex potential pathways and factors involved in cancer development. The committee sees the need for a continued and intensified focus on understanding the basic biology of breast carcinogenesis in order to gain better fundamental appreciation of the environmental factors with potential roles in the etiology of this disease.

As a crucial dimension of this research, the committee notes a growing appreciation among researchers of the important role that the timing of exposure plays in effecting changes that alter the likelihood for cancer and other diseases later in life. Observations in human studies of the effects of exposure to ionizing radiation and diethylstilbestrol (DES) in early life, as well as mechanistic and animal studies of other environmental exposures, suggest that existing assessments of the role of certain environmental factors derived from studies in adult women, such as those reviewed in Chapter 3, may have been negative or uninformative because they failed to consider critical periods of life stage and exposure—essentially asking the wrong question. As understanding grows of how the genetic, epigenetic, cellular, and tissue changes in the breast during development and over the life course influence susceptibility to breast carcinogenesis, researchers have continuing and increased appreciation for the potential for timing of exposure to make a difference in the effects of environmental agents on breast cancer risk. The committee sees the need to direct attention to the accumulating evidence that environmental exposures may have a differential impact, depending on their timing during the life course.

ENVIRONMENTAL EXPOSURES OVER THE LIFE COURSE AS DETERMINANTS OF BREAST CANCER RISK

The female breast is not static; it changes in structure and function over the life course. Breast development begins in utero and continues into adulthood, with further differentiation occurring with pregnancy and lactation and involution occurring with menopause (Russo and Russo, 2004; Polyak and Kalluri, 2010). Most breast cancers arise in and spread from the ducts or the lobules, which are the breast's main functional components (Figure 5-1).

The sections that follow consider the major life stages for women and the state of breast tissue during each stage, with indications of the potential for exposures during each stage to alter risk for breast cancer. Although evidence from human studies is limited, studies in animal models strongly indicate the potential for timing of environmental exposures to alter risk for developing cancer. Box 5-1 lists the life stages discussed by the committee and mechanisms of carcinogenesis likely to be of particular relevance or importance to breast cancer.

Early Life Exposures and Breast Cancer Risk

Preconceptional and Periconceptional Exposure

Preconception studies focus on parental exposure to environmental agents before the conception of offspring. There is no standard definition for the preconceptional period, and the term is used rather loosely. Some studies combine the time before conception with early pregnancy as the periconceptional period (Van Maele-Fabry et al., 2010). Studies examining pre- or periconceptional exposures may consider paternal or maternal exposure, or both.

To date, epidemiologic studies have not addressed parental exposure before conception and subsequent risk of breast cancer in offspring. However, childhood cancers, such as leukemia and brain tumors, have been linked to prenatal exposures such as maternal smoking, ionizing radiation,

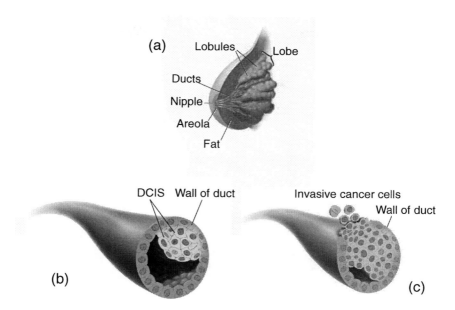

FIGURE 5-1 Schematic representation of (a) the breast, showing lobules and ducts, (b) ductal carcinoma in situ (DCIS), and (c) invasive ductal cancer.
SOURCE: NCI (2009).

and pesticides (Shim et al., 2009; Van Maele-Fabry et al., 2010). Preconceptional parental exposure to ionizing radiation was not associated with an increased risk of childhood leukemia (Wakeford, 1995), and parental exposure electromagnetic fields before conception has not been demonstrated to increase the risks of childhood cancers of any kind (Sorahan et al., 1999).

It is difficult for studies to track offspring of parents whose exposures before the child's conception are known until the children reach the ages in which breast cancer tends to manifest itself, making this a potential area for future research.

In Utero and Neonatal Exposure

The human breast begins to develop once the embryo reaches 4.5–6 mm in length (Hughes, 1950). Embryonic epidermal cells proliferate to create a breast bud, which responds to cues from the embryonic mesenchyme (Anbazhagan et al., 1998). In the newborn human, the breast is characterized by "very primitive structures, composed of ducts ending in short duct-

BOX 5-1
Life Stages Representing Potential
Windows of Susceptibility for Breast Carcinogenesis and
Hypothesized Mechanisms of Carcinogenesis

Some mechanisms are likely to be more relevant or important to breast carcinogenesis at particular life stages (e.g., epigenetic reprogramming during the in utero/perinatal period).

Life Stages as Potential Windows of Susceptibility	Hypothesized Mechanisms of Carcinogenesis
• Preconception	• Mutagenesis
• In utero/perinatal	• Nuclear hormone receptor signaling
• Early childhood	• Mitogenic signaling leading to cell proliferation
• Prepuberty	
• Puberty	• Epigenetic and developmental reprogramming
• Reproductive years	
• Menopause	• Modulation of immune function, escape from immune surveillance
• Postmenopausal years	• Alterations of tissue microenvironment

ules lined by one to two layers of epithelial and one of myoepithelial cells" (Russo and Russo, 2004, p. 3). The breast in the newborn also contains a population of stem cells that are the undifferentiated precursors of the cellular expansion that occurs as the structures of the breast develop during puberty, pregnancy, and lactation.

Strong evidence indicates that aspects of fetal growth, such as birth weight, are associated with breast cancer risk as an adult (Potischman and Troisi, 1999; Michels and Xue, 2006; dos Santos Silva et al., 2008; Park et al., 2008b). According to one postulation advanced by Trichopoulos and others (2005), the number of mammary stem cells is determined during in utero or immediate postnatal life and is under the influence of estrogens and components of the insulin-like growth factor system during pregnancy. They hold that the number of mammary tissue-specific stem cells is the core determinant of breast cancer risk. Thus, an increase in mammary stem cells associated with higher birth weight would provide more target cells for breast carcinogenesis.

This postulation also includes recognition of the increased risk of breast cancer associated with increased breast density and mass of glandular tissue (Boyd et al., 1998, 2010), which is itself likely linked to the number of stem cells (Trichopoulos et al., 2005). While this hypothesis focuses on the effects of endogenous estrogens and growth factors, it also provides a potential mechanism by which exogenous environmental agents might increase (or decrease) breast cancer risk if prenatal exposure to them promotes the formation of greater numbers of breast stem cells. There is growing evidence from rodent models to indicate that exposure to xenoestrogens during the prenatal and neonatal periods may affect mammary gland development and alter risk for cancer later in life (Soto et al., 2008). Some of these changes are thought to possibly result from developmental reprogramming as described below, but some of the mechanisms are just beginning to be elucidated.

During gestation, while maternal levels of the pregnancy hormones progesterone and estrogen soar, the developing fetus is protected from endogenous maternal hormones by steroid hormone binding proteins. Steroid hormones such as progesterone and estrogen circulate in the blood stream bound to proteins such as serum albumin and sex hormone binding globulin (SHBG), which is a glycoprotein that specifically binds testosterone and estradiol. Only a small fraction of steroid hormones is present in the circulation in the unbound "free" form. In utero, the fetal liver synthesizes sufficient amounts of steroid hormone binding proteins, including steroid-binding β-globulin (SBβG) and alpha fetoprotein (AFP), to protect the developing organism from rising levels of maternal hormones. Thus, even when maternal estrogen levels are extremely high, increased expression of AFP and other steroid hormone binding proteins (Mizejewski et al., 2004) reduces the unbound fraction of estradiol in the human fetus to 1.0 to 4.5 percent (i.e., greater than 95 percent is bound) (Pasqualini et al., 1985; vom Saal and Timms, 1999; Witorsch, 2002, citing Tulchinsky, 1973). In primates, fetal hormone levels actually decline in the face of increasing maternal hormones during pregnancy (Thau et al., 1976). Production of steroid hormone binding proteins by the fetal liver decreases dramatically after birth. Thus, some free endogenous steroid hormone is likely to come in contact with developing tissues of the fetus. In the context of environmental exposures, however, in vitro studies with human serum (e.g., Milligan et al., 1998; Jury et al., 2000) and in vivo studies in mice (Welshons et al., 1999) have found that various xenoestrogens are generally not recognized by these steroid hormone binding proteins. As a result, the mechanism that limits fetal exposure to endogenous estrogens may offer less protection from these xenoestrogens.

One of the best known examples in humans of an in utero exposure altering risk for cancer later in life is that of DES. Between 1938 and 1971, this estrogenic compound was used to prevent miscarriages, but it was taken off the market when daughters of women who took it during pregnancy developed adenocarcinomas of the vagina (Herbst et al., 1971).

Long-term follow-up studies of both mothers and their offspring have found that daughters who were exposed to pharmacological levels of DES in utero experienced important and devastating effects on adult health many years later, including an elevated risk of breast cancer (Palmer et al., 2006; Troisi et al., 2007). These studies have graphically demonstrated that the period of organogenesis during fetal life is a period when humans may be particularly sensitive to the effects of environmental agents—in this case synthetic estrogens. From an epidemiologic standpoint, these DES studies have provided a unique illustration of early development as a window of susceptibility to environmental exposures in terms of the clarity of the dose and timing of the exposure and the extremely long and costly period of follow-up.

Studies in Sprague Dawley rats exposed perinatally to 1.2 µg of DES by subcutaneous injection have demonstrated increased susceptibility to mammary carcinogenesis after postnatal treatment with the carcinogen 7,12-dimethylbenz(a)anthracene (DMBA) (Boylan and Calhoon, 1979, 1983). A higher incidence of palpable tumors and decreased tumor latency were observed compared with DMBA-treated rats without prior hormone exposure (Boylan and Calhoon, 1979, 1983); however, major differences in reproductive tracts or mammary gland structure were not observed at the DES doses used (Boylan and Calhoon, 1983). Kawaguchi et al. (2009a) also used Sprague Dawley rats to demonstrate that rats that had prenatal exposure to DES (their mothers had been fed DES [0.1 ppm] throughout pregnancy or from day 13 of pregnancy through to birth) and were then exposed to DMBA at 50 days after birth developed more mammary carcinomas than controls.

Experiments in ACI rats without additional carcinogen dosing demonstrated that mammary tumors can be induced in female rats by prenatal (0.8 µg or 8.0 µg s.c. to the pregnant mother, equivalent to 4.28 µg/kg or 42.8 µg/kg of body weight) or postnatal (2.5 mg via implanted pellet) DES exposure alone, and that prenatal and postnatal DES exposure combined yielded significantly greater tumor multiplicity and decreased tumor latency (Rothschild et al., 1987). The morphology of peripubertal mammary glands of a significant proportion of female ACI rats exposed to DES in utero were atypical; approximately 25 percent displayed hypodifferentiation and about 5 percent had hyperproliferation (Vassilacopoulou and Boylan, 1993).

In utero DES exposure is theorized to increase mammary cancer susceptibility by slowing mammary gland maturation and development (Jenkins et al., 2011). The most mature structures of the mammary gland, lobules, appear to be most resistant to carcinogenic transformation by chemical carcinogens, while terminal end buds are more susceptible (Russo and Russo, 1978; Russo et al., 1982). Prenatal DES exposure has been reported to increase the number of terminal end buds, increasing susceptibility to chemical carcinogenesis (Ninomiya et al., 2007). In contrast to the effects from fetal exposure to DES, exposure to DES at other time windows has been shown to manifest different effects (Lamartiniere and Holland, 1992; Hovey et al., 2005; Kawaguchi et al., 2009b).

Studies carried out to explore the mechanisms of uterine estrogenic effects from perinatal exposure to DES in the CD-1 mouse model might prove helpful for understanding effects in the mammary gland. Estrogenic effects include persistent expression of the lactoferrin and c-fos genes (Newbold et al., 1997; Li et al., 2003) together with a high incidence of uterine adenocarcinoma (Newbold et al., 1990). DES exposure also causes changes in the expression of several uterine genes responsible for directing tissue architecture and morphology, resulting in altered tissue structures (Ma et al., 1998; Miller et al., 1998; Block et al., 2000). Thus, altered gene expression, likely as a result of epigenetic reprogramming, may also be a contributor to increased susceptibility to mammary carcinogenesis in DES-exposed animals. The body of animal studies carried out with DES demonstrates that the in utero and neonatal periods are especially vulnerable to inappropriate xenoestrogen exposure, with these exposures inducing developmental reprogramming and potentially altering risk for cancer later in life.

Although intentional human exposures to pharmacologic doses of estrogen compounds such as DES are hopefully unlikely to recur, exposure to environmental chemicals with estrogen-like activity is common. Analysis of data from the National Health and Nutrition Examination Survey (NHANES) 2003–2004 found a large range of chemicals present in blood or urine in women who were and were not pregnant (Woodruff et al., 2011). Concentrations tended to be similar or lower in pregnant women compared to those in women who were not pregnant. The analysis demonstrates the potential opportunity for exposure of the developing fetus and the pregnant mother's breast to a wide range of chemical compounds. Whether or not the presence of these chemicals is associated with an increased risk of breast cancer in either child or parent requires additional investigation.

Early Childhood and Prepuberty

The rudimentary ductal system present in the breast at birth is under the influence of maternal hormones (Anbazhagan et al., 1991). But by age 2, the breast undergoes involution, and it has only a primitive ductal system without alveoli until the onset of puberty (Howard and Gusterson, 2000). During this period, however, the body is preparing for puberty and the next stage of mammary gland development. Various exposures during childhood appear to influence the timing of puberty. Because the timing of puberty is associated with the risk of breast cancer later in life, better understanding of the factors in childhood that influence the timing of puberty may add to understanding of pathways that can contribute to breast cancer.

Diet and the nutritional content of food consumed in childhood help set the stage for puberty by developing adequate body mass. However, prepubertal overweight and obesity are positively associated with early puberty (Kaplowitz, 2008) and late pubertal peak height velocity and early menarche (Hauspie et al., 1997). Recent national health surveillance data for the United States show that 35 percent of girls ages 6–11 are at or above the 85th percentile of body mass index (BMI) standards for their age, and 18 percent are at or over the 95th percentile (Ogden et al., 2010). Biro and colleagues (2003, 2010) have shown that girls with higher BMIs are likely to begin puberty at an earlier age. However, the mechanism underlying the association between obesity and earlier puberty is not yet clear (Jasik and Lustig, 2008). The impact of decreased physical activity, independent of energy intake, and its effects in different subgroups of children in this prepubertal period may act through decreased insulin sensitivity and are also of concern (Sorensen et al., 2009). Further investigation is needed to understand the relation between activity levels in childhood and the timing of puberty or other factors that may influence breast cancer risk in later life.

Psychosocial factors have also been found to influence pubertal timing among girls (Graber et al., 1997; Ellis and Garber, 2000; Bogaert, 2005). They are likely to operate through pathways other than diet, physical activity, and obesity. In stressful family contexts, characterized by low-quality parental investment, high levels of stress, negative relationships, and prolonged distress, reproductive maturation appears to accelerate (Romans et al., 2003). Family relationships characterized by warmth, cohesion, and stability, on the other hand, consistently predict later pubertal onset (Graber et al., 1995). One particular manifestation of disrupted family relationships exists in households with the absence of a biological father. Studies have shown an association between early pubertal maturation in both boy and girl twins in homes where the father was absent when the children were age 14. Girls were about twice as likely to experience menarche before age 12 in such households, although the mechanisms for this observation are not clear (Quinlan, 2003; Mustanski et al., 2004). One study has found this relationship confined to white girls in families with higher socioeconomic status, as measured by household income (Deardorff et al., 2011). Overall, a review of the father-absence literature suggests that girls in father-absent homes experience menarche 2 to 5 months earlier than those in homes where the father is present (Ellis, 2004). The absence of the mother or the presence of a stepfather does not appear to be related to pubertal timing (Bogaert, 2005).

Animal studies suggest that endocrine-disrupting chemicals (EDCs) may alter hormone synthesis and metabolism during this prepubertal period

(Crain et al., 2008) and may advance the onset of puberty. Individual nutrients and dietary intake of substances such as phytoestrogens, which are considered xenoestrogens, may also be factors in pubertal development. In a study of pubertal development in 9-year-old girls, higher urinary concentrations of phytoestrogens, and in particular daidzein and genistein, were associated with later age of breast development, and this effect was stronger in girls with lower BMIs (Wolff et al., 2008). The delay in pubertal development is considered protective in terms of risk for breast cancer. A limited set of case–control studies provides some support for an association between higher consumption of soy products during childhood and lower risk of breast cancer (reviewed in Hilakivi-Clarke et al., 2010).

Studies in animal models of genistein exposure have investigated the importance of the timing of exposures in altering mammary tumor susceptibility. The animal data regarding postnatal, prepubertal exposure to genistein are described as "very consistent" in showing a reduction in mammary cancer risk (Warri et al., 2008). In early studies, Lamartiniere and coworkers (1995) observed that rats treated with genistein during the early postnatal period displayed increased mammary tumor latency and decreased tumor multiplicity in classic chemical carcinogenesis models. In contrast, animal studies of in utero genistein exposure have produced conflicting results (reviewed in Warri et al., 2008).

Differences in the impact of pre- and postnatal exposures in these animal models highlight the potential for the timing of human exposure to genistein, and likely other xenoestrogens, to have differing impacts on breast cancer risk.

Puberty and Adolescence

The onset of puberty, the pubertal period, and adolescence comprise another life stage during which environmental factors may influence the development of breast cancer in adult life in unique ways. This stage spans the period from the first signs of sexual development and the external appearance of the breast to sexual maturity (Russo and Russo, 2004). Breast development in adolescent girls is characterized by branching of terminal end buds in response to hormonal cues (Russo and Russo, 2004).

The onset of puberty is clinically defined by the first signs of breast development, pubic hair, and other secondary sex characteristics (Grumbach and Styne, 2002). It coincides with the activation of the hypothalamic–pituitary–gonadal (HPG) axis, or thelarche, and the activation of the hypothalamic–pituitary–adrenal (HPA) axis, or adrenarche, which are independent events. HPG axis activation is associated with a surge of pituitary follicle-stimulating hormone (FSH), which results in the stimulation of primordial ovarian follicles to secrete estrogen. Circulating estrogen then has several important effects, including the release of pituitary luteinizing hormone (LH) and vascular proliferation and growth in the breast with the further development of the mammary ducts and the mammary stromal connective tissue (Rogol, 1998). The activation of the HPA axis stimulates the adrenal production of dehydroepiandrosterone, dehydroepiandrosterone sulfate, and androstenedione, which lead to the development of secondary sexual characteristics, including pubic hair and changing body proportions. The relative timing of thelarche and adrenache may differentially determine the onset of menarche (Biro et al., 2006).

Early age at menarche is an established risk factor for breast cancer (Kelsey and Bernstein, 1996), but its use as a marker of pubertal onset can be misleading because the relationship between the onset of puberty and menarche has not been constant over time (Euling et al., 2008; Mouritsen et al., 2010). In the United States, the correlation between the onset of puberty and menarche was greater than 0.9 for women born in the 1930s, 0.5–0.7 for those born in the 1950s, and 0.38–0.39 for those born in the 1970s (Biro et al., 2006). These results suggest that factors contributing to the onset of puberty and menarche were more similar in the past than in more recent years. Clear differentiation between the time of onset of puberty and menarche and the interval between them ("tempo") is important in studies of pubertal development (Euling et al., 2008).

It is clear that both the age when girls begin puberty and their age of menarche have declined over the past century (Euling et al., 2008). However, historical data and more recent detailed epidemiologic studies have concluded that the age of menarche declined in industrialized countries over the course of the past century, whereas the decline in the age of onset of puberty has been rapid and observed since just since the early 1990s (de Muinck Keizer-Schrama and Mul, 2001; Euling et al., 2008; Aksglaede et al., 2009; Mouritsen et al., 2010). This latter decline has not been associated with development and socioeconomic conditions and has thus raised concerns about the possible role of environmental factors such as endocrine disrupting chemicals (Mouritsen et al., 2010). Genetic regulation of puberty is unlikely to explain these rapid secular trends, but genetic factors do influence the age of pubertal onset in individual girls (Parent et al., 2005). Supporting evidence comes from studies documenting a correlation between a mother's and daughter's ages at puberty, from twin correlation studies that suggest that most (70–80 percent) of the variance between twins is explained by genetic influences, and from the observation of marked differences in pubertal timing among racial and ethnic groups (Parent et al., 2003, 2005).

At the onset of puberty, the ratio of FSH to LH favors FSH, which inhibits ovulation, and even with the onset of menarche, ovarian function can continue to be anovulatory for a time (MacMahon et al., 1982b).

The duration of anovulatory menstrual cycles after the onset of menarche varies from 1 to more than 6 years, with longer intervals to ovulation in girls with a late menarche (MacMahon et al., 1982a; Clavel-Chapelon, 2002). The shorter period of anovulation for girls with earlier menarche would suggest that the increased risk of breast cancer that is associated with earlier menarche may be related to earlier and more frequent exposure to the hormones produced during the menstrual cycle. Moreover, acceleration of menarche without a concomitant acceleration in the timing of menopause increases the duration of estrogen exposure over a lifetime, which it is widely thought to promote the development of breast cancer (de Waard and Thijssen, 2005). For each additional year of delay in menarche, the risk of breast cancer is decreased by approximately 9 percent for premenopausal cases and by approximately 4 percent for postmenopausal cases (Hsieh et al., 1990; Clavel-Chapelon and Gerber, 2002). Among women in the Nurses' Health Study II, who were followed between 1989 and 1993, a 1-year increase in age at menarche was associated with reduction in risk of 10 percent (RR = 0.90, 95% CI, 0.83–0.99) (Garland et al., 1998). In a comparison between women with an age of menarche of 13 versus 15 years or older, an older age of menarche was associated with a statistically significantly reduced risk for premenopausal breast cancer (OR = 0.72, 95% CI, 0.57–0.91), but the reduction in risk for postmenopausal breast cancer was not statistically significant (OR = 0.90, 95% CI, 0.80–1.03) (Titus-Ernstoff et al., 1998).

The contribution of timing of puberty and onset of menarche to increased breast cancer risk may be related to estrogen receptor signaling or to other mechanisms discussed later in this chapter. It is also possible that the rapidly duplicating cells of the breast during pubertal development are more susceptible to environmental insults. Studies in the rodent model show that the highest number and the greatest proliferative activity of the terminal duct lobular units (TDLUs) occurs during puberty (Rudland, 1993), and it has been suggested that this may be related to the apparent susceptibility of the breast to carcinogens during puberty (Colditz and Frazier, 1995; Knight and Sorensen, 2001).

Some of the best evidence for susceptibility of breast tissue during early-life exposures is derived from investigation of the effects of ionizing radiation from nuclear explosions and from medical diagnostic and treatment procedures. An increased risk for breast cancer has been documented among atomic bomb survivors in Japan (Tokunaga et al., 1991), and this increased risk has been related to younger age at exposure, especially during the period of puberty (Land et al., 2003). In an ecological study of the Chernobyl accident in Belarus, the areas with the highest levels of radiation contamination (estimated average cumulative doses ≥ 40 mSv) were associated with elevated breast cancer risk about 10 years after the incident, especially among women who were younger than age 45 at the time of the event (Pukkala et al., 2006).

More recently, additional follow-up of the atomic bomb survivors and analysis of both excess relative risk and excess attributable risk suggest that excess relative risks are similar across ages of exposure, for example, at ages 10, 30, or 50 (Preston et al., 2007). However, models examining excess attributable risk show a large difference by age at exposure, which the authors suggest reflects differences in factors such as reproductive history that have changed across birth cohorts and that may act multiplicatively with age at radiation exposure (Preston et al., 2007).

The potential implications of these findings are important in terms of recommendations for earlier radiographic screening among high-risk populations such as women at increased genetic risk of breast cancer. Their risks with exposure to mammographic X-rays may vary according to whether they have completed a pregnancy or other factors. Research findings such as these may influence the age at which mammographic screening is begun, reliance on other screening techniques that do not use ionizing radiation, and issues to be covered by consent documents.

An increased risk of breast cancer has also been consistently reported for exposure to ionizing radiation at young ages in conjunction with medical treatments, such as radiotherapy for Hodgkin's disease and childhood cancer, ankylosing spondylitis, tinea capitis, enlarged thymus (Shore et al., 1993), and skin hemangioma (John and Kelsey, 1993). Later in life into the reproductive years, radiation has been associated with breast cancer among women receiving radiation for postpartum mastitis (Shore et al., 1986) and during tuberculosis treatments (Boice et al., 1991). Exposures among radiologic technologists have been studied, but little impact has been found on risk for breast cancer among those employed over the past 40 years.

Reproductive Years

The reproductive period for women spans the time from sexual maturity at the end of puberty to menopause, providing an expansive time window in which exposures may influence the risk of breast cancer development. Established risk factors for breast cancer encountered during this period include later age at first full-term pregnancy and later age at menopause. The neutral or inverse association between weight or BMI and breast cancer during the reproductive period differs from the increased risk found during the years following menopause. Smoking, hormone therapy, and radiation are examples of other risk factors that may have differential effects over the life course and across the reproductive period. Indeed, pregnancy itself is associated with a short-term increased risk of breast cancer, making it a period of vulnerability of the breast for both the mother and the developing infant exposed to the in utero environment, as discussed above.

Breast tissue continues to evolve and differentiate before and during pregnancy (Russo et al., 2006), and the differentiation that occurs during pregnancy may be a factor in the reduced risk of breast cancer that is associated with childbearing. During pregnancy, the breast "attains its state of maximum development" in two distinct stages—early and late in pregnancy (Russo and Russo, 2004, p. 7). The early stage involves differentiation of ductal trees and an increase by the third month of pregnancy in the number of well-formed lobules. In the later stages of pregnancy, the continued changes in the breast are related in large part to the secretory functions that the breast tissues will perform during lactation (Russo and Russo, 2004). During pregnancy, significant changes also occur in the mammary stroma, the connective tissue in the breast. This remodeling includes changes that result in increased angiogenesis, infiltration of immune cells, and fibroblast reorganization, all of which help supply nutrients and cues to the expanding ductal and lobular structures (McCready et al., 2010). An increase in circulating hormones during pregnancy also plays a role in breast development.

Studies in rats have shown that pregnancy leads to the maximum mammary gland development. This process includes alterations in mammary stem cells (breast progenitor cells) that make them more resistant to carcinogens by virtue of primed mechanisms for metabolism of carcinogens and improved DNA repair mechanisms (Russo et al., 2006). This type of process may explain the association in humans between young age at first pregnancy and reduced breast cancer risk, and the increased risk associated with nulliparity and late age at first pregnancy. Nulliparity or late age at first pregnancy would leave the breast more vulnerable to carcinogens because of the predominance of unaltered stem cells. With late age at first pregnancy, the differentiation and proliferation of breast tissue would be more likely to involve compromised stem cells that have suffered DNA damage, which would provide a favorable environment for progression of cancer cells. Some studies have found that the increase in risk associated with a later age at first full-term birth (age 30 and older) is greater than for nulliparity (e.g., Kotsopoulos et al., 2010; Newcomb et al., 2011).

A review of the literature on breast cancer and characteristics of pregnancy found conflicting evidence for factors such as weight gain during pregnancy, fetal growth, gestational age, and gestational diabetes (Nechuta et al., 2010). Findings were more consistent that multiple births and pre-eclampsia were associated with modest reductions in risk for breast cancer (Nechuta et al., 2010). In an example of the effects of an exogenous hormone exposure during pregnancy, follow-up of women who took DES to prevent pregnancy complications from 1940 through the 1960s has found a modest association between their adult DES exposure while pregnant and subsequent breast cancer risk (RR = 1.27, 95% CI, 1.07–1.52) (Titus-Ernstoff et al., 2001), supporting the influence of hormonal factors during pregnancy and its period of rapid breast proliferation.

Other evidence that pregnancy can modify breast cancer risk comes from the literature on smoking and breast cancer. This literature is often characterized as mixed, with some studies finding associations and others not. However, a more nuanced picture emerges when the risks from smoking are analyzed separately for women who started to smoke before a first pregnancy and for those who began to smoke after their first child was born. As noted in Chapter 3, a meta-analysis of 23 studies found a weak association with increased risk of breast cancer for women who began smoking before a first pregnancy (DeRoo et al., 2011). The summary risk ratio was 1.10 (95% CI, 1.07–1.14), compared with a risk ratio 1.07 (95% CI, 0.99–1.15) for women who began smoking later (DeRoo et al., 2011).

Reynolds and colleagues (2004), for example, observed an elevated risk for women who smoked for 5 or more years before a first full-term pregnancy compared with never smokers with children (HR = 1.13, 95% CI, 1.00–1.28). Risk was not increased among the small group of women (42 cases) who began smoking after a first pregnancy (HR = 0.89, 95% CI, 0.65–1.21). This study also found no excess risk for women who initiated smoking at age 20 or later (HR = 1.03, 95% CI, 0.90–1.17), but a statistically significant increased risk for those who began before age 20 (HR = 1.17, 95% CI, 1.05–1.30). Consistent with these patterns, Ha and colleagues (2007) observed no association between breast cancer risk and pack-years of smoking after first childbirth, and no significant trend with cumulative smoking exposure. However, there was a significant trend (p <.0001) in the hazard ratios for pack-years of smoking before the birth of the first child, with the highest exposure group having a statistically significant increased risk (HR = 1.78, 95% CI, 1.27–2.49) (Figure 5-2).

As also noted in Chapter 3, recent reports from both the Nurses' Health Study (Xue et al., 2011) and the observational component of the Women's Health Initiative (Luo et al., 2011) appear to support this pattern. Xue and colleagues (2011) found that more pack-years of smoking from menarche to a first birth were associated with a statistically significant increase in risk (p for trend <.001). Luo et al. (2011) found that initiation of smoking before first full-term pregnancy was associated with a statistically significant increase in risk (HR = 1.28, 95% CI, 1.06–1.55). The risk with initiation after a first pregnancy was elevated but not statistically significant (HR = 1.17, 95% CI, 0.90–1.52).

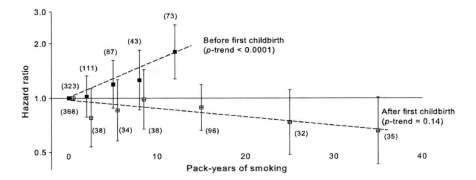

FIGURE 5-2 Breast cancer risk by pack-years of smoking before and after first childbirth among parous women, U.S. Radiologic Technologists Health Study, 1983–1998. Hazard ratios and 95 percent confidence intervals were estimated from one multivariate proportional hazards model including separate variables for pack-years of smoking before and after first childbirth, with age as the time scale, stratified for birth cohort in 5-year intervals and adjusted for alcohol intake, age at menarche, age at first childbirth, parity, family history of breast cancer, hormone replacement therapy, year that a woman first worked as a radiologic technologist, body mass index, and time-dependent menopausal status. The numbers of breast cancer cases in each category are provided in parentheses.
SOURCE: Ha et al. (2007, p. 58). Reproduced with permission from *American Journal of Epidemiology*.

Menopause and Older Ages

Most breast cancers are diagnosed in menopausal women. Menopause is characterized by a cessation of ovarian hormone production, and subsequently, the end of a woman's menstrual cycles (McCready et al., 2010). The loss of estrogen and progesterone production coincides with a process known as lobular involution, or the reduction of both number and size of lobules in the breast (Russo et al., 2001). The number of epithelial cells in the breast expressing estrogen receptors increases significantly (Shoker et al., 1999), and the interlobular stroma of the breast is increasingly replaced by adipose tissue (Howard and Gusterson, 2000). At this stage of life, breast tissue has also been influenced by the cumulative opportunity for previous exposures to endogenous and exogenous factors to have generated compromised cells.

Clear evidence from studies of menopausal hormone therapy also shows that breast cancer risks can be influenced by exposures during this life stage. Postmenopausal use of hormonal therapy that combines estrogen and progestin increases both the incidence of and mortality from breast cancer (Writing Group for the Women's Health Initiative Investigators, 2002; Million Women Study Collaborators, 2003; Chlebowski et al., 2010). The risks decline rapidly on withdrawal of combined hormone therapy in the postmenopausal period (Collaborative Group on Hormonal Factors in Breast Cancer, 1997; Clarke et al., 2006; Chlebowski et al., 2009; Beral et al., 2011). While the mechanism for the increased risk is not known in detail, it is thought that hormone therapy may be causing proliferation of cancer stem cells (Eden, 2010, 2011). In the absence of hormonal stimulus, other regulatory processes in the body may be able to inhibit the progression of tumorigenesis or expansion of an existing tumor.

Studies have also examined the risk associated with timing of initiation of hormone therapy relative to onset of menopause (Prentice et al., 2009; Beral et al., 2011). From both the Women's Health Initiative in the United States and the Million Women Study in the United Kingdom, evidence has emerged that the risk of invasive breast cancer from combination hormone therapy decreased with increasing gap time (time from menopause to initiating hormone therapy) for combined estrogen–progestin hormone therapy (Prentice et al., 2009; Beral et al., 2011; LaCroix et al., 2011). Risk was higher among women taking combined estrogen–progestin therapy within 5 years of menopause compared to those whose first use was 5 or more years after menopause (Prentice et al., 2009), and additional analyses demonstrated the risk of breast cancer increased with use within 2 years since menopause.

The Women's Health Initiative found that estrogen-only hormone therapy, which is appropriate only for women who have had hysterectomies, was not associated with increased risk of breast cancer (Women's Health Initiative Steering Committee, 2004; LaCroix et al., 2011). The Million Women Study, however, observed a small but statistically significant risk associated with estrogen-only therapy, but only when initiated before menopause or within 5 years after menopause (RR = 1.49, 95% CI, 1.40–1.58 and RR = 1.36, 95% CI, 1.27–1.46, respectively). Among estrogen-only users, there was no association with breast cancer risk if use was initiated 5 or more years after menopause (RR = 1.05, 95% CI, 0.90–1.24) (Beral et al., 2011). The findings from these studies are also discussed in Chapters 3 and 6.

MECHANISMS OF BREAST CANCER DEVELOPMENT

Several potential mechanisms may be operating in the development of breast cancer, and they may play out during a particular phase or across multiple phases of life. Researchers do not yet know the details of all of the potential mechanisms through which normal breast tissue changes to become cancerous tissue. Such changes, termed "transformation," occur through a multistage process of carcinogenesis. Multistage carcinogenesis encompasses both spontaneous and environmentally induced events that contribute to altering the cells from normal, healthy tissue into tumor tissue. Spontaneous events occur in cells by chance, or stochastically, as a by-product of normal processes. For example, reactive oxygen species (ROS) produced as a by-product of cellular respiration can cause damage to DNA, as can errors in DNA replication that result in spontaneous mutations. Environmentally induced events are those caused by an external exposure, for example, ultraviolet (UV)-induced damage from sunlight or exposure to tobacco carcinogens. In addition to physical and chemical insults, environmental contributors to cancer, as defined broadly by this committee (see Chapter 2), can include pathophysiological conditions, such as obesity.

Accumulated damage from both internal and external carcinogenic events or risk factors drives cancer development through multiple stages from normal cells through preneoplasia to metastatic disease, with many of these stages recognizable as distinct entities in the process of tumor development at the cellular (e.g., lobular hyperplasia) and/or molecular level (e.g., aberrant p16 expression). There is also a growing appreciation for the role that the tissue microenvironment can play in limiting, permitting, or potentiating the carcinogenesis. Thus, the development of breast cancer, like other adult cancers, occurs as a result of accumulated damage occurring over a person's entire life. However, susceptibility to different types of damage induced by either endogenous or exogenous carcinogens may change over the life course, so that humans may be more or less vulnerable to the effects of any given environmental agent at different stages of life.

In the remainder of this chapter, critical mechanisms of carcinogenesis are reviewed, with the recognition that mechanistic explanations of the origins of breast cancer have not commonly been situated within the life course perspective that the committee recommends.

Mutagenesis

Damage to DNA was the earliest recognized mechanism for development of cancer from environmental causes. Changes to the sequence of base pairs that make up the genome (DNA) are called mutations, and they can be caused by internal cellular processes or outside effectors such as radiation, viruses, or certain chemicals. DNA mutations can result in changes in gene (and microRNA) expression, function, or regulation that may lead to the development of tumors. Many environmental agents can induce DNA damage, which, if not properly repaired, can give rise to mutations that contribute to multistage carcinogenesis. Induction of DNA damage can be direct—for example, induction of DNA double-strand breaks by gamma irradiation, or the creation of pyrimidine dimers by UV light. DNA damage can also occur indirectly—for example, via oxidative damage resulting from depletion of ROS scavengers, or increased ROS production in the cell. The scientific community's increasingly sophisticated understanding of "mutagens as carcinogens" now includes appreciation that not all agents that induce DNA damage cause cancer. Nevertheless, environmental agents that induce mutations are still considered among the most effective types of carcinogens.

Exposure to environmental mutagens may have very different effects, depending on the target cell[1] and the extent of cell proliferation that occurs before and after repair of environmentally induced damage. If the target cells have insufficient time to repair DNA damage before they divide, mutations can be passed on to daughter cells. Therefore, those stages of life when breast epithelial cells are most proliferative (i.e., puberty and pregnancy) may be times when the breast is especially vulnerable to the effects of mutagens. If these mutations occur in genes that participate in tumor development, such exposures can be transforming and contribute to the accumulation of the necessary alterations for multistage tumorigenesis.

Estrogen, in the form of 17β-estradiol, is generally recognized as potentially carcinogenic through its promotion of cell proliferation via interactions with estrogen receptors (see below), and it may also be linked to mutagenesis through its metabolites (Yager and Davidson, 2006). In particular, a metabolite of 17β-estradiol (2,4-dihydroxy-17b-estradiol) has been found to be DNA-reactive (Cavalieri et al., 2006; Bolton and Thatcher, 2008). It is plausible for this estradiol metabolite to be produced in breast tissue, where it could contribute to the formation of DNA adducts prone to lead to changes in the gene sequence (Belous et al., 2007).

Although most studies focused on the potential mutagenic role of estradiol have been conducted in animal models, or with human tissues in vitro, there is some evidence that such estradiol-DNA adducts can occur in humans. As discussed in the International Agency for Research on Cancer (IARC) monograph update on the carcinogenicity of estrogen-related pharmaceuticals, stable DNA adducts were found in a small set of breast tissue samples from women with and without breast cancer who had used an estrogen–progestin form of hormone therapy (IARC, 2011).

Nuclear Hormone Receptor Signaling

The steroid hormones, which include estrogens and progestogens, are involved in the development and maintenance of female reproductive characteristics, via their physiological effects on a broad range of tissues (Björnström and Sjöberg, 2005). Like other hormones, the steroid

---

[1]A target cell or target tissue is the site at which an agent acts.

hormones function as signaling molecules by traveling through the blood stream to interact with cells in a variety of target tissues to affect cellular behavior. A principal point of interaction is through binding with receptors that are located within cells and referred to as nuclear hormone receptors. The receptors for the steroid hormones are part of a larger family of nuclear hormone receptors (Aranda and Pascual, 2001).

Physiologically, estrogen usually occurs in the form of estradiol, 17β-estradiol being the dominant and most potent form (Björnström and Sjöberg, 2005), or estrone. Estradiol and estrone are produced by the ovaries, and estrone can also be produced by peripheral tissues such as adipose tissue and the adrenal gland. Estrogen is especially important in the functioning of female reproductive tissues such as the uterus and breast. Estrogens are able to act on target tissues by binding to estrogen receptors. Receptors act by recognizing a molecule's structure, thereby enabling only estrogen, or molecules that closely resemble estrogen in structure, to interact with receptors and promote receptor-mediated behavior.[2]

The fact that most breast cancers express estrogen receptor(s), coupled with estrogen's potential to stimulate cell proliferation (mitogenesis), has led to the hypothesis that estrogen receptor signaling can ultimately increase susceptibility to breast cancer (Weinberg, 2007). In its classical signal transduction pathway, once estrogen is bound to its receptor, it is able to regulate gene expression, which, in turn, alters cellular function. Estrogen's biological effects are regulated by two types of estrogen receptors, ERα and ERβ (Björnström and Sjöberg, 2005). These receptors act as ligand-activated transcription factors (Björnström and Sjöberg, 2005); when no ligand (in this case, estradiol or estrone) is bound to the receptor, the receptor's ability to transactivate gene expression is greatly attenuated, although these receptors are capable of ligand-independent activity. Once receptors are activated ("turned on"), they bind to specific DNA response elements known as estrogen response elements (EREs) of target genes (Björnström and Sjöberg, 2005). Estrogen bound to its receptor also induces a change in the receptor's structure that allows it to recruit coactivator proteins (NCI, 2010). By binding to the EREs and recruiting coactivator proteins, estrogen is able to influence the transcription of messenger RNA, which allows for the production of specific proteins that can, in turn, influence cellular behavior (NCI, 2010).

One such change in cellular behavior is increased cellular proliferation,

---

[2]Receptor-mediated cellular behavior is of particular importance when considering the effects of certain environmental exposures. As discussed later in this chapter, compounds that closely resemble hormones, such as EDCs, are able to activate or block receptor function. These compounds have the potential to subsequently alter the physiological status of the whole organism, especially if it is exposed during developmental stages (Caserta et al., 2008).

a result of estrogen's mitogenic actions. Growth promotion can increase susceptibility to breast cancer in several ways: increasing the population of cells that are targets for transformation, increasing the potential for mutagenesis by shortening the cell cycle and decreasing time available for DNA repair; and selectively promoting the growth of preneoplastic and neoplastic cells of nascent tumors. As a result of the ability of 17β-estradiol and several other endogenous estrogens to promote tumor growth in the breast, estrogen is classified as a known human carcinogen by IARC (2011).

After menopause, ovarian production of 17β-estradiol diminishes, affording some protection against development of hormone-dependent tumors. Relatively little is known about how the effects of other mitogens are modulated during the life course to increase or decrease their carcinogenic potential. Some existing data suggest that promotion of growth by mitogens early in life may modulate breast cancer risk later in life.

Some evidence also indicates that progesterone has a mitogenic role in breast carcinogenesis (Bernstein, 2002; Kariagina et al., 2010). Recent studies in rats have shown that in both the normal mammary gland and mammary tumors, proliferation is stimulated more by a combination of estrogen and progesterone than by treatment with estrogen alone (Kariagina et al., 2010). This proliferation is mediated by amphiregulin, an epidermal growth factor receptor (EGFR) ligand, and intracellular signaling pathways downstream of EGFR that induce proliferation. Recently, these pathways have also received interest as potential targets for hormone-dependent cancer therapy (Kariagina et al., 2010).

Ligand-activated estrogen receptors can also affect cell growth via what is termed nongenomic signaling, which involves estrogen receptor activation of phosphoinositide 3-kinase (PI3K) and mitogen-activated protein kinase (MAPK) and other cell signaling pathways in the cytoplasm rather than the nucleus. Research in animal models indicates that during development of the mammary gland, inappropriate activation of genomic or non-genomic ER signaling by estrogens can cause changes in the developing mammary gland (see Box 5-2). For example, a study examined the effects of a single subcutaneous injection (doses of 0.0125, 0.125, 12.5, 25, or 50 µg) of DES or tamoxifen given to female BALB/c mice within 36 hours of birth (Hovey et al., 2005). At 33 days, mice with DES exposure at 12.5 µg had greater growth of the mammary ducts compared with controls. At maturity, all the mice that had received 12.5 µg of DES and remained nulliparous showed abnormalities in the ducts and alveoli of the mammary gland compared to controls. In contrast, exposure to 25 µg of tamoxifen, a selective estrogen receptor modulator, resulted in reduced ductal outgrowth relative to controls at 33 days. At 12 weeks, mice with exposure to 12.5 µg to 50 µg of tamoxifen had regressed and atrophic ducts (Hovey et al. 2005). Work by Couse et al. (2001) indicates that DES-induced developmental reprogram-

BOX 5-2
Nongenomic Estrogen Receptor Signaling
by Environmental Estrogens

The classical effects of steroid hormones are mediated by nuclear hormone receptors functioning as ligand-activated transcription factors. However, evidence now indicates that steroid hormones manifest effects by other means as well (Castoria et al., 1999; Cato et al., 2002; Losel and Wehling, 2003; Cheskis, 2004; Björnström and Sjöberg, 2005; Edwards, 2005; Levin, 2005). These other effects have been termed "nongenomic" to distinguish them from the direct, or genomic, effects of the transcription factors in the nucleus. The nongenomic effects of steroid hormones involve a subpopulation of classical receptors that associate with signaling complexes in either the cytoplasm or the plasma cell membrane (Weinberg, 2007).

Estrogens stimulate mitosis in ER-positive (ER+) breast cells, yet the precise mechanism by which estrogen is able to drive the proliferation of ER+ breast cancer cells remains unknown (Weinberg, 2007). When estrogen is added to ER+ breast cancer cells, a rapid signaling cascade ensues unrelated to ER action at a nuclear receptor (Weinberg, 2007). Genomic receptor-mediated physiological responses, which may take up to several hours, cannot explain the rapid physiological changes that occur within seconds or minutes following the administration of estrogen (Hewitt et al., 2005).

Various cell or plasma-membrane-associated-receptor signal transduction pathways have been studied to understand the non-genomic path that estrogen may take. These pathways include association with and activation of insulin-like growth factor receptors (IGF-1Rs) (Kahlert et al., 2000) or direct association with Src and Shc proteins (Migliaccio et al., 1996, 2000; Castoria et al., 2001; Kousteni et al., 2001; Song et al., 2002a,b, 2004, 2005; Wong et al., 2002), which are known to be involved in the activation of PI3K and MAPK signaling. PI3K signaling is involved in many cellular processes, including survival, proliferation, growth, and motility (Vivanco and Sawyers, 2002; Sulis and Parsons, 2003). Association of estrogen with G-protein coupled receptors such as GPCR30, which is expressed in the membranes of breast cancer cells, has also been observed (Hewitt et al., 2005; Revankar et al., 2005; Thomas et al., 2005). With further study this may explain some of estrogen's mitogenic effects (Hewitt et al., 2005). Interpreting the biological role of estrogen as a mitogen is challenging, given the myriad of proposed signal transduction pathways, and researchers urge further in vivo studies to assess the effects in mammary tissue (Hewitt et al., 2005).

ming of the murine reproductive tract requires the ERα, suggesting that this receptor has a crucial role in mediating the imprint. Further investigation is needed to assess the impact of morphologic changes in the developing mammary gland on the risk for mammary cancer later in life.

Epigenetic Reprogramming

During gestation, and in some cases postnatally, growth and differentiation of cells and tissues occurs via a well-orchestrated series of developmental processes directed by the genetic, epigenetic, and environmental milieu of the developing organism. The genetic contribution is shaped by the individual's DNA sequence. Epigenetic contributions derive from processes that modify gene expression, but do not alter the DNA sequence. Feinberg and Tycko (2004) have described three principal epigenetic mechanisms: cytosine methylation at CpG sites in the DNA (termed "DNA methylation"), genomic imprinting (silencing of an allele based on whether it was of maternal or paternal origin), and histone modifications (affecting the proteins around which DNA strands are wound).

Epigenetic methyl "marks," such as DNA methylation and histone methylation, are transmitted to daughter cells and are thus epigenetically "inherited" from the parent generation of cells (Feinberg and Tycko, 2004). Epigenetic methyl marks are the result of the addition of a methyl ($-CH_3$) group to either cytosine nucleotides in the DNA or lysine or arginine residues of histones. Methylation of cytosine residues in DNA (hypermethylation) regulates gene expression largely via gene silencing, and hypomethylation of genes is permissive for gene expression. Methylation of arginine and lysine residues of histones creates binding sites for several regulators of gene expression that recognize these site-specific marks.

During gestation, epigenomic programs that are installed in cells direct patterns of differentiation crucial for tissue and organ specification. This programming also determines how cells and tissues will respond to physiological and environmental signals throughout the life of the organism. The "developmental reprogramming" hypothesis proposes that at critical times during development, exposure of developing tissues to an adverse stimulus can result in permanent epigenetic reprogramming of normal physiological responses, and so alter risk for metabolic and hormonal disorders later in life (Frankel et al., 1996; Hattersley and Tooke, 1999; Barker, 2002; Couzin, 2002). Studies in animal models have demonstrated that perinatal exposure to xenoestrogens can reprogram the development of the mammary gland and reproductive tract, causing alterations in tissue morphology and gene expression (Muñoz-de-Toro et al., 2005; Durando et al., 2007; Murray et al., 2007; Vandenberg et al., 2008; Rudel et al., 2011).

Methylation of DNA and of histones are both thought to be targets for developmental reprogramming. A direct link between environmental exposures and changes in DNA methylation is lacking, but a direct mechanism whereby environmental exposures can interact with the epigenetic machinery and perturb histone methylation during development has been demonstrated (Bredfeldt et al., 2010; Doherty et al., 2010). Environmental estrogens such as DES can inappropriately activate nongenomic (or more aptly "pregenomic") ER signaling to perturb patterns of epigenetic histone methyl marks being installed during development. For example, DES activation of pregenomic signaling can initiate a process that lowers levels of histone H3 methylation at lysine (K) 27 (Bredfeldt et al., 2010). Reduced levels of histone H3K27 methylation, a repressive mark for gene expression, can result in increased expression of estrogen-responsive genes in response to normal physiological levels of this hormone, driving development of hormone-dependent tumors.

Although epigenetic alterations are heritable and stably maintained as cells replicate, the alterations are also potentially reversible. This has been done pharmacologically with the use of 5-azacytadine and valproic acid to modulate epigenetic methyl marks as part of cancer treatment (Yoo and Jones, 2006). It has also been found in studies in mice that maternal exposure to bisphenol A (BPA) can result in DNA hypomethylation in offspring, but that maternal folate and genistein supplementation blocked the BPA-induced DNA hypomethylation (Dolinoy et al., 2007). Elucidating the role of environmental agents in inducing epigenetic alterations and the potential contribution of these alterations in the development of breast cancer will broaden understanding of the mechanisms of action of these agents, and potentially open avenues to reverse or prevent their adverse health effects.

Modulation of Immune Function

"Immunoediting" is a term used to describe the process by which the immune system influences cancer development and progression (Bui and Schreiber, 2007). The immune system functions to eliminate cancer cells by immunosurveillance, that is, controlling the growth of transformed cells by mechanisms of adaptive immunity. However, cancer cells may escape detection by the body's immune cells (immune escape), and thus evade being killed by the immune system.

Environmental exposures may modulate the effectiveness of the immune system in detecting or eliminating cancer cells. For example, the environmental estrogen genistein, present at levels associated with soy consumption, has been shown in studies using human breast cancer cell lines grown in mice to block the ability of cytotoxic T-cells and natural killer cells to recognize and destroy breast cancer cells during immunosurveillance (Jiang et al., 2008). Chemical immunotoxicity is an active area of research, with investigation of effects from lead, cigarette smoke, endocrine disruptors, and ambient air pollution, among others (Svensson et al., 1994; Weisglas-Kuperus et al., 2000; Bunn et al., 2001; Hertz-Picciotto et al., 2005; Ng et al., 2006; Park et al., 2008a). Conversely, pregnancy can increase the expression of immunosurveillance genes in breast epithelial cells, suggesting that one of the protective effects of pregnancy may be to enhance elimination of nascent tumor cells by immunosurveillance (Balogh et al., 2007). More needs to be learned about how environmental factors influence immunoediting and how modulation of immune function by environmental exposures over the life course influences breast cancer risk.

The Microenvironment Model of Carcinogenesis

In the multistage model of carcinogenesis, cancer progression results from the stimulation of abnormal growth of an initiated cell with altered genetic material. However, mutations that may initiate carcinogenesis can occur frequently in the body and do not necessarily result in cancer. Rodent studies have demonstrated that the number of initiated cells far exceeds the number of tumors that develop in vivo (as reviewed in Barcellos-Hoff and Ravani, 2000). The multistage model of carcinogenesis alone does not fully explain the complex series of transitions from mutation to a cancerous state. It is now appreciated that in addition to the genetic and epigenetic alterations that occur within transformed cells, the process of carcinogenesis and tumor progression includes alterations of the dialog between cells and their surroundings (Bissell and Hines, 2011).

Cells do not exist as isolated units, but are part of a complex environment of various cell types and tissues, as well as extracellular space. The extracellular space is filled with a network of macromolecules called the extracellular matrix (ECM), which serves as a structural anchor for cells. Once viewed as an inert scaffold, it is now accepted that the ECM is dynamic and influences cellular processes through initiation of intracellular signal transduction (Divoux and Clement, 2011). A sheet-like deposition of ECM, called the basement membrane, separates epithelial cells from supportive connective tissue cells called the stroma. The stroma consists of fibroblasts, immune cells, fat cells, and blood vessel cells (Mueller and Fusenig, 2004). Because all cells in the body have the same genetic code, molecular signals are necessary to direct the cells destined to become an eye to become an eye, and the nose to become a nose, and so on.

The stroma is responsible for some of the regulatory signals necessary to determine proper organ development and function in an embryo, and later to maintain homeostasis in an adult organism (Shekhar et al., 2003). Occasionally, such regulatory signals are disturbed, resulting in an alteration of normal external context for cells. Usually the effects are transient, but sustained inflammation may result in an up-regulation of enzymes such as matrix metalloproteinases (Bissell and Radisky, 2001). These enzymes normally promote healing, but their prolonged presence can contribute to the degradation of the ECM and allow for invading immune cells to over-produce factors that may lead to unregulated cellular proliferation. In this way, it is thought that the cellular environment, including the extracellular matrix and stroma, may play an important role in progression from nor-malcy to neoplasia. It has been proposed that disruption of the stromal–cell interactions in early stages of carcinogenesis can "provide the stimulus for initiated cells to move further down the neoplastic pathway" (Barcellos-Hoff and Ravani, 2000, p. 1254).

Microenvironments are specialized systems, composed of the ECM and soluble growth factors. They mediate epithelial–stromal interactions and thereby play an important role in normal growth and tissue development (Barcellos-Hoff and Ravani, 2000). The microenvironment theory comple-ments the multistage model of carcinogenesis, with the microenvironment and stroma in the breast interacting with transformed epithelial cells to constrain or permit invasive tumor growth (Shekhar et al., 2003; Bissell, 2007). Based on this model, the destruction of tissue itself could be a car-cinogenic event; an event such as wounding or tissue damage could trigger an inflammatory cascade leading to unregulated proliferation.

Alternatively, exposure to a carcinogen may disrupt the interactions between a cell and its external environment in a manner that permits transformation. Ionizing radiation has carcinogenic potential in both the breast and the rodent mammary gland, and it has been demonstrated to elicit rapid remodeling of the mammary gland ECM (Barcellos-Hoff, 1993). When genomically unstable cells harboring mutations in both alleles of p53 (COMMA-D cells) were transplanted into epithelial-free mammary stroma in irradiated host mice, tumors arose more quickly and were able to grow larger than in unirradiated hosts (Barcellos-Hoff and Ravani, 2000).

Since carcinogenic environments are not necessarily mutagenic or mito-genic, it is likely that changes in cell–cell contact or cell–ECM interactions promote a malignant phenotype (Barcellos-Hoff and Ravani, 2000). Stud-ies in a mouse mammary chimera model showed that when an irradiated host received oncogenic tissue, the development of aggressive tumors was accelerated and the molecular signatures of the tumors differed from those arising in nonirradiated hosts (Nguyen et al., 2011). Maffini et al. (2004) also demonstrated the stroma to be a critical target for carcinogenesis using a rat mammary tissue recombination model and the chemical carcinogen N-methyl-N-nitrosourea (MNU). They found that mammary epithelial cells were neoplastically transformed only when the stroma was exposed to MNU, regardless of MNU exposure of the epithelial cells.

The microenvironment model is especially relevant to breast cancer because of both the nature of the disease and the tissue structure of the organ. Most breast cancers originate from the epithelial cells of the terminal lobular ducts, and stroma accounts for more than 80 percent of breast volume (Shekhar et al., 2003). Furthermore, the breast is an organ that undergoes substantial development after birth and is a dynamic tissue that continually undergoes changes throughout a woman's lifetime, most notably during puberty, pregnancy, and menopause (McCready et al., 2010). Changes in the stromal and hormonal environments are part of these age- and event-related processes.

One reflection of stromal differences is mammographic density. Breasts with a greater proportion of fat produce a darker radiographic image that is described as less dense, and breasts with greater amounts of connective and epithelial tissue produce lighter images that are considered more dense (Boyd et al., 2010). Greater breast density is strongly associated with a higher risk of breast cancer (reviewed in Boyd et al., 2010). Women with higher mammographic density have been shown to have an altered ECM composition, including an increase in fibrillar collagen (McCready et al., 2010). As discussed above, it is hypothesized that changes to ECM architecture may play an early role in the path of tumorigenesis (e.g., McCready et al., 2010) or may reflect the effects of processes promoting tumorigenesis (e.g., Martin and Boyd, 2008). Further research is required to elucidate the mechanisms underlying the association between mammographic density and breast cancer risk, including the influence of environmental factors in altering breast tissue composition.

POSSIBLE BIOLOGIC MECHANISMS FOR ALTERATIONS IN BREAST CANCER RISK ASSOCIATED WITH OBESITY

As summarized in Chapter 3, obesity or excess body fatness has become a major public health issue in the United States. Obesity is associated with a myriad of adverse health outcomes, including many forms of cancer (ACS, 2011).[3] The picture for breast cancer is mixed. Obesity is associated with a lower risk for premenopausal breast cancer, while it is associated with an increased risk for postmenopausal breast cancer (Carmichael and Bates, 2004; WCRF/AICR, 2007). Pooled analysis of cohort and case–control studies suggests that greater abdominal fatness (measured by waist circumference or waist-to-hip ratio) is a "probable cause" of breast cancer diagnosed in postmenopausal women (WCRF/AICR, 2007, 2010).

---

[3]Overweight and obesity are commonly defined on the basis of body mass index (BMI), which is an approximate measure of body fat based on height and weight. BMI is calculated as body weight (kilograms) divided by height (meters) squared. The following BMI categories are used: Underweight, <18.5; Normal weight, 18.5–24.9; Overweight, 25.0–29.0; and Obese, ≥30.0.

Being obese is associated with a poorer prognosis and higher mortality rates for women diagnosed with breast cancer (Calle et al., 2003; Carmichael, 2006; Protani et al., 2010). The American Cancer Society's Cancer Prevention Study II, a prospective cohort study that included 2,852 breast cancer deaths, indicated that breast cancer mortality rates increased continually with increasing BMI at entry into the study. For women with the highest BMI (≥40) compared to women with normal weight (BMI 18.5–20.49), the relative risk (RR) of death from breast cancer was 3.08 (95% CI, 2.09–4.51) (Petrelli et al., 2002).

Overweight and obesity have been found to be associated with diagnosis of advanced-stage disease; however, the experience of African American women differs from that of white women (Hunter et al., 1993; Jones et al., 1997). Jones et al. (1997), in a retrospective study in Connecticut, found that a greater prevalence of severe obesity (BMI ≥ 32.3) among black women could account for approximately 33 percent of their excess risk of diagnosis of later-stage cancer compared to white women. In the National Cancer Institute survival study by Hunter et al. (1993), no differences between blacks and whites were observed.

Obesity probably affects breast cancer risk through several different and overlapping pathways (Fletcher et al., 2005; Slattery et al., 2007; WCRF/AICR, 2007; Cleary and Grossmann, 2009). Obesity-associated carcinogenesis has been explained by three main candidate mechanisms involving (1) insulin and insulin-like growth factor axis, (2) steroid hormones, and (3) circulating levels of cytokines and adipokines (Roberts et al., 2010).

Insulin plays an integral role in short-term metabolism, signaling muscle, liver, and adipose tissues to convert the glucose in the bloodstream into glycogen for storage. Circulating insulin levels have been shown to correlate positively with an increasing BMI (Roberts et al., 2010). Insulin-like growth factors (IGFs) mediate metabolism in the longer term and affect cell growth. Insulin contributes to inhibiting IGF by reducing hepatic secretion of IGF binding proteins (IGFBPs) that bind IGFs with high affinity (Giovanucci, 2001).

Obese individuals are often insulin resistant, a state of reduced responsiveness of muscle, liver, and adipose tissues to insulin. As a result, obese individuals are likely to have higher blood glucose and circulating insulin levels. The insulin resistance–cancer hypothesis is based on hyperinsulinemia, or elevated insulin levels, reducing production of IGFBPs and leaving free "bioactive" IGFs, which are then able to promote cell proliferation (Roberts et al., 2010).

Insulin itself has also been shown to suppress apoptosis (programmed cell death) and contribute to cell proliferation by acting as a mitogen for breast epithelial cells via insulin and IGF-1 receptors (Ish-Shalom et al., 1997; Chappell et al., 2001). In animal models, administration of insulin promotes mammary tumor growth (Heuson et al., 1972; Shafie and Grantham, 1981; Shafie and Hilf, 1981). Insulin also has the ability to affect sex hormone levels (Poretsky and Kalin, 1987) and to lower levels of SHBG, which binds sex hormones and renders them temporarily biologically "inactive" (Pugeat et al., 1991).

Several clinical trials are under way to evaluate potential benefits of the common diabetes drug metformin in treatment of women who have breast cancer (NIH, 2010). Early studies also suggested that metformin, but not other diabetes drugs, decreased the incidence of breast and other cancers in diabetics (Bodmer et al., 2010; Zakikhani et al., 2010). The mechanism of the protective effect of metformin is thought to be mediated through the action of this drug on AMP-activated protein kinase (AMPK), an enzyme that acts as a calorie restriction mimetic and that inhibits mitogenic signaling via the PI3K pathway in tumor cells (Zakikhani et al., 2006; Wysocki and Wierusz-Wysocka, 2010), as well as perhaps from direct reduction in levels of HER2 protein levels through inhibition of specific kinase activity (Vazquez-Martin et al., 2009).

Obesity may also be linked with carcinogenesis through its impact on estrogen and other steroid hormones and circulating growth factors. Adipose tissue is an important source of the estrogen estrone in postmenopausal women, and production of estrone by aromatization of androstenedione in the breast is thought to be one of the major factors in obesity's association with increased risk of breast cancer in postmenopausal women. Estrogens such as estrone can induce IGF-1 and increase cellular proliferation via IGF-1 (Suenson et al., 1984; Stewart et al., 1990; Owens et al., 1993; Ruan et al., 1995). As noted, the IGFs are key mitogens in regulating cell proliferation and differentiation, and they are anti-apoptotic (Pollak, 1998; Lee et al., 1999; Yu and Rohan, 2000). Studies have produced conflicting results regarding the association between higher IGF-1 levels and increased breast cancer risk (Yu and Rohan, 2000; Renehan et al., 2004; Shi et al., 2004; Sugumar et al., 2004; Fletcher et al., 2005). A collaborative group formed to carry out pooled analyses of individual data from prospective studies explored this relationship (Endogenous Hormones and Breast Cancer Collaborative Group, 2010). The group's analysis of data from 17 studies showed that IGF-1 concentrations, adjusted for age, were higher in moderately overweight women than in other women. A positive association was found between circulating IGF-1 and breast cancer risk; the odds ratio for breast cancer for women in the highest versus the lowest fifth of IGF-1 concentration was 1.28 (95% CI, 1.14–1.44). The association was not influenced by menopausal status at the time blood samples were collected or by adjustment for other risk factors. In addition, it seemed to be limited to estrogen receptor–positive tumors (Endogenous Hormones and Breast Cancer Collaborative Group, 2010).

Another aspect of obesity that may relate to carcinogenesis is its induction of a chronic state of low-grade inflammation. Inflammation promotes an increase in cell proliferation and differentiation, inhibition of apoptosis, generation of new blood vessels, and induction of epigenetic events (Federico et al., 2007; WCRF/AICR, 2007). Obesity is characterized by an increased production of inflammatory factors by adipocytes, or fat cells; depending on the degree of obesity, up to 40 percent of fat tissue may be composed of macrophages, a type of white blood cell that plays a role in adaptive immunity and acts as a chemoattractant (WCRF/AICR, 2007, p. 39). These macrophages in turn recruit proinflammatory cells such as monocytes (McCready et al., 2010). Compared with lean people, obese individuals also have elevated concentrations of circulating leptin, which can function as an inflammatory cytokine (Zhang et al., 2002); tumor necrosis factor (TNF)-alpha (Jarvinen et al., 2001); interleukin (IL)-6; and C-reactive protein (Lunn et al., 1997). Human studies of obesity, inflammation, and breast cancer are limited (Kundu and Surh, 2008; Aggarwal and Gehlot, 2009), but studies in animal models suggest a connection between chronic inflammation and mammary tumors (McCready et al., 2010).

The mechanisms proposed to explain obesity-related carcinogenesis have many notable shortcomings, including, but not limited to, a focus on endocrine mechanisms that may ignore the role of paracrine signaling; complex isoforms of insulin, and IGF receptors that may alter their roles in carcinogenesis; and some inconsistency between animal and epidemiologic evidence (e.g., IGF-1 increases with increasing fatness in mice but declines in humans above a BMI of about 27) (Roberts et al., 2010). Other proposed mechanisms of obesity-induced carcinogenesis include obesity-related hypoxia, migrating stromal cells, and shared genetic susceptibility (Roberts et al., 2010). Adipose tissue hypoxia (ATH), or reduced levels of oxygen in adipose tissue, is seen in the white adipose tissue of obese mice in comparison to lean mice (Ye et al., 2007). ATH also plays a role in increased insulin resistance and chronic inflammation, potentially leading to an increased cancer risk (Roberts et al., 2010).

Migrating stromal cells have also been proposed to play a role in obesity-induced carcinogenesis. Studies have shown that adipose stromal cells from fat deposits in obese mice can promote tumor vascularization by migrating to the tumor site upon recruitment (Zhang et al., 2009). Collagen, a component of connective tissue, is also up-regulated in the adipose tissue of mice during the progression of mammary tumors, due to promotion from adipocyte-derived factors (Iyengar et al., 2003).

Shared genetic susceptibility and "obesity genetics" are a third novel area of research, in which genetic maps are used to explore overlapping "mutual candidate genes" for both obesity and cancer. Such approaches have found promising patterns of polymorphisms for diseases such as colorectal cancers (Roberts et al., 2010).

A NEED TO CONSIDER TIMING OF EXPOSURE ACROSS THE LIFE COURSE

Thus, substantial evidence supports varying effects of environmental exposures at different stages along the life course. As science in this field has matured, it now seems clear that to examine the effect of exposure to an environmental chemical or other factor without consideration for the stage in life when that factor may be acting is shortsighted. The currently modest amount of evidence on the impact of environmental factors on breast cancer may actually stem from a lack of clear, biologically based hypotheses about the relevant timing of exposures and implementation of designs and protocols that address such hypotheses. Although studies that take into account the timing of exposure will be challenging to conduct, the committee strongly recommends that both epidemiologic and laboratory investigations of breast cancer risk factors specify the stage of the life course being studied (see Chapter 7).

REFERENCES

ACS (American Cancer Society). 2011. *Cancer facts and figures 2011*. http://www.cancer. org/Research/CancerFactsFigures/CancerFactsFigures/cancer-facts-figures-2011 (accessed June 22, 2011).

Aggarwal, B. B., and P. Gehlot. 2009. Inflammation and cancer: How friendly is the relationship for cancer patients? *Curr Opin Pharmacol* 9(4):351–369.

Aksglaede, L., K. Sorensen, J. H. Petersen, N. E. Skakkebaek, and A. Juul. 2009. Recent decline in age at breast development: The Copenhagen Puberty Study. *Pediatrics* 123(5): e932–e939.

Anbazhagan, R., J. Bartek, P. Monaghan, and B. A. Gusterson. 1991. Growth and development of the human infant breast. *Am J Anat* 192(4):407–417.

Anbazhagan, R., P. P. Osin, J. Bartkova, B. Nathan, E. B. Lane, and B. A. Gusterson. 1998. The development of epithelial phenotypes in the human fetal and infant breast. *J Pathol* 184(2):197–206.

Aranda, A., and A. Pascual. 2001. Nuclear hormone receptors and gene expression. *Physiol Rev* 81(3):1269–1304.

Balogh, G. A., I. H. Russo, C. Spittle, R. Heulings, and J. Russo. 2007. Immune-surveillance and programmed cell death-related genes are significantly overexpressed in the normal breast epithelium of postmenopausal parous women. *Int J Oncol* 31(2):303–312.

Barcellos-Hoff, M. H. 1993. Radiation-induced transforming growth factor β and subsequent extracellular matrix reorganization in murine mammary gland. *Cancer Res* 53(17):3880–3886.

Barcellos-Hoff, M. H., and S. A. Ravani. 2000. Irradiated mammary gland stroma promotes the expression of tumorigenic potential by unirradiated epithelial cells. *Cancer Res* 60(5):1254–1260.

Barker, D. J. 2002. Fetal programming of coronary heart disease. *Trends Endocrinol Metab* 13(9):364–368.

Belous, A. R., D. L. Hachey, S. Dawling, N. Roodi, and F. F. Parl. 2007. Cytochrome P450 1B1-mediated estrogen metabolism results in estrogen-deoxyribonucleoside adduct formation. *Cancer Res* 67(2):812–817.

Beral, V., G. Reeves, D. Bull, and J. Green. 2011. Breast cancer risk in relation to the interval between menopause and starting hormone therapy. *J Natl Cancer Inst* 103(4):296–305.

Bernstein, L. 2002. Epidemiology of endocrine-related risk factors for breast cancer. *J Mammary Gland Biol Neoplasia* 7(1):3–15.

Biro, F. M., A. W. Lucky, L. A. Simbartl, B. A. Barton, S. R. Daniels, R. Striegel-Moore, S. S. Kronsberg, and J. A. Morrison. 2003. Pubertal maturation in girls and the relationship to anthropometric changes: Pathways through puberty. *J Pediatr* 142(6):643–646.

Biro, F. M., P. Khoury, and J. A. Morrison. 2006. Influence of obesity on timing of puberty. *Int J Androl* 29(1):272–277; discussion 286–290.

Biro, F. M., M. P. Galvez, L. C. Greenspan, P. A. Succop, N. Vangeepuram, S. M. Pinney, S. Teitelbaum, G. C. Windham, et al. 2010. Pubertal assessment method and baseline characteristics in a mixed longitudinal study of girls. *Pediatrics* 126(3):e583–e590.

Bissell, M. J. 2007. Modelling molecular mechanisms of breast cancer and invasion: Lessons from the normal gland. *Biochem Soc Trans* 35(Pt 1):18–22.

Bissell, M. J., and W. C. Hines. 2011. Why don't we get more cancer? A proposed role of the microenvironment in restraining cancer progression. *Nat Med* 17(3):320–329.

Bissell, M. J., and D. Radisky. 2001. Putting tumours in context. *Nat Rev Cancer* 1(1):46–54.

Björnström, L., and M. Sjöberg. 2005. Mechanisms of estrogen receptor signaling: Convergence of genomic and nongenomic actions on target genes. *Mol Endocrinol* 19(4):833–842.

Block, K., A. Kardana, P. Igarashi, and H. S. Taylor. 2000. In utero diethylstilbestrol (DES) exposure alters Hox gene expression in the developing mullerian system. *FASEB J* 14(9):1101–1108.

Bodmer, M., C. Meier, S. Krahenbuhl, S. S. Jick, and C. R. Meier. 2010. Long-term metformin use is associated with decreased risk of breast cancer. *Diabetes Care* 33(6):1304–1308.

Bogaert, A. F. 2005. Age at puberty and father absence in a national probability sample. *J Adolesc* 28(4):541–546.

Boice, J. D., Jr., D. Preston, F. G. Davis, and R. R. Monson. 1991. Frequent chest X-ray fluoroscopy and breast cancer incidence among tuberculosis patients in Massachusetts. *Radiat Res* 125(2):214–222.

Bolton, J. L., and G. R. Thatcher. 2008. Potential mechanisms of estrogen quinone carcinogenesis. *Chem Res Toxicol* 21(1):93–101.

Boyd, N. F., G. A. Lockwood, J. W. Byng, D. L. Tritchler, and M. J. Yaffe. 1998. Mammographic densities and breast cancer risk. *Cancer Epidemiol Biomarkers Prev* 7(12): 1133–1144.

Boyd, N. F., L. J. Martin, M. Bronskill, M. J. Yaffe, N. Duric, and S. Minkin. 2010. Breast tissue composition and susceptibility to breast cancer. *J Natl Cancer Inst* 102(16): 1224–1237.

Boylan, E. S., and R. E. Calhoon. 1979. Mammary tumorigenesis in the rat following prenatal exposure to diethylstilbestrol and postnatal treatment with 7,12-dimethylbenz[a]anthracene. *J Toxicol Environ Health* 5(6):1059–1071.

Boylan, E. S., and R. E. Calhoon. 1983. Transplacental action of diethylstilbestrol on mammary carcinogenesis in female rats given one or two doses of 7,12-dimethylbenz(a) anthracene. *Cancer Res* 43(10):4879–4884.

Bredfeldt, T. G., K. L. Greathouse, S. H. Safe, M. C. Hung, M. T. Bedford, and C. L. Walker. 2010. Xenoestrogen-induced regulation of EZH2 and histone methylation via estrogen receptor signaling to PI3K/AKT. *Mol Endocrinol* 24(5):993–1006.

Bui, J. D., and R. D. Schreiber. 2007. Cancer immunosurveillance, immunoediting and inflammation: Independent or interdependent processes? *Curr Opin Immunol* 19(2):203–208.

Bunn, T. L., P. J. Parsons, E. Kao, and R. R. Dietert. 2001. Exposure to lead during critical windows of embryonic development: Differential immunotoxic outcome based on stage of exposure and gender. *Toxicol Sci* 64(1):57–66.

Calle, E. E., C. Rodriguez, K. Walker-Thurmond, and M. J. Thun. 2003. Overweight, obesity, and mortality from cancer in a prospectively studied cohort of U.S. adults. *N Engl J Med* 348(17):1625–1638.

Carmichael, A. R. 2006. Obesity and prognosis of breast cancer. *Obes Rev* 7(4):333–340.

Carmichael, A. R., and T. Bates. 2004. Obesity and breast cancer: A review of the literature. *Breast* 13(2):85–92.

Caserta, D., L. Maranghi, A. Mantovani, R. Marci, F. Maranghi, and M. Moscarini. 2008. Impact of endocrine disruptor chemicals in gynaecology. *Hum Reprod Update* 14(1):59–72.

Castoria, G., M. V. Barone, M. Di Domenico, A. Bilancio, D. Ametrano, A. Migliaccio, and F. Auricchio. 1999. Non-transcriptional action of oestradiol and progestin triggers DNA synthesis. *EMBO J* 18(9):2500–2510.

Castoria, G., A. Migliaccio, A. Bilancio, M. Di Domenico, A. de Falco, M. Lombardi, R. Fiorentino, L. Varricchio, et al. 2001. PI3-kinase in concert with Src promotes the S-phase entry of oestradiol-stimulated MCF-7 cells. *EMBO J* 20(21):6050–6059.

Cato, A. C., A. Nestl, and S. Mink. 2002. Rapid actions of steroid receptors in cellular signaling pathways. *Sci STKE* 2002(138):re9.

Cavalieri, E., D. Chakravarti, J. Guttenplan, E. Hart, J. Ingle, R. Jankowiak, P. Muti, E. Rogan, et al. 2006. Catechol estrogen quinones as initiators of breast and other human cancers: Implications for biomarkers of susceptibility and cancer prevention. *Biochim Biophys Acta* 1766(1):63–78.

Chappell, J., J. W. Leitner, S. Solomon, I. Golovchenko, M. L. Goalstone, and B. Draznin. 2001. Effect of insulin on cell cycle progression in MCF-7 breast cancer cells: Direct and potentiating influence. *J Biol Chem* 276(41):38023–38028.

Cheskis, B. J. 2004. Regulation of cell signalling cascades by steroid hormones. *J Cell Biochem* 93(1):20–27.

Chlebowski, R. T., L. H. Kuller, R. L. Prentice, M. L. Stefanick, J. E. Manson, M. Gass, A. K. Aragaki, J. K. Ockene, et al. 2009. Breast cancer after use of estrogen plus progestin in postmenopausal women. *N Engl J Med* 360(6):573–587.

Chlebowski, R. T., G. L. Anderson, M. Gass, D. S. Lane, A. K. Aragaki, L. H. Kuller, J. E. Manson, M. L. Stefanick, et al. 2010. Estrogen plus progestin and breast cancer incidence and mortality in postmenopausal women. *JAMA* 304(15):1684–1692.

Clarke, C. A., S. L. Glaser, C. S. Uratsu, J. V. Selby, L. H. Kushi, and L. J. Herrinton. 2006. Recent declines in hormone therapy utilization and breast cancer incidence: Clinical and population-based evidence. *J Clin Oncol* 24(33):e49–e50.

Clavel-Chapelon, F. 2002. Evolution of age at menarche and at onset of regular cycling in a large cohort of French women. *Hum Reprod* 17(1):228–232.

Clavel-Chapelon, F., and M. Gerber. 2002. Reproductive factors and breast cancer risk. Do they differ according to age at diagnosis? *Breast Cancer Res Treat* 72(2):107–115.

Cleary, M. P., and M. E. Grossmann. 2009. Minireview: Obesity and breast cancer: The estrogen connection. *Endocrinology* 150(6):2537–2542.

Colditz, G. A., and A. L. Frazier. 1995. Models of breast cancer show that risk is set by events of early life: Prevention efforts must shift focus. *Cancer Epidemiol Biomarkers Prev* 4(5):567–571.

Collaborative Group on Hormonal Factors in Breast Cancer. 1997. Breast cancer and hormone replacement therapy: Collaborative reanalysis of data from 51 epidemiological studies of 52,705 women with breast cancer and 108,411 women without breast cancer. *Lancet* 350(9084):1047–1059.

Couse, J. F., D. Dixon, M. Yates, A. B. Moore, L. Ma, R. Maas, and K. S. Korach. 2001. Estrogen receptor-alpha knockout mice exhibit resistance to the developmental effects of neonatal diethylstilbestrol exposure on the female reproductive tract. *Dev Biol* 238(2):224–238.

Couzin, J. 2002. Quirks of fetal environment felt decades later. *Science* 296(5576):2167–2169.

Crain, D. A., S. J. Janssen, T. M. Edwards, J. Heindel, S. M. Ho, P. Hunt, T. Iguchi, A. Juul, et al. 2008. Female reproductive disorders: The roles of endocrine-disrupting compounds and developmental timing. *Fertil Steril* 90(4):911–940.

de Muinck Keizer-Schrama, S. M., and D. Mul. 2001. Trends in pubertal development in Europe. *Hum Reprod Update* 7(3):287–291.

de Waard, F., and J. H. Thijssen. 2005. Hormonal aspects in the causation of human breast cancer: Epidemiological hypotheses reviewed, with special reference to nutritional status and first pregnancy. *J Steroid Biochem Mol Biol* 97(5):451–458.

Deardorff, J., J. P. Ekwaru, L. H. Kushi, B. J. Ellis, L. C. Greenspan, A. Mirabedi, E. G. Landaverde, and R. A. Hiatt. 2011. Father absence, body mass index, and pubertal timing in girls: Differential effects by family income and ethnicity. *J Adolesc Health* 48(5):441–447.

DeRoo, L. A., P. Cummings, and B. A. Mueller. 2011. Smoking before the first pregnancy and the risk of breast cancer: A meta-analysis. *Am J Epidemiol* 174(4):390–402.

Divoux, A., and K. Clement. 2011. Architecture and the extracellular matrix: The still unappreciated components of the adipose tissue. *Obes Rev* 12(5):e494–e503.

Doherty, L. F., J. G. Bromer, Y. Zhou, T. S. Aldad, and H. S. Taylor. 2010. In utero exposure to diethylstilbestrol (DES) or bisphenol-A (BPA) increases EZH2 expression in the mammary gland: An epigenetic mechanism linking endocrine disruptors to breast cancer. *Horm Cancer* 1(3):146–155.

Dolinoy, D. C., D. Huang, and R. L. Jirtle. 2007. Maternal nutrient supplementation counteracts bisphenol A-induced DNA hypomethylation in early development. *Proc Natl Acad Sci U S A* 104(32):13056–13061.

dos Santos Silva, I., B. De Stavola, and V. McCormack. 2008. Birth size and breast cancer risk: Reanalysis of individual participant data from 32 studies. *PLoS Med* 5(9):e193.

Durando, M., L. Kass, J. Piva, C. Sonnenschein, A. M. Soto, E. H. Luque, and M. Muñoz-de-Toro. 2007. Prenatal bisphenol A exposure induces preneoplastic lesions in the mammary gland in Wistar rats. *Environ Health Perspect* 115(1):80–86.

Eden, J. A. 2010. Breast cancer, stem cells and sex hormones. Part 2: The impact of the reproductive years and pregnancy. *Maturitas* 67(3):215–218.

Eden, J. A. 2011. Breast cancer, stem cells and sex hormones. Part 3: The impact of the menopause and hormone replacement. *Maturitas* 68(2):129–136.

Edwards, D. P. 2005. Regulation of signal transduction pathways by estrogen and progesterone. *Annu Rev Physiol* 67:335–376.

Ellis, B. J. 2004. Timing of pubertal maturation in girls: An integrated life history approach. *Psychol Bull* 130(6):920–958.

Ellis, B. J., and J. Garber. 2000. Psychosocial antecedents of variation in girls' pubertal timing: Maternal depression, stepfather presence, and marital and family stress. *Child Dev* 71(2):485–501.

Endogenous Hormones and Breast Cancer Collaborative Group. 2010. Insulin-like growth factor 1 (IGF1), IGF binding protein 3 (IGFBP3), and breast cancer risk: Pooled individual data analysis of 17 prospective studies. *Lancet Oncol* 11(6):530–542.

Euling, S. Y., M. E. Herman-Giddens, P. A. Lee, S. G. Selevan, A. Juul, T. I. Sorensen, L. Dunkel, J. H. Himes, et al. 2008. Examination of U.S. puberty timing data from 1940 to 1994 for secular trends: Panel findings. *Pediatrics* 121(Suppl 3):S172–S191.

Federico, A., F. Morgillo, C. Tuccillo, F. Ciardiello, and C. Loguercio. 2007. Chronic inflammation and oxidative stress in human carcinogenesis. *Int J Cancer* 121(11):2381–2386.

Feinberg, A. P., and B. Tycko. 2004. The history of cancer epigenetics. *Nat Rev Cancer* 4(2):143–153.

Fletcher, O., L. Gibson, N. Johnson, D. R. Altmann, J. M. Holly, A. Ashworth, J. Peto, and S. Silva Idos. 2005. Polymorphisms and circulating levels in the insulin-like growth factor system and risk of breast cancer: A systematic review. *Cancer Epidemiol Biomarkers Prev* 14(1):2–19.

Frankel, S., P. Elwood, P. Sweetnam, J. Yarnell, and G. D. Smith. 1996. Birthweight, body-mass index in middle age, and incident coronary heart disease. *Lancet* 348(9040):1478–1480.

Garland, M., D. J. Hunter, G. A. Colditz, J. E. Manson, M. J. Stampfer, D. Spiegelman, F. Speizer, and W. C. Willett. 1998. Menstrual cycle characteristics and history of ovulatory infertility in relation to breast cancer risk in a large cohort of U.S. women. *Am J Epidemiol* 147(7):636–643.

Giovannucci, E. 2001. Insulin, insulin-like growth factors and colon cancer: A review of the evidence. *J Nutr* 131(11 Suppl):3109S–3120S.

Graber, J. A., J. Brooks-Gunn, and M. P. Warren. 1995. The antecedents of menarcheal age: Heredity, family environment, and stressful life events. *Child Dev* 66(2):346–359.

Graber, J. A., P. M. Lewinsohn, J. R. Seeley, and J. Brooks-Gunn. 1997. Is psychopathology associated with the timing of pubertal development? *J Am Acad Child Adolesc Psychiatry* 36(12):1768–1776.

Grumbach, M. M., and D. M. Styne. 2002. Puberty: Ontogeny, neuroendocrinology, physiology, and disorders. In *Williams textbook of endocrinology*, 10th ed. Edited by P. R. Larsen, H. M. Kronenberg, S. Melmed, and K. S. Polonsky. Philadelphia, PA: Saunders. Pp. 1115–1286.

Ha, M., K. Mabuchi, A. J. Sigurdson, D. M. Freedman, M. S. Linet, M. M. Doody, and M. Hauptmann. 2007. Smoking cigarettes before first childbirth and risk of breast cancer. *Am J Epidemiol* 166(1):55–61.

Hattersley, A. T., and J. E. Tooke. 1999. The fetal insulin hypothesis: An alternative explanation of the association of low birthweight with diabetes and vascular disease. *Lancet* 353(9166):1789–1792.

Hauspie, R. C., M. Vercauteren, and C. Susanne. 1997. Secular changes in growth and maturation: An update. *Acta Paediatr Suppl* 423:20–27.

Herbst, A. L., H. Ulfelder, and D. C. Poskanzer. 1971. Adenocarcinoma of the vagina. Association of maternal stilbestrol therapy with tumor appearance in young women. *N Engl J Med* 284(15):878–881.

Hertz-Picciotto, I., C. E. Herr, P. S. Yap, M. Dostal, R. H. Shumway, P. Ashwood, M. Lipsett, J. P. Joad, et al. 2005. Air pollution and lymphocyte phenotype proportions in cord blood. *Environ Health Perspect* 113(10):1391–1398.

Heuson, J. C., N. Legros, and R. Heimann. 1972. Influence of insulin administration on growth of the 7,12-dimethylbenz(a)anthracene-induced mammary carcinoma in intact, oophorectomized, and hypophysectomized rats. *Cancer Res* 32(2):233–238.

Hewitt, S. C., J. C. Harrell, and K. S. Korach. 2005. Lessons in estrogen biology from knockout and transgenic animals. *Annu Rev Physiol* 67:285–308.

Hilakivi-Clarke, L., J. E. Andrade, and W. Helferich. 2010. Is soy consumption good or bad for the breast? *J Nutr* 140(12):2326S–2334S.

Hovey, R. C., M. Asai-Sato, A. Warri, B. Terry-Koroma, N. Colyn, E. Ginsburg, and B. K. Vonderhaar. 2005. Effects of neonatal exposure to diethylstilbestrol, tamoxifen, and toremifene on the BALB/c mouse mammary gland. *Biol Reprod* 72(2):423–435.

Howard, B. A., and B. A. Gusterson. 2000. Human breast development. *J Mammary Gland Biol Neoplasia* 5(2):119–137.

Hsieh, C. C., D. Trichopoulos, K. Katsouyanni, and S. Yuasa. 1990. Age at menarche, age at menopause, height and obesity as risk factors for breast cancer: Associations and interactions in an international case–control study. *Int J Cancer* 46(5):796–800.

Hughes, E. S. 1950. The development of the mammary gland: Arris and Gale Lecture, delivered at the Royal College of Surgeons of England on 25 October, 1949. *Ann R Coll Surg Engl* 6(2):99–119.

Hunter, C. P., C. K. Redmond, V. W. Chen, D. F. Austin, R. S. Greenberg, P. Correa, H. B. Muss, M. R. Forman, et al. 1993. Breast cancer: Factors associated with stage at diagnosis in black and white women. Black/White Cancer Survival Study Group. *J Natl Cancer Inst* 85(14):1129–1137.

IARC (International Agency for Research on Cancer). 2011. *IARC monographs on the evaluation of carcinogenic risks to humans.* Vol. 100, Part A: Pharmaceuticals. Lyon, France: IARC.

Ish-Shalom, D., C. T. Christoffersen, P. Vorwerk, N. Sacerdoti-Sierra, R. M. Shymko, D. Naor, and P. De Meyts. 1997. Mitogenic properties of insulin and insulin analogues mediated by the insulin receptor. *Diabetologia* 40(Suppl 2):S25–S31.

Iyengar, P., T. P. Combs, S. J. Shah, V. Gouon-Evans, J. W. Pollard, C. Albanese, L. Flanagan, M. P. Tenniswood, et al. 2003. Adipocyte-secreted factors synergistically promote mammary tumorigenesis through induction of anti-apoptotic transcriptional programs and proto-oncogene stabilization. *Oncogene* 22(41):6408–6423.

Jarvinen, R., P. Knekt, T. Hakulinen, H. Rissanen, and M. Heliovaara. 2001. Dietary fat, cholesterol and colorectal cancer in a prospective study. *Br J Cancer* 85(3):357–361.

Jasik, C. B., and R. H. Lustig. 2008. Adolescent obesity and puberty: The "perfect storm." *Ann N Y Acad Sci* 1135:265–279.

Jenkins, S., A. M. Betancourt, J. Wang, and C. A. Lamartiniere. 2011. Endocrine-active chemicals in mammary cancer causation and prevention. *J Steroid Biochem Mol Biol* June 23. [Epub ahead of print]

Jiang, X., N. M. Patterson, Y. Ling, J. Xie, W. G. Helferich, and D. J. Shapiro. 2008. Low concentrations of the soy phytoestrogen genistein induce proteinase inhibitor 9 and block killing of breast cancer cells by immune cells. *Endocrinology* 149(11):5366–5373.

John, E. M., and J. L. Kelsey. 1993. Radiation and other environmental exposures and breast cancer. *Epidemiol Rev* 15(1):157–162.

Jones, B. A., S. V. Kasi, M. G. Curnen, P. H. Owens, and R. Dubrow. 1997. Severe obesity as an explanatory factor for the black/white difference in stage at diagnosis of breast cancer. *Am J Epidemiol* 146(5):394–404.

Jury, H. H., T. R. Zachzrewski, and G. L. Hammond. 2000. Interactions between human plasma sex hormone-binding globulin and xenobiotic ligands. *J Steroid Biochem Mol Biol* 75:167–176.

Kahlert, S., S. Nuedling, M. van Eickels, H. Vetter, R. Meyer, and C. Grohe. 2000. Estrogen receptor alpha rapidly activates the IGF-1 receptor pathway. *J Biol Chem* 275(24):18447–18453.

Kaplowitz, P. B. 2008. Link between body fat and the timing of puberty. *Pediatrics* 121(Suppl 3):S208–S217.

Kariagina, A., J. Xie, J. R. Leipprandt, and S. Z. Haslam. 2010. Amphiregulin mediates estrogen, progesterone, and EGFR signaling in the normal rat mammary gland and in hormone-dependent rat mammary cancers. *Horm Cancer* 1(5):229–244.

Kawaguchi, H., N. Miyoshi, Y. Miyamoto, M. Souda, Y. Umekita, N. Yasuda, and H. Yoshida. 2009a. Effects of fetal exposure to diethylstilbestrol on mammary tumorigenesis in rats. *J Vet Med Sci* 71(12):1599–1608.

Kawaguchi, H., Y. Umekita, M. Souda, K. Gejima, H. Kawashima, T. Yoshikawa, and H. Yoshida. 2009b. Effects of neonatally administered high-dose diethylstilbestrol on the induction of mammary tumors induced by 7,12-dimethylbenz[a]anthracene in female rats. *Vet Pathol* 46(1):142–150.

Kelsey, J. L., and L. Bernstein. 1996. Epidemiology and prevention of breast cancer. *Annu Rev Public Health* 17:47–67.

Knight, C. H., and A. Sorensen. 2001. Windows in early mammary development: Critical or not? *Reproduction* 122(3):337–345.

Kotsopoulos, J., W. Y. Chen, M. A. Gates, S. S. Tworoger, S. E. Hankinson, and B. A. Rosner. 2010. Risk factors for ductal and lobular breast cancer: Results from the Nurses' Health Study. *Breast Cancer Res* 12(6):R106.

Kousteni, S., T. Bellido, L. I. Plotkin, C. A. O'Brien, D. L. Bodenner, L. Han, K. Han, G. B. DiGregorio, et al. 2001. Nongenotropic, sex-nonspecific signaling through the estrogen or androgen receptors: Dissociation from transcriptional activity. *Cell* 104(5):719–730.

Kundu, J. K., and Y. J. Surh. 2008. Inflammation: Gearing the journey to cancer. *Mutat Res* 659(1–2):15–30.

LaCroix, A. Z., R. T. Chlebowski, J. E. Manson, A. K. Aragaki, K. C. Johnson, L. Martin, K. L. Margolis, M. L. Stefanick, et al. 2011. Health outcomes after stopping conjugated equine estrogens among postmenopausal women with prior hysterectomy: A randomized controlled trial. *JAMA* 305(13):1305–1314.

Lamartiniere, C. A., and M. B. Holland. 1992. Neonatal diethylstilbestrol prevents spontaneously developing mammary tumors. In *Hormonal carcinogenesis: Proceedings of the first international symposium*. Edited by J. Li, S. Nandi, and S. A. Li. New York: Springer-Verlag.

Lamartiniere, C. A., J. B. Moore, N. M. Brown, R. Thompson, M. J. Hardin, and S. Barnes. 1995. Genistein suppresses mammary cancer in rats. *Carcinogenesis* 16(11):2833–2840.

Land, C. E., M. Tokunaga, K. Koyama, M. Soda, D. L. Preston, I. Nishimori, and S. Tokuoka. 2003. Incidence of female breast cancer among atomic bomb survivors, Hiroshima and Nagasaki, 1950–1990. *Radiat Res* 160(6):707–717.

Lee, A. V., J. G. Jackson, J. L. Gooch, S. G. Hilsenbeck, E. Coronado-Heinsohn, C. K. Osborne, and D. Yee. 1999. Enhancement of insulin-like growth factor signaling in human breast cancer: Estrogen regulation of insulin receptor substrate-1 expression in vitro and in vivo. *Mol Endocrinol* 13(5):787–796.

Levin, E. R. 2005. Integration of the extranuclear and nuclear actions of estrogen. *Mol Endocrinol* 19(8):1951–1959.

Li, S., R. Hansman, R. Newbold, B. Davis, J. A. McLachlan, and J. C. Barrett. 2003. Neonatal diethylstilbestrol exposure induces persistent elevation of c-fos expression and hypomethylation in its exon-4 in mouse uterus. *Mol Carcinog* 38(2):78–84.

Losel, R., and M. Wehling. 2003. Nongenomic actions of steroid hormones. *Nat Rev Mol Cell Biol* 4(1):46–56.

Lunn, R. M., Y. J. Zhang, L. Y. Wang, C. J. Chen, P. H. Lee, C. S. Lee, W. Y. Tsai, and R. M. Santella. 1997. p53 mutations, chronic hepatitis B virus infection, and aflatoxin exposure in hepatocellular carcinoma in Taiwan. *Cancer Res* 57(16):3471–3477.

Luo, J., K. L. Margolis, J. Wactawski-Wende, K. Horn, C. Messina, M. L. Stefanick, H. A. Tindle, E. Tong, et al. 2011. Association of active and passive smoking with risk of breast cancer among postmenopausal women: A prospective cohort study. *BMJ* 342:d1016.

Ma, L., G. V. Benson, H. Lim, S. K. Dey, and R. L. Maas. 1998. *Abdominal B (AbdB) Hoxa* genes: Regulation in adult uterus by estrogen and progesterone and repression in mullerian duct by the synthetic estrogen diethylstilbestrol (DES). *Dev Biol* 197(2):141–154.

MacMahon, B., D. Trichopoulos, J. Brown, A. P. Andersen, K. Aoki, P. Cole, F. deWaard, T. Kauraniemi, et al. 1982a. Age at menarche, probability of ovulation and breast cancer risk. *Int J Cancer* 29(1):13–16.

MacMahon, B., D. Trichopoulos, J. Brown, A. P. Andersen, P. Cole, F. deWaard, T. Kauraniemi, A. Polychronopoulou, et al. 1982b. Age at menarche, urine estrogens and breast cancer risk. *Int J Cancer* 30(4):427–431.

Maffini, M. V., A. M. Soto, J. M. Calabro, A. A. Ucci, and C. Sonnenschein. 2004. The stroma as a crucial target in rat mammary gland carcinogenesis. *J Cell Sci* 117(Pt 8):1495–1502.

Martin, L. J., and N. F. Boyd. 2008. Mammographic density. Potential mechanisms of breast cancer risk associated with mammographic density: Hypotheses based on epidemiological evidence. *Breast Cancer Res* 10(1):201.

McCready, J., L. M. Arendt, J. A. Rudnick, and C. Kuperwasser. 2010. The contribution of dynamic stromal remodeling during mammary development to breast carcinogenesis. *Breast Cancer Res* 12(3):205.

Michels, K. B., and F. Xue. 2006. Role of birthweight in the etiology of breast cancer. *Int J Cancer* 119(9):2007–2025.

Migliaccio, A., M. Di Domenico, G. Castoria, A. de Falco, P. Bontempo, E. Nola, and F. Auricchio. 1996. Tyrosine kinase/p21ras/MAP-kinase pathway activation by estradiol-receptor complex in MCF-7 cells. *EMBO J* 15(6):1292–1300.

Migliaccio, A., G. Castoria, M. Di Domenico, A. de Falco, A. Bilancio, M. Lombardi, M. V. Barone, D. Ametrano, et al. 2000. Steroid-induced androgen receptor–oestradiol receptor β-Src complex triggers prostate cancer cell proliferation. *EMBO J* 19(20):5406–5417.

Miller, C., K. Degenhardt, and D. A. Sassoon. 1998. Fetal exposure to DES results in deregulation of Wnt7a during uterine morphogenesis. *Nat Genet* 20(3):228–230.

Milligan, S. R., O. Khan, and M. Nash. 1998. Competitive binding of xenobiotic oestrogens to rat alpha-fetoprotein and to sex steroid binding proteins in human and rainbow trout (*Oncorhynchus mykiss*) plasma. *Gen Comp Endocrinol* 112(1):89–95.

Million Women Study Collaborators. 2003. Breast cancer and hormone-replacement therapy in the Million Women Study. *Lancet* 362(9382):419–427.

Mizejewski, G., G. Smith, and G. Butterstein. 2004. Review and proposed action of alpha-fetoprotein growth inhibitory peptides as estrogen and cytoskeleton-associated factors. *Cell Biol Int* 28(12):913–933.

Mouritsen, A., L. Aksglaede, K. Sorensen, S. S. Mogensen, H. Leffers, K. M. Main, H. Frederiksen, A. M. Andersson, et al. 2010. Hypothesis: Exposure to endocrine-disrupting chemicals may interfere with timing of puberty. *Int J Androl* 33(2):346–359.

Mueller, M. M., and N. E. Fusenig. 2004. Friends or foes—bipolar effects of the tumour stroma in cancer. *Nat Rev Cancer* 4(11):839–849.

Muñoz-de-Toro, M., C. M. Markey, P. R. Wadia, E. H. Luque, B. S. Rubin, C. Sonnenschein, and A. M. Soto. 2005. Perinatal exposure to bisphenol-A alters peripubertal mammary gland development in mice. *Endocrinology* 146(9):4138–4147.

Murray, T. J., M. V. Maffini, A. A. Ucci, C. Sonnenschein, and A. M. Soto. 2007. Induction of mammary gland ductal hyperplasias and carcinoma in situ following fetal bisphenol A exposure. *Reprod Toxicol* 23(3):383–390.

Mustanski, B. S., R. J. Viken, J. Kaprio, L. Pulkkinen, and R. J. Rose. 2004. Genetic and environmental influences on pubertal development: Longitudinal data from Finnish twins at ages 11 and 14. *Dev Psychol* 40(6):1188–1198.

NCI (National Cancer Institute). 2009. *What you need to know about™ breast cancer.* http://www.cancer.gov/cancertopics/wyntk/breast/page2 (accessed August 1, 2011).

NCI. 2010. *Understanding cancer series: Estrogen receptors/SERMs.* http://www.cancer.gov/cancertopics/understandingcancer/estrogenreceptors (accessed November 2, 2011).

Nechuta, S., N. Paneth, and E. M. Velie. 2010. Pregnancy characteristics and maternal breast cancer risk: A review of the epidemiologic literature. *Cancer Causes Control* 21(7):967–989.

Newbold, R. R., B. C. Bullock, and J. A. McLachlan. 1990. Uterine adenocarcinoma in mice following developmental treatment with estrogens: A model for hormonal carcinogenesis. *Cancer Res* 50(23):7677–7681.

Newbold, R. R., R. B. Hanson, and W. N. Jefferson. 1997. Ontogeny of lactoferrin in the developing mouse uterus: A marker of early hormone response. *Biol Reprod* 56(5):1147–1157.

Newcomb, P. A., A. Trentham-Dietz, J. M. Hampton, K. M. Egan, L. Titus-Ernstoff, S. Warren Andersen, E. R. Greenberg, and W. C. Willett. 2011. Late age at first full term birth is strongly associated with lobular breast cancer. *Cancer* 117(9):1946–1956.

Ng, S. P., A. E. Silverstone, Z. W. Lai, and J. T. Zelikoff. 2006. Effects of prenatal exposure to cigarette smoke on offspring tumor susceptibility and associated immune mechanisms. *Toxicol Sci* 89(1):135–144.

Nguyen, D. H., H. Martinez-Ruiz, and M. H. Barcellos-Hoff. 2011. Consequences of epithelial or stromal TGFβ1 depletion in the mammary gland. *J Mammary Gland Biol Neoplasia* 16(2):147–155.

NIH (National Institutes of Health). 2011. *Clinical trials.gov.* http://clinicaltrials.gov/ct2/results?term=breast+cancer+and+metformin (accessed November 4, 2011).

Ninomiya, K., H. Kawaguchi, M. Souda, S. Taguchi, M. Funato, Y. Umekita, and H. Yoshida. 2007. Effects of neonatally administered diethylstilbestrol on induction of mammary carcinomas induced by 7,12-dimethylbenz(a)anthracene in female rats. *Toxicol Pathol* 35(6):813–818.

Ogden, C. L., M. D. Carroll, L. R. Curtin, M. M. Lamb, and K. M. Flegal. 2010. Prevalence of high body mass index in U.S. children and adolescents, 2007–2008. *JAMA* 303(3):242–249.

Owens, P. C., P. G. Gill, N. J. De Young, M. A. Weger, S. E. Knowles, and K. J. Moyse. 1993. Estrogen and progesterone regulate secretion of insulin-like growth factor binding proteins by human breast cancer cells. *Biochem Biophys Res Commun* 193(2):467–473.

Palmer, J. R., L. A. Wise, E. E. Hatch, R. Troisi, L. Titus-Ernstoff, W. Strohsnitter, R. Kaufman, A. L. Herbst, et al. 2006. Prenatal diethylstilbestrol exposure and risk of breast cancer. *Cancer Epidemiol Biomarkers Prev* 15(8):1509–1514.

Parent, A. S., G. Teilmann, A. Juul, N. E. Skakkebaek, J. Toppari, and J. P. Bourguignon. 2003. The timing of normal puberty and the age limits of sexual precocity: Variations around the world, secular trends, and changes after migration. *Endocr Rev* 24(5):668–693.

Parent, A. S., G. Rasier, A. Gerard, S. Heger, C. Roth, C. Mastronardi, H. Jung, S. R. Ojeda, et al. 2005. Early onset of puberty: Tracking genetic and environmental factors. *Horm Res* 64 (Suppl 2):41–47.

Park, H. Y., I. Hertz-Picciotto, J. Petrik, L. Palkovicova, A. Kocan, and T. Trnovec. 2008a. Prenatal PCB exposure and thymus size at birth in neonates in Eastern Slovakia. *Environ Health Perspect* 116(1):104–109.

Park, S. K., D. Kang, K. A. McGlynn, M. Garcia-Closas, Y. Kim, K. Y. Yoo, and L. A. Brinton. 2008b. Intrauterine environments and breast cancer risk: Meta-analysis and systematic review. *Breast Cancer Res* 10(1):R8.

Pasqualini, J. R., F. A. Kincl, and C. Sumida. 1985. *Hormones and the fetus.* New York: Pergamon Press.

Petrelli, J. M., E. E. Calle, C. Rodriguez, and M. J. Thun. 2002. Body mass index, height, and postmenopausal breast cancer mortality in a prospective cohort of U.S. women. *Cancer Causes Control* 13(4):325–332.

Pollak, M. N. 1998. Endocrine effects of IGF-I on normal and transformed breast epithelial cells: Potential relevance to strategies for breast cancer treatment and prevention. *Breast Cancer Res Treat* 47(3):209–217.

Polyak, K., and R. Kalluri. 2010. The role of the microenvironment in mammary gland development and cancer. *Cold Spring Harb Perspect Biol* 2(11):a003244.

Poretsky, L., and M. F. Kalin. 1987. The gonadotropic function of insulin. *Endocr Rev* 8(2):132–141.

Potischman, N., and R. Troisi. 1999. In-utero and early life exposures in relation to risk of breast cancer. *Cancer Causes Control* 10(6):561–573.

Prentice, R. L., J. E. Manson, R. D. Langer, G. L. Anderson, M. Pettinger, R. D. Jackson, K. C. Johnson, L. H. Kuller, et al. 2009. Benefits and risks of postmenopausal hormone therapy when it is initiated soon after menopause. *Am J Epidemiol* 170(1):12–23.

Preston, D. L., E. Ron, S. Tokuoka, S. Funamoto, N. Nishi, M. Soda, K. Mabuchi, and K. Kodama. 2007. Solid cancer incidence in atomic bomb survivors: 1958–1998. *Radiat Res* 168(1):1–64.

Protani, M., M. Coory, and J. H. Martin. 2010. Effect of obesity on survival of women with breast cancer: Systematic review and meta-analysis. *Breast Cancer Res Treat* 123(3): 627–635.

Pugeat, M., J. C. Crave, M. Elmidani, M. H. Nicolas, M. Garoscio-Cholet, H. Lejeune, H. Dechaud, and J. Tourniaire. 1991. Pathophysiology of sex hormone binding globulin (SHBG): Relation to insulin. *J Steroid Biochem Mol Biol* 40(4–6):841–849.

Pukkala, E., A. Kesminiene, S. Poliakov, A. Ryzhov, V. Drozdovitch, L. Kovgan, P. Kyyronen, I. V. Malakhova, et al. 2006. Breast cancer in Belarus and Ukraine after the Chernobyl accident. *Int J Cancer* 119(3):651–658.

Quinlan, R. J. 2003. Father absence, parental care, and female reproductive development. *Evol Hum Behav* 24(6):376–390.

Renehan, A. G., M. Zwahlen, C. Minder, S. T. O'Dwyer, S. M. Shalet, and M. Egger. 2004. Insulin-like growth factor (IGF)-I, IGF binding protein-3, and cancer risk: Systematic review and meta-regression analysis. *Lancet* 363(9418):1346–1353.

Revankar, C. M., D. F. Cimino, L. A. Sklar, J. B. Arterburn, and E. R. Prossnitz. 2005. A transmembrane intracellular estrogen receptor mediates rapid cell signaling. *Science* 307(5715):1625–1630.

Reynolds, P., S. E. Hurley, K. Hoggatt, H. Anton-Culver, L. Bernstein, D. Deapen, D. Peel, R. Pinder, et al. 2004. Correlates of active and passive smoking in the California Teachers Study cohort. *J Womens Health (Larchmt)* 13(7):778–790.

Roberts, D. L., C. Dive, and A. G. Renehan. 2010. Biological mechanisms linking obesity and cancer risk: New perspectives. *Annu Rev Med* 61:301–316.

Rogol, A. D. 1998. Leptin and puberty. *J Clin Endocrinol Metab* 83(4):1089–1090.

Romans, S. E., J. M. Martin, K. Gendall, and G. P. Herbison. 2003. Age of menarche: The role of some psychosocial factors. *Psychol Med* 33(5):933–939.

Rothschild, T. C., E. S. Boylan, R. E. Calhoon, and B. K. Vonderhaar. 1987. Transplacental effects of diethylstilbestrol on mammary development and tumorigenesis in female ACI rats. *Cancer Res* 47(16):4508–4516.

Ruan, W., V. Catanese, R. Wieczorek, M. Feldman, and D. L. Kleinberg. 1995. Estradiol enhances the stimulatory effect of insulin-like growth factor-I (IGF-I) on mammary development and growth hormone-induced IGF-I messenger ribonucleic acid. *Endocrinology* 136(3):1296–1302.

Rudel, R. A., S. E. Fenton, J. M. Ackerman, S. Y. Euling, and S. L. Makris. 2011. Environmental exposures and mammary gland development: State of the science, public health implications, and research recommendations. *Environ Health Perspect* 119(8):1053–1061.

Rudland, P. S. 1993. Epithelial stem cells and their possible role in the development of the normal and diseased human breast. *Histol Histopathol* 8(2):385–404.

Russo, J., and I. H. Russo. 1978. DNA labeling index and structure of the rat mammary gland as determinants of its susceptibility to carcinogenesis. *J Natl Cancer Inst* 61(6): 1451–1459.

Russo, J., and I. H. Russo. 2004. Development of the human breast. *Maturitas* 49(1):2–15.

Russo, J., L. K. Tay, and I. H. Russo. 1982. Differentiation of the mammary gland and susceptibility to carcinogenesis. *Breast Cancer Res Treat* 2 (1):5–73.

Russo, J., H. Lynch, and I. H. Russo. 2001. Mammary gland architecture as a determining factor in the susceptibility of the human breast to cancer. *Breast J* 7(5):278–291.

Russo, J., G. A. Balogh, J. Chen, S. V. Fernandez, R. Fernbaugh, R. Heulings, D. A. Mailo, R. Moral, et al. 2006. The concept of stem cell in the mammary gland and its implication in morphogenesis, cancer and prevention. *Front Biosci* 11:151–172.

Shafie, S. M., and F. H. Grantham. 1981. Role of hormones in the growth and regression of human breast cancer cells (MCF-7) transplanted into athymic nude mice. *J Natl Cancer Inst* 67(1):51–56.

Shafie, S. M., and R. Hilf. 1981. Insulin receptor levels and magnitude of insulin-induced responses in 7,12-dimethylbenz(a)anthracene-induced mammary tumors in rats. *Cancer Res* 41(3):826–829.

Shekhar, M. P., R. Pauley, and G. Heppner. 2003. Host microenvironment in breast cancer development: Extracellular matrix-stromal cell contribution to neoplastic phenotype of epithelial cells in the breast. *Breast Cancer Res* 5(3):130–135.

Shi, R., H. Yu, J. McLarty, and J. Glass. 2004. IGF-I and breast cancer: A meta-analysis. *Int J Cancer* 111(3):418–423.

Shim, Y. K., S. P. Mlynarek, and E. van Wijngaarden. 2009. Parental exposure to pesticides and childhood brain cancer: U.S. Atlantic Coast Childhood Brain Cancer Study. *Environ Health Perspect* 117(6):1002–1006.

Shoker, B. S., C. Jarvis, R. B. Clarke, E. Anderson, J. Hewlett, M. P. Davies, D. R. Sibson, and J. P. Sloane. 1999. Estrogen receptor-positive proliferating cells in the normal and precancerous breast. *Am J Pathol* 155(6):1811–1815.

Shore, R. E., N. Hildreth, E. Woodard, P. Dvoretsky, L. Hempelmann, and B. Pasternack. 1986. Breast cancer among women given X-ray therapy for acute postpartum mastitis. *J Natl Cancer Inst* 77(3):689–696.

Shore, R. E., N. Hildreth, P. Dvoretsky, E. Andresen, M. Moseson, and B. Pasternack. 1993. Thyroid cancer among persons given X-ray treatment in infancy for an enlarged thymus gland. *Am J Epidemiol* 137(10):1068–1080.

Slattery, M. L., C. Sweeney, R. Wolff, J. Herrick, K. Baumgartner, A. Giuliano, and T. Byers. 2007. Genetic variation in IGF1, IGFBP3, IRS1, IRS2 and risk of breast cancer in women living in Southwestern United States. *Breast Cancer Res Treat* 104(2):197–209.

Song, R. X., R. A. McPherson, L. Adam, Y. Bao, M. Shupnik, R. Kumar, and R. J. Santen. 2002a. Linkage of rapid estrogen action to MAPK activation by ERα-Shc association and Shc pathway activation. *Mol Endocrinol* 16(1):116–127.

Song, R. X., R. J. Santen, R. Kumar, L. Adam, M. H. Jeng, S. Masamura, and W. Yue. 2002b. Adaptive mechanisms induced by long-term estrogen deprivation in breast cancer cells. *Mol Cell Endocrinol* 193(1–2):29–42.

Song, R. X., C. J. Barnes, Z. Zhang, Y. Bao, R. Kumar, and R. J. Santen. 2004. The role of Shc and insulin-like growth factor 1 receptor in mediating the translocation of estrogen receptor α to the plasma membrane. *Proc Natl Acad Sci U S A* 101(7):2076–2081.

Song, R. X., Z. Zhang, and R. J. Santen. 2005. Estrogen rapid action via protein complex formation involving ERα and Src. *Trends Endocrinol Metab* 16(8):347–353.

Sorahan, T., L. Hamilton, K. Gardiner, J. T. Hodgson, and J. M. Harrington. 1999. Maternal occupational exposure to electromagnetic fields before, during, and after pregnancy in relation to risks of childhood cancers: Findings from the Oxford Survey of Childhood Cancers, 1953–1981 deaths. *Am J Ind Med* 35(4):348–357.

Sorensen, K., L. Aksglaede, T. Munch-Andersen, N. J. Aachmann-Andersen, J. H. Petersen, L. Hilsted, J. W. Helge, and A. Juul. 2009. Sex hormone-binding globulin levels predict insulin sensitivity, disposition index, and cardiovascular risk during puberty. *Diabetes Care* 32(5):909–914.

Soto, A. M., L. N. Vandenberg, M. V. Maffini, and C. Sonnenschein. 2008. Does breast cancer start in the womb? *Basic Clin Pharmacol Toxicol* 102(2):125–133.

Stewart, A. J., M. D. Johnson, F. E. May, and B. R. Westley. 1990. Role of insulin-like growth factors and the type I insulin-like growth factor receptor in the estrogen-stimulated proliferation of human breast cancer cells. *J Biol Chem* 265(34):21172–21178.

Suenson, E., O. Lutzen, and S. Thorsen. 1984. Initial plasmin-degradation of fibrin as the basis of a positive feed-back mechanism in fibrinolysis. *Eur J Biochem* 140(3):513–522.

Sugumar, A., Y. C. Liu, Q. Xia, Y. S. Koh, and K. Matsuo. 2004. Insulin-like growth factor (IGF)-I and IGF-binding protein 3 and the risk of premenopausal breast cancer: A meta-analysis of literature. *Int J Cancer* 111(2):293–297.

Sulis, M. L., and R. Parsons. 2003. PTEN: From pathology to biology. *Trends Cell Biol* 13(9):478–483.

Svensson, B. G., T. Hallberg, A. Nilsson, A. Schutz, and L. Hagmar. 1994. Parameters of immunological competence in subjects with high consumption of fish contaminated with persistent organochlorine compounds. *Int Arch Occup Environ Health* 65(6):351–358.

Thau, R., J. T. Lanman, and A. Brinson. 1976. Declining plasma progesterone concentration with advancing gestation in blood from umbilical and uterine veins and fetal heart in monkeys. *Biol Reprod* 14(4):507–509.

Thomas, P., Y. Pang, E. J. Filardo, and J. Dong. 2005. Identity of an estrogen membrane receptor coupled to a G protein in human breast cancer cells. *Endocrinology* 146(2):624–632.

Titus-Ernstoff, L., M. P. Longnecker, P. A. Newcomb, B. Dain, E. R. Greenberg, R. Mittendorf, M. Stampfer, and W. Willett. 1998. Menstrual factors in relation to breast cancer risk. *Cancer Epidemiol Biomarkers Prev* 7(9):783–789.

Titus-Ernstoff, L., E. E. Hatch, R. N. Hoover, J. Palmer, E. R. Greenberg, W. Ricker, R. Kaufman, K. Noller, et al. 2001. Long-term cancer risk in women given diethylstilbestrol (DES) during pregnancy. *Br J Cancer* 84(1):126–133.

Tokunaga, M., C. E. Land, and S. Tokuoka. 1991. Follow-up studies of breast cancer incidence among atomic bomb survivors. *J Radiat Res (Tokyo)* 32 (Suppl):201–211.

Trichopoulos, D., P. Lagiou, and H. O. Adami. 2005. Towards an integrated model for breast cancer etiology: The crucial role of the number of mammary tissue-specific stem cells. *Breast Cancer Res* 7(1):13–17.

Troisi, R., E. E. Hatch, L. Titus-Ernstoff, M. Hyer, J. R. Palmer, S. J. Robboy, W. C. Strohsnitter, R. Kaufman, et al. 2007. Cancer risk in women prenatally exposed to diethylstilbestrol. *Int J Cancer* 121(2):356–360.

Tulchinsky, D. 1973. Placental secretion of unconjugated estrone, estradiol and estriol into the maternal and the fetal circulation. *J Clin Endocrinol Metab* 36(6):1079–1087.

Van Maele-Fabry, G., A. C. Lantin, P. Hoet, and D. Lison. 2010. Childhood leukaemia and parental occupational exposure to pesticides: A systematic review and meta-analysis. *Cancer Causes Control* 21(6):787–809.

Vandenberg, L. N., M. V. Maffini, C. M. Schaeberle, A. A. Ucci, C. Sonnenschein, B. S. Rubin, and A. M. Soto. 2008. Perinatal exposure to the xenoestrogen bisphenol-A induces mammary intraductal hyperplasias in adult CD-1 mice. *Reprod Toxicol* 26(3–4):210–219.

Vassilacopoulou, D., and E. S. Boylan. 1993. Mammary gland morphology and responsiveness to regulatory molecules following prenatal exposure to diethylstilbestrol. *Teratog Carcinog Mutagen* 13(2):59–74.

Vazquez-Martin, A., C. Oliveras-Ferraros, and J. A. Menendez. 2009. The antidiabetic drug metformin suppresses HER2 (erbB-2) oncoprotein overexpression via inhibition of the mTOR effector p70S6K1 in human breast carcinoma cells. *Cell Cycle* 8(1):88–96.

Vivanco, I., and C. L. Sawyers. 2002. The phosphatidylinositol 3-kinase AKT pathway in human cancer. *Nat Rev Cancer* 2(7):489–501.

vom Saal, F. S., and B. G. Timms. 1999. The role of natural and manmade estrogens in prostate development. In *Endocrine disruptors: Effects on male and female reproductive systems*. Edited by R. K. Naz. Boca Raton, FL: CRC Press. Pp. 307–327.

Wakeford, R. 1995. The risk of childhood cancer from intrauterine and preconceptional exposure to ionizing radiation. *Environ Health Perspect* 103(11):1018–1025.

Warri, A., N. M. Saarinen, S. Makela, and L. Hilakivi-Clarke. 2008. The role of early life genistein exposures in modifying breast cancer risk. *Br J Cancer* 98(9):1485–1493.

WCRF/AICR (World Cancer Research Fund/American Institute for Cancer Research). 2007. *Food, nutrition, physical activity, and the prevention of cancer: A global perspective.* Washington, DC: AICR.

WCRF/AICR. 2010. Food, nutrition and physical activity and the prevention of breast cancer. *WCRF/AICR Continuous update report summary*. http://dietandcancerreport.org/downloads/cu/cu_breast_cancer_summary_2008.pdf (accessed October 29, 2011).

Weinberg, R. A. 2007. *The biology of cancer*. New York: Garland Science.

Weisglas-Kuperus, N., S. Patandin, G. A. Berbers, T. C. Sas, P. G. Mulder, P. J. Sauer, and H. Hooijkaas. 2000. Immunologic effects of background exposure to polychlorinated biphenyls and dioxins in Dutch preschool children. *Environ Health Perspect* 108(12):1203–1207.

Welshons, W. V., S. C. Nagel, K. A. Thayer, B. M. Judy, and F. S. vom Saal. 1999. Low-dose bioactivity of xenoestrogens in animals: Fetal exposure to low doses of methoxychlor and other xenoestrogens increases adult prostate size in mice. *Toxicol Ind Health* 15(1–2):12–25.

Witorsch, R. J. 2002. Low-dose in utero effects of xenoestrogens in mice and their relevance to humans: An analytical review of the literature. *Food Chem Toxicol* 40(7):905–912.

Wolff, M. S., J. A. Britton, L. Boguski, S. Hochman, N. Maloney, N. Serra, Z. Liu, G. Berkowitz, et al. 2008. Environmental exposures and puberty in inner-city girls. *Environ Res* 107(3):393–400.

Women's Health Initiative Steering Committee. 2004. Effects of conjugated equine estrogen in postmenopausal women with hysterectomy: The Women's Health Initiative randomized controlled trial. *JAMA* 291(14):1701–1712.

Wong, C. W., C. McNally, E. Nickbarg, B. S. Komm, and B. J. Cheskis. 2002. Estrogen receptor-interacting protein that modulates its nongenomic activity-crosstalk with Src/Erk phosphorylation cascade. *Proc Natl Acad Sci U S A* 99(23):14783–14788.

Woodruff, T. J., A. R. Zota, and J. M. Schwartz. 2011. Environmental chemicals in pregnant women in the United States: NHANES 2003–2004. *Environ Health Perspect* 119(6):878–885.

Writing Group for the Women's Health Initiative Investigators. 2002. Risks and benefits of estrogen plus progestin in healthy postmenopausal women: Principal results from the Women's Health Initiative randomized controlled trial. *JAMA* 288(3):321–333.

Wysocki, P. J., and B. Wierusz-Wysocka. 2010. Obesity, hyperinsulinemia and breast cancer: Novel targets and a novel role for metformin. *Expert Rev Mol Diagn* 10(4):509–519.

Xue, F., W. C. Willett, B. A. Rosner, S. E. Hankinson, and K. B. Michels. 2011. Cigarette smoking and the incidence of breast cancer. *Arch Intern Med* 171(2):125–133.

Yager, J. D., and N. E. Davidson. 2006. Estrogen carcinogenesis in breast cancer. *N Engl J Med* 354(3):270–282.

Ye, J., Z. Gao, J. Yin, and Q. He. 2007. Hypoxia is a potential risk factor for chronic inflammation and adiponectin reduction in adipose tissue of ob/ob and dietary obese mice. *Am J Physiol Endocrinol Metab* 293(4):E1118–E1128.

Yoo, C. B., and P. A. Jones. 2006. Epigenetic therapy of cancer: Past, present and future. *Nat Rev Drug Discov* 5(1):37–50.

Yu, H., and T. Rohan. 2000. Role of the insulin-like growth factor family in cancer development and progression. *J Natl Cancer Inst* 92(18):1472–1489.

Zakikhani, M., R. Dowling, I. G. Fantus, N. Sonenberg, and M. Pollak. 2006. Metformin is an AMP kinase-dependent growth inhibitor for breast cancer cells. *Cancer Res* 66(21):10269–10273.

Zakikhani, M., M. J. Blouin, E. Piura, and M. N. Pollak. 2010. Metformin and rapamycin have distinct effects on the AKT pathway and proliferation in breast cancer cells. *Breast Cancer Res Treat* 123(1):271–279.

Zhang, Y., M. Matheny, S. Zolotukhin, N. Tumer, and P. J. Scarpace. 2002. Regulation of adiponectin and leptin gene expression in white and brown adipose tissues: Influence of β3-adrenergic agonists, retinoic acid, leptin and fasting. *Biochim Biophys Acta* 1584(2–3):115–122.

Zhang, Y., A. Daquinag, D. O. Traktuev, F. Amaya-Manzanares, P. J. Simmons, K. L. March, R. Pasqualini, W. Arap, et al. 2009. White adipose tissue cells are recruited by experimental tumors and promote cancer progression in mouse models. *Cancer Res* 69(12):5259–5266.

6

Opportunities for Action to Reduce Environmental Risks for Breast Cancer

The committee was asked to consider the potential for evidence-based actions to reduce the risk of breast cancer. Individual women, health care providers, advocacy organizations, and many other stakeholders are all eager to know what concrete steps can be taken to reduce the risk of breast cancer for an individual or the population, and when during the life course those actions might be most effective. This chapter outlines several evidence-based actions that women can take. However, the scientific community still has only limited understanding of which exposures might best be avoided and when, and which actions might have a long-term positive benefit in reducing risk for breast cancer.

Even when research strongly supports classifying an exposure as a risk factor for breast cancer, that research does not necessarily provide the information needed to determine the appropriate response to reduce risk. Should exposure be avoided completely? Will reducing or eliminating exposure in adulthood reduce a risk that has accrued from exposure at younger ages? Will the presence or absence of other risk factors for breast cancer influence the likely benefit or harm from a change in exposure to a given risk factor? Will changing one type of exposure lead to another that carries new and possibly as yet unrecognized risks for breast cancer, other diseases, or perhaps some other adverse economic or environmental outcome?

Finding ways to reduce risk and avert cases of breast cancer is a high priority for everyone concerned about this disease. Although some definite actions can be taken to reduce risk, the committee found overall that evidence-based options are limited because few studies have been done to test the effectiveness of actions that may be hypothesized to reduce risk. In this chapter, the committee discusses some specific areas where action appears warranted, but it first summarizes the significance of the uncertainty around preventive action.

RECOGNIZING UNCERTAINTY OF BENEFITS AND RISKS

Potential for Introducing New Hazards or Risks

A key concept to remember when evaluating a particular risk associated with a particular factor is that an action that is aimed at eliminating the specific risk of concern may result in a substitution of one risk for another, or perhaps shifting risk from one group to another. Any risk of the alternative action thus needs to be considered and weighed against the risk that the change is intended to reduce or eliminate.

The complexity of trade-offs from substitutions can be illustrated with the case of contamination of potable ground water sources with pesticides or industrial chemicals shown to be carcinogenic in experimental animals or humans. Reducing exposures to potentially carcinogenic substances in drinking water from groundwater sources seems to be a logical, health-protective action, even if the actual or perceived risk from the contaminants is small. A typical action to reduce the potential cancer risk from using the contaminated ground water is to switch the consumer to an alternative source of potable water, such as a public water supply system. However, such systems require disinfection, usually by chlorination, and chlorination of surface water introduces trace levels of disinfection by-products (DBPs). Several DBPs have been found to be carcinogenic in animal bioassays (e.g., NTP, 2007a,b), and some epidemiologic studies have suggested that long-term exposure to DBPs is associated with an increase in bladder cancer (reviewed in Richardson et al., 2007), especially in a subset of the population with specific genetic polymorphisms (Cantor et al., 2010).

In this scenario, one would have to consider many factors, including (1) the relative carcinogenic potency of the groundwater contaminant(s) versus that of the DBPs, (2) the concentrations of the groundwater contaminants or the DBPs in the drinking water and indoor air following use, and (3) the duration and frequency of likely exposure to a given drinking water source over a lifetime. Depending on these values, it is possible that a comparative risk assessment would show that switching from the contaminated groundwater supply to the uncontaminated but disinfected surface water supply actually increased, rather than decreased, potential cancer risks to the exposed population.

This example, however, also illustrates the challenges in assessing trade-offs in population- and individual-level risks and benefits. There are certainly no clear benefits, at least to the individual consumer, of drinking groundwater contaminated with low levels of pesticides. But the potential cancer risk associated with the presence of DBPs in a public water system must be assessed against the very real risks of widespread acute illness from microbial contamination that could result in the absence of disinfection (Gibbons and Laha, 1999; Schoeny, 2010).

This pattern of trading one hazardous substance for another is not uncommon. Although federal agencies evaluate the toxicity and carcinogenicity of new pesticides and prescription drugs before they are approved for sale, the United States does not have a comprehensive program to evaluate the safety of chemicals before their widespread use in consumer products. In the face of consumer concern about bisphenol A (BPA), for example, BPA-free plastics are now available, but new research appears to show that BPA-free plastics may leach other chemicals with estrogenic activity comparable to that of BPA (Yang et al., 2011). As noted in Chapter 2, the European Union has adopted a program (Registration, Evaluation, Authorisation and Restriction of Chemical Substances, or REACH) for broader safety testing by manufacturers of their products before they are approved for use. In the United States, the Government Accountability Office (GAO, 2009a,b) has recommended changes to improve the effectiveness of federal regulation of chemicals.

Risk trade-offs may also be hard to judge because a given factor can have both positive and negative health effects. For example, there is fairly compelling evidence that moderate alcohol consumption is associated with a small but consistently observed increase in the risk of breast cancer.[1] However, there is also compelling evidence that consumption of the same moderate amounts of alcohol is associated with a reduction in mortality from cardiovascular disease (Maskarinec et al., 1998; Gunzerath et al., 2004; Klatsky, 2009; Ronksley et al., 2011). The risks associated with any specific environmental exposure occur against the background of a woman's genetic susceptibility, reproductive history, and lifestyle.

Challenges in Public Health Policy Aimed at Risk Reduction

A significant challenge—relevant to the discussion of environmental risk factors for breast cancer and frequently faced by regulators of environmental pollutants and public health officials—is a lack of information about the nature of the effects of many exposures on risks for breast cancer. Chapter 4 reviewed the diverse challenges in trying to generate and interpret relevant information. Noted here are a few specific areas where substantial uncertainty faces policy makers.

Assessing the net effect of environmental exposures is one challenge.

---

[1]The trade-offs associated with alcohol use are discussed further later in this chapter.

Individuals and populations are never exposed to only one risk or protective factor at a time, but complex combinations of exposures are rarely the subject of laboratory or epidemiologic studies. The U.S. Environmental Protection Agency (EPA) and other environmental regulatory agency scientists often assume that cancer risks in a population are simply the sum of risks estimated for each individual chemical in the absence of data on risks of co-exposures. In practice, this means that one might be confident that a lifetime of exposure to a single chemical that causes a theoretical increase in cancer risk of 1 additional case per 1 million exposed people is essentially lost in the background and that the risk from that exposure may be considered minimal. But what if a population is exposed to 1,000 of these types of "1 in 1 million" lifetime risks? Are the risks simply additive (e.g., the increase in lifetime risk becomes 1,000 per 1 million, or 1 in 1,000), or is it possible that the chemicals can interact to alter risk in some nonadditive manner, either by reducing each other's effects (e.g., competition for receptor binding) or by mutually enhancing each other's effects? For example, smoking and asbestos exposure are each well-recognized risk factors for lung cancer, but exposure to both multiplies the risk of lung cancer, making the risk far greater than the addition of the individual effects of these two exposures. A study of asbestos workers found that the lung cancer mortality rate was 122.6 per 100,000 among men with a history of smoking and no asbestos exposure; 58.4 per 100,000 among those with asbestos exposure but no history of smoking; and 601.6 per 100,000 with exposure to both smoking and asbestos (Hammond et al., 1979). As a qualitative example, cigarette smoke is a mixture of relatively low levels of numerous carcinogens, and both direct and passive exposure to cigarette smoke are associated with a variety of cancers, including breast cancer. On the other hand, some exposures may increase risks for breast cancer, but reduce them for other cancers. In the vast majority of instances, the scientific information is typically not sufficiently developed to calculate the joint effects of multiple exposures with confidence.

Another area of uncertainty is whether risks occur at very low levels of exposures and whether those risks can be estimated from information on hazards and risks that are determined for high-dose exposures. The EPA and other regulatory agencies have made specific assumptions about the shape of the dose–response curve in relation to certain mechanisms of action. For example, a linear extrapolation of cancer risk from high doses to low doses is used for carcinogenic agents determined to be mutagenic. Although linear extrapolation may not apply in some circumstances, it is considered a protective approach in the absence of evidence to guide the selection of an alternative model (EPA, 2005). Directly detecting small differences in risk in human or animal studies may be difficult, if not impossible, because of challenges that include the need for very large study populations, the potential for errors in measuring exposure, and the possibility of unrecognized confounding.

Finally, as highlighted in earlier chapters, the risk from a given exposure may depend on the age at which it occurs, although current knowledge about susceptible windows pertains to few exposures. Perhaps most salient is the difficulty in assessing whether reduction or elimination of an exposure will alter long-term risk of breast cancer, and if so, by how much.

EVIDENCE-BASED OPPORTUNITIES FOR ACTION TO REDUCE RISK

Identifying evidence-based opportunities for action to reduce risk of breast cancer depends, ideally, on a convergence of several elements, including

- sufficient evidence to demonstrate that a specific factor is associated with increasing or decreasing breast cancer risk;
- a means by which to modify exposure to the risk factor;
- an understanding of whether effective changes can be made by an individual or would require instead, or in addition, changes at governmental, social, or cultural levels;
- evidence that a specific action to modify exposure will result in the desired impact on breast cancer risk, the characteristics of women who could be expected to benefit, and when the intervention needs to occur; and
- awareness of the trade-offs (potentially as yet unrecognized) that may occur in terms of other health outcomes, personal preferences, or economic consequences.

As illustrated in the reviews in Chapter 3, the evidence on many of the environmental factors that have been investigated as potential risk factors for breast cancer remains inconclusive. But for a modest set, the evidence is relatively strong and points to likely opportunities for prevention when these factors are modifiable. What, then, is the distinction between "modifiable" and "nonmodifiable" risk factors? For example, the age at which women have a first full-term pregnancy is known to influence the risk of breast cancer, with later age at first birth generally associated with higher risk. At the individual level, women can make decisions as to when they will have their first pregnancy, but changes at the population level are influenced by a range of social, economic, educational, cultural, and personal forces, and any effort to influence personal choices could have unexpected consequences. Furthermore, opportunities for modification of some factors may be limited to certain ages—a woman's age at a first full-term pregnancy may be modifiable before menopause, but obviously it is not after menopause.

Another question is whether particular actions to change a modifiable risk factor will actually translate into lowered risk. For example, although the evidence is relatively strong that greater body fatness is associated with increased risk of breast cancer for postmenopausal women (WCRF/AICR, 2007), it is less clear whether these women can reduce their risk if they lose weight during the postmenopausal period. It is possible that the adverse effects of being overweight are hard to reverse at older ages and can best be prevented by avoiding overweight and obesity throughout life.

Overweight and weight reduction also illustrate the complexity of framing guidance on action when the consequences of an exposure differ among groups in a population. Whereas evidence indicates that greater weight is associated with an increased risk for breast cancer for postmenopausal women, it also indicates that greater weight is associated with a lower risk of breast cancer for premenopausal women (WCRF/AICR, 2007). Therefore, avoiding overweight is not a reasonable strategy for reducing the low, but still present, risk of premenopausal breast cancer, although avoiding overweight has many other important health benefits for women of all ages.

The association between shift work and increased risk of breast cancer concerns the committee, but it does not see a sound basis, at this time, for proposing action. More research is needed to understand the mechanisms underlying the association between shift work and breast cancer and to develop a clearer, more consistent characterization of the kind of work or work schedule that is associated with increased risk. This deeper understanding is needed to guide any effort to frame and test interventions in a realm with significant socioeconomic ramifications. A specific call for research on shift work is included in the recommendations in Chapter 7.

In some cases, the available evidence from animal or mechanistic studies suggests that a chemical or other factor may be a hazard, but evidence to directly assess the breast cancer risk for women is lacking (or perhaps not possible to obtain). In such circumstances, policy makers may use formal risk assessments to gauge the magnitude of possible risk and the appropriateness of actions to mitigate it. The identification of hazards—factors that have the ability to cause adverse effects—is an essential element in risk assessment and is often based on laboratory studies of biological mechanisms and effects of exposures on laboratory animals. Estimates of risk represent the probability that a particular adverse outcome—breast cancer in this case—will occur in an individual person or a population as a result of defined exposures to a hazard. A risk assessment considers not only the hazard of the substance, but also its potency (roughly speaking, how strong its effect is for a given dose) and the magnitude, nature, and timing of expected human exposure. A highly potent carcinogen may pose substantial risk to an exposed individual, but if exposure to the general public is very low or extremely uncommon, the *population risk* will tend to be low. Alternatively, a low-potency carcinogen may pose risks that are low, but if exposures are common, it may be associated with a measureable effect in the population as a whole.

The committee did not undertake formal risk assessments for the environmental chemicals it found to be biologically plausible or possible contributors to breast cancer. Critical pieces of information were lacking, particularly robust data for estimating the magnitude of human breast cancer risk for a given dose (potency) at different life stages, and the prevalence and magnitude of the exposures across the population at different life stages. These data gaps were an obstacle to proposing evidence-based action that women could take to reduce risks from exposure to any particular chemical.

LIKELY OPPORTUNITIES TO ACT TO REDUCE RISK OF BREAST CANCER

With these limitations to the evidence in mind, the committee highlights here the areas where it sees the clearest indications of opportunities for actions to reduce breast cancer risk. These actions are reviewed in this section and summarized in Table 6-1. It is important to recognize that the evidence is generally more extensive and therefore stronger for postmenopausal women than for premenopausal women, and for white, non-Hispanic women than for women of other races and ethnicities. In addition, some of the prevention opportunities that the committee points to appear more likely to apply to the prevention of the more common estrogen receptor–positive (ER+) tumors than estrogen receptor–negative (ER–) tumors. Younger women, however, tend to have ER– forms of breast cancer, as do women who have strong inherited susceptibility to breast cancer, such as carrying a mutation in *BRCA1*. All women should know their personal risk factors for breast cancer and seek clinical guidance from their health care providers regarding their breast cancer risk and how to modify it.

Medical Radiation

Among the strongest evidence reviewed by the committee regarding environmental exposures that have been causally linked to breast cancer was the evidence on ionizing radiation. Based on standard models developed from the radiation exposures of the Japanese atomic bomb survivors, it is commonly assumed that the breasts are most sensitive to carcinogenic effects of radiation at early ages (e.g., below ages 20–30). Nevertheless, models also predict elevated risks after exposure, even in middle age (Berrington

TABLE 6-1 Summary of Committee Assessment of Opportunities for Actions by Women That May Reduce Risk of Breast Cancer

Opportunity for Action	Strength of Evidence That Exposure Is Associated with Breast Cancer Risk[a]	Modification of Exposure	
		Personal Action Possible	Requires Action by Others
Avoid inappropriate medical radiation exposure[d]	+++	Yes	Yes
Avoid combination menopausal hormone therapy, unless medically appropriate[e]	+++	Yes	Confer with physician
Avoid or end active smoking	+	Yes	Others can facilitate
Avoid passive smoking	(no committee consensus)	Varies	Yes
Limit or eliminate alcohol consumption	++	Yes	Others can facilitate
Maintain or increase physical activity	– –[f]	Yes	Others can facilitate
Maintain healthy weight or reduce overweight or obesity to reduce postmenopausal risk	+++	Yes	Others can facilitate

Action			
Target Population Defined	Effective Form and Timing Established[b]	Affects Risk for Specific Subtype	Other Prominent Known Risks or Benefits from Taking Action[c]
All ages	Yes, especially at younger ages	?	May result in loss of clinically useful information in some instances Likely to decrease risk for other cancers
Postmenopausal women	Yes	ER+	May experience moderate to severe menopausal symptoms, continued menopausal associated bone loss
All ages, especially before first pregnancy	Yes (form) No (timing)	?	Likely to reduce risk for other cancers, heart disease, stroke
All ages	Yes	?	Likely to reduce risk for other cancers, heart disease
All women	Yes (form) No (timing)	ER?	May increase risk for cardiovascular disease No known benefit of high alcohol consumption
All ages	No	?	Likely to reduce risk for cardiovascular disease, diabetes May increase risk for injury
Unclear	No	ER+?	Likely to reduce risk for cardiovascular disease, diabetes, other cancers

continued

TABLE 6-1 Continued

Opportunity for Action	Strength of Evidence That Exposure Is Associated with Breast Cancer Risk[a]	Modification of Exposure	
		Personal Action Possible	Requires Action by Others
Limit or eliminate workplace, consumer, and environmental exposure to chemicals that are plausible contributors to breast cancer risk while considering risks of substitutes[g]	Varies by chemical	Varies	Yes
If at high risk for breast cancer, consider use of chemoprevention	– – –[b]	Yes	Confer with physician

[a]The assessments of the evidence of an association between an exposure and risk of breast cancer are qualitative representations of the committee's conclusions from its review of available evidence: strong conclusion of increased risk, +++; moderately strong conclusion of increased risk, ++; conclusion of increased risk, +; unclear, ?; conclusion of reduced risk, –; moderately strong conclusion of reduced risk, – –; strong conclusion of reduced risk, – – –.

[b]Actions to address risk factors can take various forms, some of which may be more effective than others. For example, increasing physical activity might be based on amount of time spent in any one exercise opportunity, on increasing specific types of exercise, or increasing the frequency of exercise, or perhaps some combination of any of these. Studies have not been done that provide evidence that a specific form of physical activity is optimal for reducing breast cancer risk.

[c]The committee's comments on other benefits or risks highlight major considerations, but are not intended to be exhaustive.

de Gonzalez et al., 2009; Shuryak et al., 2010). Recent models suggest the possibility that exposures to ionizing radiation across the age range of 10 to 50 years may result in excess relative risks of tumors in breast tissue that are more similar than previously estimated (Preston et al., 2007).

For the U.S. population, about half of the exposure to ionizing radiation comes from medical radiation, primarily in the diagnostic setting and

| Action | | | |
| | | | |

Target Population Defined	Effective Form and Timing Established[b]	Affects Risk for Specific Subtype	Other Prominent Known Risks or Benefits from Taking Action[c]
Varies	No	?	May reduce risk for other forms of cancer. May result in replacement with products that have health or other risks not yet identified
High-risk women	Yes	ER+	Depending on the agent, increased risk of endometrial cancer, stroke, deep-vein thrombosis among others

[d]While recognizing the risks of ionizing radiation exposure, particularly for certain higher dose methods (e.g., CT scans), it was not the committee's intent to dissuade women from routine mammography screening, which aids in detecting early-stage tumors.

[e]Combination hormone therapy with estrogen and progestin increases the risk of breast cancer and the associated risk is reduced upon stopping therapy. Oral contraceptives are also associated with an increased risk of breast cancer while they are being used. This risk is superimposed on a low background risk for younger women, who are most likely to use oral contraceptives. These contraceptives are associated with long-term risk reduction for ovarian and endometrial cancer.

[f]Reflects reduced risk of breast cancer associated with greater physical activity.

[g]Plausibility may be indicated by epidemiologic evidence, animal bioassays, or mechanistic studies.

[h]Reflects reduced risk of breast cancer associated with use of chemopreventive agents.

especially from computed tomography (CT) scans and myocardial perfusion imaging (Fazel et al., 2009). As outlined in a paper commissioned by the committee (see Appendix F), the average annual dose of radiation from medical diagnostic sources in the U.S. population approximately doubled from 1985 to 2006 (Smith-Bindman, 2011). As further elaborated in that paper, there is evidence that exposure doses for the same imaging tests vary widely among institutions. Superimposed on this variability is the element of human error, which has resulted in very high doses of radiation inadvertently delivered to patients by inadequately trained or supervised technicians and poorly designed equipment (e.g., Bogdanich and Rebelo, 2010; Smith-Bindman, 2010; Bogdanich, 2011). Extrapolating from estimates based on CT scans (Berrington de Gonzalez et al., 2009), Smith-Bindman (2011) estimated that among women in the United States in 2007, about 2,800 future breast cancers would result over their remaining lifetimes from exposure to all sources of medical diagnostic radiation delivered in 2007.[2]

There have been successful efforts through the Mammography Quality Standards Act (MQSA) to standardize and minimize radiation doses received from mammography, but the United States has no federal oversight for other imaging examinations and no guidelines on optimal doses. Evidence shows that physicians are insufficiently informed about radiation doses or the cancer risks attributable to the medical imaging they order (Lee et al., 2004). Evidence from patient surveys also shows that the public has little appreciation that just two to three abdominal CT scans can deliver radiation doses in the range of exposure experienced by Hiroshima survivors, doses that have been associated with elevated risks of breast and other cancers (Baumann et al., 2011).

The committee is encouraged that in 2010, the Food and Drug Administration (FDA) launched an Initiative to Reduce Unnecessary Radiation Exposure from Medical Imaging that addresses many of these areas (FDA, 2010).[3] Sufficient resources and staff are needed for the development of detailed programs, full implementation of all components, and prospective evaluation of outcomes. The patient perspectives need to be incorporated into the planning of the programs, and breast cancer and other patient advocacy groups could provide important contributions to the development and evaluation of the programs. Radiology and imaging professionals are also seeking improvements. The Image Wisely program (http://www.image wisely.org/) is an effort to inform patients and professional colleagues of the importance of minimizing unnecessary exposure to ionizing radiation in imaging procedures. The Image Gently program (Alliance for Radiation Safety in Pediatric Imaging, 2011) specifically focuses on maximizing safety for children (Don, 2011; Moreno, 2011).

Used properly, medical imaging, including mammography, is a valu-

---

[2]Ionizing radiation is also an important tool for treatment of breast and other cancers and other conditions. For individuals who have been diagnosed with cancer, the benefits and risks of exposure to ionizing radiation are different from those for an individual who does not have cancer. However, even in treatment settings, patients can be exposed to excessively high doses of radiation because of errors or equipment malfunctions.

[3]A broad description of this initiative is available at http://www.fda.gov/Radiation-EmittingProducts/RadiationSafety/RadiationDoseReduction/ucm199994.htm#_Toc253092879.

able tool in diagnosing illness and guiding treatment, but unnecessary or improper use may increase risks because of the exposure to ionizing radiation. The committee sees important opportunities for actions at several levels that may contribute to lowering the risk of breast cancer due to exposure to ionizing radiation by improving medical imaging procedures.

Patients and Families: Individuals can question health providers specifically about what is known regarding the health benefits and harms associated with proposed diagnostic tests that involve exposure to ionizing radiation. They can request information about the relative doses of radiation associated with each type of procedure they undergo. Informed women may be able to avoid unnecessary tests for themselves and their families.

Health Professionals: Medical education programs and training can be created to enhance health care providers' and students' understanding of the doses of radiation involved in diagnostic imaging tests, and the health risks associated with those doses. Continuing education opportunities can include evaluation of the medical literature on the health benefits and harms of diagnostic imaging. Technicians can be trained in the avoidance of radiation overdoses and in methods to minimize dose while maintaining image quality.

Hospitals and Medical Practices: Hospitals and medical practices can make every effort to obtain previous imaging tests that have recently been done in other settings and avoid repeating imaging studies only for convenience. Expanding use of electronic medical records and ensuring compatibility and interoperability of records and digital films may facilitate the transmission of images between facilities.

Industry: Manufacturers could be encouraged to engineer diagnostic imaging devices to maximize safety and minimize human error. Through organizations and other collaborative mechanisms that promote the development of industry standards, manufacturers could adopt design standards that promote safe operation of the equipment. They might also take steps such as increasing the similarity of the "look and feel" of imaging equipment so that technicians can better transfer operating skills across manufacturers' machines, thereby reducing errors.

Public Health: Public education campaigns could help inform consumers about levels of exposure to ionizing radiation that result from relevant medical procedures and the cumulative risks of such exposures. Educational campaigns could also encourage consumers to work with their health care providers to minimize exposures and might provide consumer tools to help track those exposures.

Regulation: Relevant agencies can work with the appropriate professional societies and scientists trained in evidence-based decision making to develop standardization of dosimetry and evidence-based guidelines for appropriate use of tests for screening, diagnosis, and follow-up.

Menopausal and Contraceptive Hormone Use

Menopausal Hormone Therapy

The Women's Health Initiative (WHI), which included a randomized clinical trial to assess the health effects of combination hormone therapy (i.e., a product with estrogen and progestin), demonstrated an increased risk of breast cancer among postmenopausal women taking combination hormone therapy (Writing Group for the Women's Health Initiative Investigators, 2002). The annualized incidence rate of invasive breast cancer over the intervention period was 0.38 percent for women taking combination estrogen–progestin hormone therapy and 0.30 percent for women on placebo (HR = 1.26, 95% CI, 1.00–1.59). The association between combination hormone therapy and increased risk of breast cancer confirmed prior results from observational studies (Collaborative Group on Hormonal Factors in Breast Cancer, 1997).

The increased risk associated with current use of combination hormone therapy has been found to decline when use stops (e.g., Chlebowski et al., 2009; Beral et al., 2011). Follow-up of the WHI clinical trial participants demonstrated that the risk of breast cancer declined rapidly after combination hormone treatment ended (Chlebowski et al., 2009). Two or more years after the end of the intervention period, the annualized incidence of breast cancer among those assigned to the combination hormone therapy group was 0.49 percent compared to 0.42 percent among women assigned to placebo (HR = 1.19, 95% CI, 0.59–2.42). In the observational Million Women Study in the United Kingdom (Beral et al., 2011), former users' risk became comparable to that of never users within 4 years. After publication of the WHI results in 2002, rates of hormone therapy use declined rapidly, and coincident with that, U.S. breast cancer rates among women ages of 50 to 69 years were observed to decline by 11.8 percent (95% CI, 9.2–14.5) between 2001 and 2004 (Ravdin et al., 2007).

The committee is confident in urging that women avoid or minimize use of combination hormone therapy, thereby avoiding increasing their risk of breast cancer. The WHI considered other outcomes in addition to breast cancer, including heart disease, fractures, stroke, and colorectal cancer. Overall, the increased risks for breast cancer, heart disease, and stroke were considered to outweigh the reduction in risk for hip fractures and colorectal cancer (Writing Group for the Women's Health Initiative Investigators, 2002). Managing menopausal symptoms is a common reason for women to consider taking hormone therapy, but women should confer with their health care providers to determine the most appropriate way to manage these or other symptoms.

This position is consistent with the guidance from the U.S. Preventive Services Task Force (USPSTF, 2005). The USPSTF guidelines recommend against the routine use of combination hormone therapy for the prevention of chronic conditions because the increased risks of breast cancer, stroke, and other conditions are considered to outweigh the potential benefits of reduced risks for fractures and colorectal cancer. However, this guidance specifically excludes consideration of management of menopausal symptoms. The USPSTF advises women and their health care providers to give individualized consideration to personal risk factors and preferences in deciding whether use of hormone therapy is medically appropriate.

The Endocrine Society (Santen et al., 2010) issued a scientific statement that provides a detailed review and grading of evidence on benefits and harms associated with use of postmenopausal hormone therapy (HT). The highest-quality evidence concerning combination HT and breast health is that it increases mammographic density, which is associated with increased risk of breast cancer (e.g., Cuzick, 2008; Boyd et al., 2009, 2010). Other evidence concerning breast cancer risks was considered less strong. The statement advocates individualized assessments of women's potential risks and benefits but suggests that some form of HT may be appropriate for menopausal women younger than age 60.

The WHI also included a trial of estrogen-only hormone therapy among women who had a hysterectomy. Women with a hysterectomy who took estrogen-only therapy were less likely to develop invasive breast cancer during the intervention period and the subsequent period after the intervention ended than women who took the placebo, but the absolute differences were small (LaCroix et al., 2011). Over the entire follow-up period (during the intervention and afterward), the incidence of breast cancer in the group that took conjugated estrogen was 0.27 percent, compared with 0.35 percent in the placebo group (HR = 0.77, 95% CI, 0.62–0.95). Although this WHI intervention trial observed no excess risk with estrogen-only therapy among women who had a hysterectomy, observational studies have found a small increased risk of breast cancer (Million Women Study Collaborators, 2003; Beral et al., 2011). One reason for this discrepancy between observational studies and the randomized clinical trials may be that observational studies are more likely to have misclassification of exposure between combination hormone therapy and estrogen-only therapy. Observational studies also suggest that these risks are higher in lean compared with obese women (Huang et al., 1997; Reeves et al., 2006; Brinton et al., 2008; Beral et al., 2011).

A criticism of the combination hormone therapy products that are prescribed most often and used in clinical trials of menopausal hormone therapy is that they are synthetic. The concern is that they therefore may be associated with risk not present for products that are considered more "natural," often referred to as bioidentical hormones. Bioidentical hormones are derived from plants and treated to have the same chemical structure as endogenous human hormones (Cirigliano, 2007). They may be commercially available or individually compounded in pharmacies. The Endocrine Society (2006) has issued a position statement on bioidentical hormones, emphasizing the lack of data about their safety and effectiveness, expressing concern about potentially misleading or inaccurate claims, and supporting FDA regulation of all hormone products. A review of bioidentical hormone therapy (Cirigliano, 2007) supports the concerns raised by The Endocrine Society, citing a lack of evidence to support claims of improved safety with bioidentical hormones. Because observational studies have consistently shown an increased risk of breast cancer among women with higher endogenous estrogen and androgen serum concentrations (Key et al., 2002; Hankinson, 2005–2006), there is little to suggest that use of bioidentical hormones would be a safe alternative to other forms of hormone therapy.

Oral Contraceptives

Oral contraceptives with combination estrogen and progestin are also associated with an increased risk of breast cancer while women are using them (Collaborative Group on Hormonal Factors in Breast Cancer, 1996; Marchbanks et al., 2002; Hunter et al., 2010), and the risk may be greater for certain product formulations (Hunter et al., 2010). The increased risks associated with oral contraceptives are short term, and they decline after use ends, as is the case with hormone therapy after menopause. This increased risk occurs against a background of low risk for the younger women who are most likely to be taking oral contraceptives. The result is that the overall impact on the incidence of breast cancer is small. In addition, oral contraceptive use is associated with a long-term reduction in the risk of both ovarian and endometrial cancers (reviewed in La Vecchia, 2001). Use of oral contraceptives in the perimenopausal period would be expected to be associated with risks similar to hormone therapy use during the same window, although data on this practice are limited (Davidson and Helzlsouer, 2002).

Chemoprevention for Women at Increased Risk of Breast Cancer

The committee noted that for women at increased risk of breast cancer, chemoprevention with tamoxifen or raloxifene has been shown in clinical trials to reduce the risk of developing breast cancer. The clinical trials evaluating the ability of these medications to reduce the risk of breast cancer considered women with an estimated risk of being diagnosed with breast cancer within 5 years of 1.7 percent or greater to be at increased risk and eligible for the trials.[4]

Research has demonstrated that drugs that alter responses to estrogen (e.g., selective estrogen receptor modulators or SERMs) or production of estrogen (e.g., aromatase inhibitors) can substantially reduce risk of ER+ breast cancer (e.g., Cummings et al., 2009; Nelson et al., 2009; Goss et al., 2011). Tamoxifen and raloxifene, both SERMS, are two of the best known and best studied products of this type. The FDA has approved their use for this purpose by women who are considered at increased risk of breast cancer and are not at increased risk for cerebrovascular disease. Raloxifene is approved for use only after menopause. Other medications not currently approved for use for breast cancer risk reduction are also being evaluated. One category of drugs is aromatase inhibitors, designed to inhibit the conversion of androgen to estrogen, and early results from a study of the aromatase inhibitor exemestane show a reduced risk of breast cancer among high-risk women (Goss et al., 2011). Other medications being studied include other SERMs and aromatase inhibitors, bisphosphonates, and metformin (Cuzick et al., 2011).

A meta-analysis of randomized controlled trials of breast cancer prevention reported that with 5 years of tamoxifen use, women at high risk had a statistically significant reduction in the risk of invasive ER+ breast cancer (meta-analysis risk ratio 0.70, 95% CI, 0.59–0.82) compared with women who had not used the drug (Nelson et al., 2009). The same authors also reported a meta-analysis of randomized controlled trials for raloxifene, finding a statistically significant reduction in invasive ER+ breast cancer for women who used the drug compared to those who did not (meta-analysis risk ratio 0.44, 95% CI, 0.27–0.71). However, they noted that the mean age at entry into the tamoxifen studies ranged from 47 to 51 years, compared with a mean ranging from 67 to 68 years for the raloxifene studies. The authors estimated that use of tamoxifen or raloxifene for 5 years would be expected to result in 7 to 10 fewer breast cancer cases per 1,000 women per year (Nelson et al., 2009).

The Study of Tamoxifen and Raloxifene (STAR) trial was designed to provide a direct comparison of the two products. The study population consisted of postmenopausal women who were at increased risk of breast cancer, but did not have a history of cancer or various other condi-

---

[4]The risk of developing breast cancer can be estimated from statistical models that consider factors such as age, reproductive history, and personal and family history of breast cancer. In the United States, a commonly used model is the Breast Cancer Risk Assessment Tool (Gail et al., 1989; NCI, 2011) (available at http://www.cancer.gov/bcrisktool/). It is discussed further later in this chapter.

tions, including stroke, uncontrolled diabetes, or uncontrolled hypertension (Vogel et al., 2010). With 6.75 years of follow-up, the ratio of risk for invasive breast cancer with use of raloxifene to that with use of tamoxifen was 1.24 (95% CI, 1.05–1.47) (Vogel et al., 2010). Although the use of raloxifene reduced risk less than use of tamoxifen, raloxifene had fewer adverse effects than tamoxifen. Relatively few eligible women have chosen to use tamoxifen or raloxifene, at least in part because of their association with increased risk for serious adverse health effects, including endometrial cancer (tamoxifen) and stroke (Fisher et al., 2005; Vogel et al., 2010).

The committee endorses recommendations of the USPSTF (2002) that women have their breast cancer risk assessed and discuss with their health care providers whether use of tamoxifen or raloxifene as chemoprevention to reduce their risk of breast cancer is appropriate for them. Risk assessment to weigh the potential benefits and risks should be available to all women. Use may be appropriate for women who are at increased risk of breast cancer (a 5-year risk of at least 1.7 percent) and who have low risk for the adverse effects associated with these medications. The adverse effects can include menopausal symptoms, risk of deep vein thrombosis (blood clots), endometrial hyperplasia and cancer (for tamoxifen), and stroke. Benefits for menopausal women, in addition to breast cancer risk reduction, include lower fracture risks. Statistical models are available to help guide decision making regarding the use of chemoprevention for premenopausal women ages 35 and older (Gail et al., 1989) and for menopausal women ages 50 and older who are at increased risk for breast cancer (Freedman et al., 2011).

Active and Passive Smoking

Accumulating evidence points to active smoking being associated with an increase in risk for breast cancer (Reynolds et al., 2004; CalEPA, 2005; Ha et al., 2007; Collishaw et al., 2009; Secretan et al., 2009). Some evidence indicates the most consistent findings are for earlier initiation of smoking and smoking before a first full-term pregnancy (DeRoo et al., 2011; Luo et al., 2011; Xue et al., 2011). In addition, some expert reviews have concluded that the evidence is consistent with a causal association between passive smoking and increased risk for premenopausal breast cancer (CalEPA, 2005; Collishaw et al., 2009), while the evidence regarding passive smoking is described as inconclusive in the most recent review by the International Agency for Research on Cancer (IARC) review (Secretan et al., 2009). Some evidence also suggests a possible association between high levels of exposure to passive smoking and postmenopausal breast cancer (Reynolds et al., 2009; Luo et al., 2011).

Smoking poses substantial health risks in addition to any contribution it may make to increased risk of breast cancer, and the committee has no hesitation in urging women not to begin smoking, to stop smoking if they are current smokers, and to protect themselves and their children from exposure to secondhand smoke. Women who have become smokers are certainly likely to gain health benefits by ceasing to smoke.

Exposure to secondhand tobacco smoke increases the risk of several diseases (HHS, 2006), and so it should be avoided. Public and private policies that call for smoke-free environments in public spaces and workplaces help reduce exposure to secondhand smoke, especially for adults. However, children and nonsmoking adults who live with smokers may still be exposed within the home and in private cars. The evidence of increased potency of smoking before pregnancy suggests the possibility that a similar window of greater vulnerability may exist at younger ages for exposure to secondhand smoke, although exposure to secondhand smoke only during childhood does not appear to increase the risk of breast cancer (HHS, 2006; Chuang et al., 2011; Luo et al., 2011).

There is opportunity for improving health at both the individual and societal level through reduction in both active and passive exposure to tobacco smoke.

Individuals: Girls and women can avoid beginning to smoke, and those who smoke can quit. Individuals can also avoid exposing themselves and their children to secondhand smoke.

Public and private sectors: Efforts can be made to expand smoke-free environments in workplaces and public spaces. Educational programs can inform smokers and nonsmokers of the dangers that secondhand smoke presents. Efforts can also be made to encourage smoke-free homes and cars. Given that initiation of smoking generally occurs in adolescence or earlier, the evidence linking smoking before a first full-term pregnancy with increased risk of breast cancer underscores the need for effective programs geared towards smoking prevention in preteen and teenage girls.

Alcohol Consumption

Alcohol consumption has been shown to modestly increase risk for both pre- and postmenopausal breast cancer (IARC, 2010), with the largest studies suggesting a linear relation between intake and risk. Risk was estimated to increase approximately 7 percent (Collaborative Group on Hormonal Factors in Breast Cancer, 2002) to 9 percent (Smith-Warner et al., 1998) for each additional 10 grams of alcohol consumed per day. (In the United States one drink is considered to contain approximately 14 grams of alcohol [CDC, 2011].) An analysis of data from 53 studies found that women who had substantial levels of daily alcohol consumption (≥ 45 g per day) had a relative risk of breast cancer of 1.46 (95% CI, 1.3–1.6), compared to those who reported drinking no alcohol (Collaborative Group on Hormonal Factors in Breast Cancer, 2002). For those whose consumption was approximately one to two drinks per day (15–24 g), the relative risk was 1.13 (95% CI, 1.08–1.19) (Collaborative Group on Hormonal Factors in Breast Cancer, 2002).

Questions remain unresolved about whether the association between breast cancer and alcohol consumption is cumulative over years of exposure, or a time-limited and reversible association (IARC, 2010). Some studies also suggest that the increased risk associated with higher alcohol consumption (> 20 g/day) is primarily among women who use menopausal hormone therapy (Gapstur et al., 1992; Chen et al., 2002; Horn-Ross et al., 2004); however, an IARC (2010) review reported no significant variation. The health risks and potential benefits of moderate alcohol intake were evaluated and published as a formal position paper by the National Institutes of Health (Gunzerath et al., 2004). In addition to breast cancer, consumption of more than one to two drinks per day for women (more than two to three drinks per day for men) is associated with an increased risk for a variety of other cancers and other adverse health conditions (e.g., WCRF/AICR, 2007; Gronbaek, 2009; IARC, 2010). However, the moderate levels of consumption that are associated with an increased risk of breast cancer are also associated with positive outcomes such as lower mortality from cardiovascular disease (Maskarinec et al., 1998; Klatsky, 2009; Ronksley et al., 2011), which is a much larger contributor to morbidity and mortality among women than breast cancer. A meta-analysis found that for those who consumed an average of one drink or less (2.5–14.9 grams) per day, the relative risk of cardiovascular disease mortality was 0.77 (95% CI, 0.71–0.83) compared with those who consumed no alcohol (Ronksley et al., 2011). A similar reduction in risk was seen for the incidence of coronary heart disease, one specific type of cardiovascular disease.

With respect to balancing alcohol's risk of breast cancer with potential benefits, Gunzenrath and colleagues advised in their position paper, "individual women, with the help of their physicians, must weight their potential increased risk for breast cancer against their potential reduced risk for CHD [coronary heart disease] in determining whether alcohol consumption should be reduced" (Gunzerath et al., 2004, p. 833).

The committee concluded that the consistent evidence that even moderate consumption of alcohol is associated with an increased risk of breast cancer warranted note in this discussion of modifiable risk factors. However, with the lack of evidence regarding the impact on breast cancer risk of *changes* in consumption and the evidence supporting beneficial effects of moderate alcohol consumption related to cardiovascular disease, the merits of restricting or eliminating moderate alcohol consumption as a breast cancer risk reduction strategy are hard to judge for individual women. The committee urges women to confer with their health care providers about the potential benefits and risks of reducing their alcohol consumption.

Physical Activity

Reviews by the World Cancer Research Fund/American Institute for Cancer Research (WCRF/AICR, 2007, 2010) characterized as probable an association between greater physical activity and a reduction in risk for postmenopausal breast cancer. The evidence regarding reduction in risk for premenopausal breast cancer is described as limited. Other reviews (e.g., Monninkhof et al., 2007; Physical Activity Guidelines Advisory Committee, 2008; Friedenreich, 2010) have produced similar assessments of the available evidence. The beneficial effects of physical activity appear to be stronger for women of normal weight and without a family history of breast cancer; they are, however, observed in women of all races and ethnicities (Friedenreich, 2010).

Additional research is needed to clarify the type of activity, the amount, and the timing of physical activity over the life course that can produce a reduction of breast cancer risk. Three primary prevention studies (McTiernan et al., 2004a,b; Monninkhopf et al., 2009; Friedenreich et al., 2010, 2011; also reviewed in Winzer et al., 2011) offer some initial insight into the feasibility of exercise interventions to reduce risk among inactive postmenopausal women, most of whom were overweight. In these 1-year trials, it was not possible to measure changes in breast cancer risk directly. Outcomes were assessed on the basis of a variety of biomarkers considered relevant to breast cancer risk. Among the biomarkers were weight, body mass index (BMI), sex hormone concentrations, mammographic density, and insulin concentrations. For example, moderate to vigorous aerobic exercise of approximately 3 hours per week (3 to 4 days per week) among previously sedentary women ages 50–74 resulted in statistically significant decreases in weight, BMI, and abdominal fat (Friedenreich et al., 2011). In another study (McTiernan et al., 2004b), an average of nearly 3 hours per week of moderate intensity exercise among postmenopausal women resulted after a year in a statistically significant decline in serum estrogen levels, but only among the women whose percent body fat decreased by at least 2 percentage points. But a study that tested a program of 2.5 hours per week of combined aerobic exercise and strength training did not detect a significant change in serum estrogen levels, even among the women whose percent body fat declined (Monnihkhopf et al., 2009). The results of studies such as these suggest that changes considered likely to be indicative of reduced risk for breast cancer can be achieved, but only with greater frequency and duration of exercise.

Physical activity throughout the life course is generally recognized as having wide-ranging health benefits, which include the likely reduction in risk for postmenopausal breast cancer among women who are more active. The committee endorses the guidance of the Department of Health and Human Services (HHS, 2008) for regular physical activity at all ages.

Excess Weight and Weight Gain

As discussed in Chapter 3, data from 2007–2008 indicate that approximately 36 percent of adult women of all ages can be considered obese and another 29 percent as overweight (Flegal et al., 2010). The systematic review by the WCRF/AICR (2007) and subsequent updates (WCRF/AICR, 2008, 2010) classified greater body fatness[5] as convincingly associated with greater risk for postmenopausal breast cancer and adult weight gain as probably associated with increased risk. Some studies have found that the increased risk associated with weight gain is stronger for women who have not used HT (Eliassen et al., 2006; Ahn et al., 2007).

For younger women, however, greater body fatness is probably associated with reduced risk of premenopausal breast cancer (WCRF/AICR, 2007), although this is a time of life for which breast cancer risk is much lower than for older women. But weight gain earlier in life may be difficult to reverse later in life when the increase in risk caused by body fatness may have a greater effect because of the higher breast cancer risk at this life stage. It also appears that the association of greater weight and adult weight gain with increased postmenopausal breast cancer risk is dominated by the experience of white women and may not hold for African American women (Palmer et al., 2007).

The committee is persuaded that maintaining weight within what is considered a normal range (a BMI of 18.5–24.9) is appropriate guidance for all women. Overweight and obesity are associated with increased risk for a wide range of adverse health consequences beyond the specific relation to breast cancer. Preventing weight gain may be especially important because it is less clear whether overweight and obese women can reduce their risk of postmenopausal breast cancer by losing weight. The Nurses' Health Study (Eliassen et al., 2006) and the Iowa Women's Health Study (Harvie et al., 2005) found evidence of reduced risk for women who lost weight compared with those who maintained a stable weight. Other studies (Ahn et al., 2007; Teras et al., 2011), however, failed to find reduced risk among women who lost weight. For African American women, and perhaps other population

---

[5]Body fatness and overweight and obesity are commonly measured using body mass index (BMI). BMI is defined as body weight in kilograms divided by height in meters squared. The following weight categories are based on BMI values: underweight, <18.5; normal weight, 18.5–24.9; overweight, 25–29.0; and obese, ≥30.

groups for which data on weight-related breast cancer risk factor patterns are still limited, different or additional prevention strategies may be needed.

Chemicals and Consumer Products

The committee evaluated the potential role that some individual exogenously produced chemicals found in the diet, air, water, household products, and workplaces may play in the development of breast cancer in humans. Because there are vast numbers of such chemicals and often very limited evidence regarding breast cancer, the committee chose to examine the evidence for only a selected set (see Chapter 3). The comments that follow are typically specific to the chemicals that the committee reviewed. It is not possible for the committee to comment on the chemicals that it did not review. For some chemicals, relevant information may be available from other sources (e.g., Brody et al., 2007; California Breast Cancer Research Program, 2007; WCRF/AICR, 2007; EPA, 2011; IARC, 2011; NTP, 2011).

Ethylene Oxide, Benzene, and 1,3-Butadiene

Among the chemicals considered, the evidence for an association with increased risk of breast cancer was clearest for ethylene oxide. Benzene and 1,3-butadiene are also probably human breast carcinogens. Cigarette smoke is a source of exposure, either through active or passive smoking, to these chemicals (Fennell et al., 2000). All three substances are raw materials used in the production of numerous industrial chemicals. Ethylene oxide is also used for sterilization in industrial and medical settings. Benzene has been used as a fuel additive and is a natural constituent of crude oil. Vehicular emissions and gasoline vapors at filling stations are a source of exposure to both benzene and 1,3-butadiene. Although ambient air levels have been substantially curtailed through regulatory actions, widespread, low-level environmental exposure, especially to benzene, continues. While recognizing potential hazards, the committee did not have the capacity to estimate breast cancer risks at these low doses because the information necessary to do so is insufficient.

Because these chemicals are recognized carcinogens (NTP, 2011), steps are taken to reduce occupational and public exposures. However, there is limited awareness of the possible association between these chemicals and increased risk for breast cancer, and federal occupational health standards do not call for medical surveillance for breast cancer for exposed workers.[6] Women whose work involves the potential for exposure to these chemicals

---

[6]The medical surveillance guidelines for workers exposed to these chemicals are available at 29 CFR 1910.1028 App C, 29 CFR 1910.1051 App C, and 29 CFR 1910.1047 App C.

may be able to take additional steps to minimize their exposure. Accomplishing that goal, however, will require awareness of the possibility of exposure and access to appropriate resources, procedures, and policies to make minimizing exposure possible. Although women can accomplish some of this on their own, they will also have to depend on actions by employers, equipment manufacturers, and agencies responsible for ensuring workplace and environmental safety to limit or eliminate exposure. The general public can minimize exposure through avoidance of tobacco smoke and by limiting exposure to gasoline vapors and vehicular exhaust.

Other Environmental Agents

For many of the other chemicals that the committee considered, as well as those discussed in reviews by others (e.g., Brody et al., 2007; Rudel et al., 2007; Gray, 2010), little or no epidemiologic evidence on breast cancer risk is available. However, evidence from in vivo cancer bioassays, mechanistic studies, or both may suggest the potential for exposure to contribute to breast cancer in humans. Where these indications exist, the committee recommends further research to improve understanding of the relevance of the findings for humans (see Chapter 7). Many of these chemicals have been identified as probable or likely carcinogenic hazards by authoritative organizations (e.g., IARC, EPA, National Toxicology Program), but the findings are not specific to breast cancer hazard.

For two of the agents—hair dyes and non-ionizing radiation—substantial epidemiologic evidence from large populations studied over long periods of time has consistently failed to identify a significant increase in risk of breast cancer associated with exposure. The committee concluded that avoiding exposure to either hair dyes or non-ionizing radiation has little potential to contribute to a substantial reduction in breast cancer risk for individuals or the population. For certain other compounds, such as dioxins, polychlorinated biphenyls (PCBs), some metals, and vinyl chloride, regulatory actions taken many years ago have greatly reduced exposures. However, low-level exposure continues because these chemicals persist in the environment or some sources are difficult to eliminate even if they are subject to regulatory controls. Although individuals may be able to control some sources of exposure to some of these persistent chemicals (e.g., by avoiding certain types of fish known to have high levels of PCBs or dioxins), it may be difficult for individuals to act on their own to avoid or limit many of these low-level exposures.

For many of the reviewed compounds, including those discussed above, evidence of hazard may be present, but information to assess the magnitude of risk, particularly at environmentally relevant doses is lacking or inadequate, posing a substantial challenge for gauging the extent to which an individual's actions may reduce risk (Table 6-1). For example, for BPA, epidemiologic data are largely lacking, and the available studies on timing of exposure do not adequately address potentially important windows such as fetal and early life that may influence adult disease. Avoidance or reduction of exposures to such substances at the individual level may be difficult or infeasible in some cases, but eminently possible in others. For example, a small study demonstrated substantial reduction in urinary levels of BPA when participants shifted to use of minimally packaged foods (Rudel et al., 2011). Levels of BPA increased when the study participants resumed eating packaged foods. Determining the sustainability of such changes, their acceptability to a broader population, and whether such reductions would actually decrease breast cancer risk would require further investigation.

The committee recognizes, however, that existing data indicate that BPA and some other substances may be hazards to human health and may well warrant consideration of actions by regulatory agencies that are aimed at reducing future population-based exposures. Other considerations for regulators may include the possibility that exposure to multiple chemicals that contribute to mechanisms involved in breast cancer (e.g., mutagens, endocrine disruptors, etc.) may present a cumulative risk that could be controlled in part through regulatory actions on individual substances. Even where evidence regarding breast cancer is limited, evidence related to other health effects (e.g., developmental effects or other types of cancer) may provide a stronger basis for regulatory action or individual efforts to avoid exposure.

Such policy action would be based on many factors, including taking into account the impact of foreseeable substitutions for a regulated substance and the likely prospect of unanticipated substitutions of substances with as yet unknown properties. Given the limits in the evidence base regarding breast cancer, and the complexity of the analysis it would entail, it is beyond the charge and capacity of this committee to make specific recommendations for regulatory action. However, it notes that GAO (2005, 2006, 2007, 2009a,b) has called several times for improvements in monitoring and regulation of toxic chemicals, citing both constraints resulting from the 1976 Toxic Substances Control Act (TSCA) and a need for better use of the authority it does provide. Under TSCA, EPA has limited authority to require that manufacturers test products for carcinogenicity (or other health hazards), and its authority to share information that may be provided is also limited. Interested organizations can help inform the public about the current provisions for testing chemicals and encourage manufacturers to improve testing and make existing information on their products more readily available.

Dietary Supplements and Cosmetics

Dietary supplements and cosmetics are widely used products, but the FDA has limited authority to test their safety before they are marketed. Rules regarding FDA regulation of dietary supplements and cosmetics differ from those covering pharmaceutical agents. Since the passage of the 1994 Dietary Supplement Health and Education Act (DSHEA), manufacturers are responsible for ensuring that their supplement products are safe and that product label information is truthful and not misleading. No FDA approval or proof of safety or efficacy is required before dietary supplements are marketed (FDA, 2009). Similarly for cosmetics, the products are not subject to premarket approval by the FDA under the laws governing the sale and use of cosmetics (FDA, 2005).

Data from the National Health Interview Survey indicate that about 114 million Americans, or more than half of the U.S. adult population, consume dietary supplements (Cohen, 2009). Vitamin and some nutrient supplements have been well studied, but many supplements marketed as alternatives to prescription hormone therapies (e.g., for control of peri-menopausal symptoms or weakness in old age), or for improvement of athletic performance or weight loss, have not been tested for safety or effectiveness. Interest in such supplements could be amplified by messages that hormone therapies, physical inactivity, and overweight are risk factors for breast cancer. Similarly, cosmetics that are widely used by girls and women of all ages may also contain hormonally active ingredients that are intended to produce a more youthful appearance.

The limited role for the FDA in the marketing of dietary supplements and cosmetics is poorly understood. In a 2002 Harris poll, a majority of respondents believed that dietary supplements are approved by a federal regulatory agency (Taylor and Leftman, 2002). Moreover, in an online questionnaire completed by medical residents affiliated with 15 internal medicine programs, baseline knowledge about regulation of dietary supplements was poor and did not vary by training year of residency (Ashar et al., 2007). A third of the residents were not aware that the FDA does not require premarketing submission of safety or efficacy data.

The FDA does have the authority to withdraw dietary supplements and cosmetics from the market if product adulteration is discovered or if cosmetics are found to be misbranded. The FDA can declare such products adulterated when they present an unreasonable risk of illness or injury under the conditions of use. For example, a number of side effects—breast enlargement, loss of libido, cardiovascular side effects, thromboembolism, and bleeding—were found in men with prostate cancer who were taking the herbal dietary supplement PC-SPES. When chemical and bioassays of PC-SPES lots were performed, pharmacologic levels of diethylstilbestrol

(DES), warfarin, and indomethacin were found, leading to product withdrawal (Sovak et al., 2002; White, 2002). Currently, no prospective system is in place to routinely detect product adulteration prior to marketing.

The committee sees a need for better means for the FDA to prospectively survey or detect contaminants or ingredients in cosmetics and dietary supplements, including estrogenic substances that are known or possible causes of breast cancer, or otherwise monitor products designed to have pharmacologically active levels of such substances. It also urges consumer organizations and other interested groups to develop educational programs for the public and the health professions to enhance awareness of the rules governing marketing of dietary supplements and cosmetics, including that manufacturers are responsible for establishing the safety of these products and that the FDA has limited authority to act before they are marketed. Consumers and interested organizations can also urge manufacturers to provide consumers with more information regarding the presence of potentially hormonally active ingredients in dietary supplements and cosmetics, ideally by identifying such ingredients on product labels.

ASSESSING THE POTENTIAL IMPACT OF RISK REDUCTION EFFORTS

The committee has proposed several actions that women could take that may reduce their risk of breast cancer, but based on the existing literature, found it difficult to estimate the magnitude of the potential impact of these actions for either individuals or population groups. Although numerous studies have established associations between risk factors and breast cancer incidence, those associations may or may not be causal. If a risk factor is not causally linked to breast cancer, then changing exposure to that factor will not have a direct impact on breast cancer risk. In addition, there is limited research demonstrating that the effect of an exposure on breast cancer risk can be reversed by removing the exposure or, if it could be reversed, the magnitude of risk reduction that could be achieved by modifying or preventing the exposure. For example, for combination hormone therapy, for which a clinical trial has been conducted, the magnitude of breast cancer risk has been quantified, and therefore the excess risk that can be avoided by refraining from use of combination HT can be quantified.

Here, the committee offers some perspective on levels of breast cancer risks, interrelationships among risk factors, and what we know about risk reduction.

Average Risk for Breast Cancer

Estimates of risk and changes in risk should be viewed with an understanding of the incidence of the disease in the population. For breast cancer, incidence is very low until women reach their thirties, when it begins to rise steadily into older ages. But even among women in their seventies, only a small minority will be diagnosed with breast cancer. The data in Table 6-2 show the percentage of women who on average would be expected to receive a diagnosis of invasive breast cancer within 10 years of a given age. For example, for a group of 50-year-old white women at average risk, this 10-year risk is 2.43 percent Thus, out of 100 white women aged 50 years who are followed for 10 years, 2 to 3 will be diagnosed with breast cancer and 97 to 98 will not. It is also possible to calculate risk for longer periods, or even for a lifetime: the cumulative risk from birth to the end of life is 12.57 percent in white women (not shown in the table). This is how the familiar statistic of about "one in eight" white women expected to be diagnosed with breast cancer over a lifetime is derived (NCI, 2010, Table 4-18).

Because each number in Table 6-2 is an average among women in that age and race/ethnic group, there are obviously women whose risk is higher than the average and women whose risk is lower. The concept of relative risk (see Chapter 2) can be illustrated in this context. In a hypothetical 10-year study in a group of 50-year-old white women who have the average risk shown in Table 6-2, a certain risk factor might have a *relative* risk of 1.5—which is a 50 percent increase in risk. If half of the women have that risk factor and half do not, then the overall 2.43 percent 10-year risk of breast cancer would be just the middle ground between a risk of approximately 2.9 percent for those with the risk factor, and approximately 1.9 percent for those without it. In other words, for 50-year-old white women, the risk of being diagnosed with breast cancer in the next 10 years is about 3 out of 100 women in those who have the risk factor, compared with 2 of 100 women who do not.

Risk Estimates for Individuals

As in the hypothetical example just described, observational studies or controlled trials in groups of women produce estimates of the risk of breast cancer associated with given exposures that are based on the experience of the overall study population. A separate but related question is what this means for the individual who is exposed. Because many factors can increase or decrease an individual's risk of cancer, the risk associated with a single factor has to be put into context.

The Breast Cancer Risk Assessment Tool (Gail et al., 1989; NCI, 2011a) and the Tyrer-Cuzick breast cancer risk assessment model (Tyrer et

TABLE 6-2 Absolute Risk, Expressed as a Percentage of Women at a Specified Age Expected to Be Diagnosed with Invasive Breast Cancer Within the Next 10 Years

| Current Age | Race/Ethnicity | | | | |
	White	Black	Asian/ Pacific Islander	American Indian/ Alaska Native	Hispanic (any race)
0	0.00	0.00	0.00	0.00	0.00
10	0.00	0.00	0.00	0.00	0.00
20	0.06	0.08	0.06	0.03	0.05
30	0.43	0.47	0.39	0.32	0.33
40	1.46	1.41	1.33	1.05	1.08
50	2.43	2.24	1.96	1.41	1.73
60	3.59	3.08	2.41	2.09	2.44
70	3.93	3.23	2.35	1.76	2.54
80	3.12	2.72	1.86	1.65	1.95

NOTES: A percent of 0.00 represents a value that is less than 0.005. Incidence data are from the SEER 17 areas (San Francisco, Connecticut, Detroit, Hawaii, Iowa, New Mexico, Seattle, Utah, Atlanta, San Jose-Monterey, Los Angeles, Alaska Native Registry, Rural Georgia, California excluding SF/SJM/LA, Kentucky, Louisiana, and New Jersey).
SOURCE: NCI (2010).

al., 2004) are designed to generate estimates of absolute risk for individuals in conjunction with absolute risk estimates for the general population as a reference.[7] The Breast Cancer Risk Assessment Tool (NCI, 2011a) uses a limited set of characteristics (e.g., age, reproductive history, family history) to assess risk for an individual who is a member of a group with these characteristics and generates an estimate of the absolute risk over the next 5 years for women in that group. For comparison, the model generates an estimate of average risk for women of the same age. An individual woman's characteristics may put her in a group at higher or lower risk than the average. The performance of this model is better for white women than women of other races or ethnicities and is intended for women who are at least 35 years old (NCI, 2011b).

These tools are used primarily to guide decisions about medical care in clinical practice, including whether a woman's risk of breast cancer is high enough to make her eligible for chemopreventive medications, such as

---

[7]The Breast Cancer Risk Assessment Tool is based on breast cancer rates in the U.S. population, and the Tyrer-Cuzick model is based on breast cancer rates from the United Kingdom.

tamoxifen or raloxifene. Because these risk assessment tools do not make use of information about environmental exposures, the committee did not examine them in detail.

Population Attributable Risk for Understanding the Relative Contributions of Different Factors

Table 6-2 provides not only estimates of risk for individual women, but also provides a framework for understanding the concept of population attributable risk (PAR), which was introduced in Chapter 2 and discussed in Chapter 4. The PAR represents a population-based measure of the percentage of excess cases associated with the exposure of interest (i.e., among the exposed in comparison with the unexposed) that also takes into account the distribution of the exposure within the population. While it has sometimes been defined as the proportion of all cases that would not have occurred if exposure to a causal factor was removed from the population (Rothman and Greenland, 1998), this definition represents the ideal: it assumes that all observed associations are actually causal. In reality, many associations observed in epidemiologic studies are confounded by other factors, and when those studies are used to estimate the PAR, the PAR becomes confounded. The PAR is useful for summarizing current understanding of the relative contributions of different factors to the overall "burden" or incidence of breast cancer in the population, when confounding has been adequately controlled. It is helpful for researchers and policy makers in assessing possible opportunities to reduce disease burden through public health interventions that target specific modifiable risk factors in the population as a whole, but has limited applicability for an individual.

For instance, the PAR can be used to estimate how many cases of breast cancer might be prevented if half of the women offered unnecessarily high levels of medical radiation were able to avoid those high levels, or if an additional 25 percent of women did not gain weight and become obese or overweight by the time they reached menopause. The PAR itself is an estimate of the maximum potential benefit of eliminating a risk factor; it is not an anticipated outcome, partly because most risk factors are unlikely to be completely eliminated and partly because some risk factors are proxies for others that are causal. In other words, the PAR estimates assume that the studies identified truly causal associations and that any remaining confounding or other biases would have little impact on the estimated role of the factor under study. Finally, the PAR may be different in other populations with a different combination of characteristics, even if the proportion with a specific risk factor is the same.

Unlike the risk estimates of Table 6-2, which are for women in the population *before* any of them develop breast cancer, the calculation of the PAR is based on all cases of the disease observed in a specific population of women *after* they have been diagnosed. These cases of disease then represent 100 percent, and the PAR for a given risk factor is the proportion or percentage of these women who have breast cancer in whom that factor may play a causal role. Sometimes the PAR is calculated for a group of factors rather than a single one, but it is always calculated as a percentage of all persons in a specific population with the disease (e.g., women with breast cancer). Values calculated for individual risk factors *cannot* be summed to generate an estimate of their combined contribution to risk. This is because many cases of breast cancer are the result of multiple risk factors that interact with each other. Therefore, an estimate of the combined contribution of several individual risk factors to the total number of cases in the population must allow for the nature of the interaction of those factors among women who have breast cancer.

Estimates of PARs vary across studies. (A table summarizing estimates from several studies appears in Appendix D.) For example, alcohol ranges from 2 percent (Tseng et al., 1999) to 11 percent (Mezzetti et al., 1998), whereas hormone therapy has PAR estimates from approximately 4 percent in the United States in about 2001 (Clarke et al., 2006) to 27 percent in Norway in the 1990s (Bakken et al., 2004), and physical inactivity, from 6 percent in Canada in 2006 (Neutel and Morrison, 2010) to 20 percent in a combination of several European countries in 2002 (Friedenreich et al., 2010). The variation in PAR estimates arises from differences across studies in the prevalence of the risk factors in the study populations as well as in other characteristics of the study populations and in the design and quality of the studies. For example, if combination hormone therapy is widely used in a study population, its PAR would tend to higher than the PAR for hormone therapy in a study population with relatively limited use. PARs should at best be viewed as ballpark estimates of potential impact on breast cancer risk on a population level, under the assumption that the associations are causal.

Implications for Breast Cancer Reduction

As indicated above, the PAR estimates the percentage reduction in disease burden that can be achieved on a population level with a reduction in the prevalence of a risk factor. The estimated benefit on an absolute scale either for the population or for an individual may be small (Petracci et al., 2011). For a woman with a diagnosis of breast cancer who is of normal weight, the contribution of being overweight or obese to her breast cancer has to be zero. Alternatively, for a woman with breast cancer who is overweight or obese, the chances that her weight contributed to the development of her breast cancer might be higher than the contribution of overweight and obesity to breast cancer cases in the population as a whole. An uncommon exposure will usually have a small PAR because the percentage of all cases that is attributable to the exposure will be small. However, for a woman who has that exposure, reducing or eliminating it could substantially lower her risk and be very important to her individually. That is, a rare, high-risk exposure may have little impact on population rates of cancer, but it may be a quite important determinant of an exposed woman's personal risk.

Recent efforts have tried to further clarify risks for both individual women and populations by developing models that estimate the absolute risk of breast cancer from relative risks and estimates of attributable risk. In a study by Petracci et al. (2011), the authors used data on Italian women to develop a model to predict breast cancer risk, making use of both nonmodifiable risk factors and the modifiable risk factors of BMI, alcohol consumption, and physical activity. Data from a cohort study were used to assess the potential impact on absolute breast cancer risk of reducing exposures to the modifiable risk factors. The projected 20-year absolute risk of breast cancer for 65-year-old women, for example, ranged from 6.5 to 18.6 percent, depending on their risk profiles. If these women optimized their BMI, alcohol consumption, and physical activity, the estimated 20-year absolute risks would be reduced to 4.9 and 14.1 percent, respectively (Petracci et al., 2011). Presentation of the absolute risk reductions along with estimates of relative risk and the PAR reduction that could maximally be achieved may be a useful approach to both individual counseling and public health decision making (Schwartz et al., 2006; Akl et al., 2011; Helzlsouer, 2011). It illustrates the well-known concept that small changes at the individual level can have a large impact at the population level (Rose, 1992).

SUMMARY

Many of the established risk factors for breast cancer—age, sex, age at menarche and menopause, age at first full-term pregnancy—offer little or no opportunity to intervene. For a limited set of other risk factors, evidence suggests that action can be taken in ways that that have the potential to reduce risk for breast cancer for many women: eliminating unnecessary medical radiation throughout life, avoiding use of postmenopausal hormone therapy, avoiding active and passive smoking, reducing alcohol consumption, increasing physical activity, and minimizing weight gain. Chemoprevention may be an appropriate choice for some women.

For the many chemicals that are manufactured or generated as by-products of other processes, the committee found little basis in the human evidence it examined to point to avoiding or eliminating exposure as a specific strategy for reducing breast cancer risk. Exceptions were benzene,

1,3-butadiene, ethylene oxide, for which certain measures to control occupational exposures are already be in place. However, these chemicals can also be encountered by the general public (although likely at much lower exposure levels) through exposure to facilities emissions, tobacco smoke, and gasoline vapors and vehicular exhaust (benzene and 1,3-butadiene). While for other compounds that were reviewed, such as BPA, animal and mechanistic evidence may indicate breast cancer hazard is biologically plausible, given sufficient dosing, information to assess the magnitude of risk in humans is lacking or inadequate in human studies, posing a substantial challenge for gauging the extent to which an individual's actions may reduce risk.

Even when action appears possible, most approaches to risk reduction come with potentially complex trade-offs. These trade-offs may be social or economic (e.g., the potential influence of earlier age at first birth on a woman's education or employment), or they may be health related (e.g., moderate alcohol consumption increases breast cancer risk, but it may reduce risk of heart disease; tamoxifen reduces risk for breast cancer but increases risk for stroke and endometrial cancer). It is also important to keep in mind that what the committee has outlined in this chapter are areas where the evidence indicates that action is likely to reduce risk in an average *population*. The actual change in risk for any *individual* woman who takes such actions might range from very small to moderate.

Chapter 7 outlines the committee's recommendations for further research to strengthen the knowledge base on breast cancer and, hopefully, to point to more and better opportunities to reduce risk for this disease.

REFERENCES

Ahn, J., A. Schatzkin, J. V. Lacey, Jr., D. Albanes, R. Ballard-Barbash, K. F. Adams, V. Kipnis, T. Mouw, et al. 2007. Adiposity, adult weight change, and postmenopausal breast cancer risk. *Arch Intern Med* 167(19):2091–2102.

Akl, E. A., A. D. Oxman, J. Herrin, G. E. Vist, I. Terrenato, F. Sperati, C. Costiniuk, D. Blank, et al. 2011. Using alternative statistical formats for presenting risks and risk reductions. *Cochrane Database Syst Rev* (3):CD006776.

Alliance for Radiation Safety in Pediatric Imaging. 2008. *Image gently.* http://www.pedrad. org/associations/5364/ig/ (accessed June 23, 2011).

Ashar, B. H., T. N. Rice, and S. D. Sisson. 2007. Physicians' understanding of the regulation of dietary supplements. *Arch Intern Med* 167(9):966–969.

Bakken, K., E. Alsaker, A. E. Eggen, and E. Lund. 2004. Hormone replacement therapy and incidence of hormone-dependent cancers in the Norwegian Women and Cancer study. *Int J Cancer* 112(1):130–134.

Baumann, B. M., E. H. Chen, A. M. Mills, L. Glaspey, N. M. Thompson, M. K. Jones, and M. C. Farner. 2011. Patient perceptions of computed tomographic imaging and their understanding of radiation risk and exposure. *Ann Emerg Med* 58(1):1–7.

Beral, V., G. Reeves, D. Bull, and J. Green. 2011. Breast cancer risk in relation to the interval between menopause and starting hormone therapy. *J Natl Cancer Inst* 103(4):296–305.

Berrington de Gonzalez, A., M. Mahesh, K. P. Kim, M. Bhargavan, R. Lewis, F. Mettler, and C. Land. 2009. Projected cancer risks from computed tomographic scans performed in the United States in 2007. *Arch Intern Med* 169(22):2071–2077.

Bogdanich, W. 2011. West Virginia hospital overradiated brain scan patients, records show. *New York Times*, March 5.

Bogdanich, W., and K. Rebelo. 2010. A pinpoint beam strays invisibly, harming instead of healing. *New York Times*, December 28.

Boyd, N. F., L. J. Martin, M. Yaffe, and S. Minkin. 2009. Mammographic density. *Breast Cancer Res* 11(Suppl 3):S4.

Boyd, N. F., L. J. Martin, M. Bronskill, M. J. Yaffe, N. Duric, and S. Minkin. 2010. Breast tissue composition and susceptibility to breast cancer. *J Natl Cancer Inst* 102(16): 1224–1237.

Brinton, L. A., D. Richesson, M. F. Leitzmann, G. L. Gierach, A. Schatzkin, T. Mouw, A. R. Hollenbeck, and J. V. Lacey, Jr. 2008. Menopausal hormone therapy and breast cancer risk in the NIH–AARP Diet and Health Study Cohort. *Cancer Epidemiol Biomarkers Prev* 17(11):3150–3160.

Brody, J. G., K. B. Moysich, O. Humblet, K. R. Attfield, G. P. Beehler, and R. A. Rudel. 2007. Environmental pollutants and breast cancer: Epidemiologic studies. *Cancer* 109(12 Suppl):2667–2711.

CalEPA (California Environmental Protection Agency). 2005. *Proposed identification of environmental tobacco smoke as a toxic air contaminant.* http://www.arb.ca.gov/regact/ets2006/ets2006.htm (accessed November 9, 2011).

California Breast Cancer Research Program. 2007. *Identifying gaps in breast cancer research: Addressing disparities and the roles of the physical and social environment.* http://cbcrp.org/sri/reports/identifyingGaps/index.php (accessed October 25, 2011).

Cantor, K. P., C. M. Villanueva, D. T. Silverman, J. D. Figueroa, F. X. Real, M. Garcia-Closas, N. Malats, S. Chanock, et al. 2010. Polymorphisms in GSTT1, GSTZ1, and CYP2E1, disinfection by-products, and risk of bladder cancer in Spain. *Environ Health Perspect* 118(11):1545–1550.

CDC (Centers for Disease Control and Prevention). 2011. *Alcohol and public health—frequently asked questions.* http://www.cdc.gov/alcohol/faqs.htm#heavyDrinking (accessed November 9, 2011).

Chen, W. Y., G. A. Colditz, B. Rosner, S. E. Hankinson, D. J. Hunter, J. E. Manson, M. J. Stampfer, W. C. Willett, et al. 2002. Use of postmenopausal hormones, alcohol, and risk for invasive breast cancer. *Ann Intern Med* 137(10):798–804.

Chlebowski, R. T., L. H. Kuller, R. L. Prentice, M. L. Stefanick, J. E. Manson, M. Gass, A. K. Aragaki, J. K. Ockene, et al. 2009. Breast cancer after use of estrogen plus progestin in postmenopausal women. *N Engl J Med* 360(6):573–587.

Chuang, S. C., V. Gallo, D. Michaud, K. Overvad, A. Tjonneland, F. Clavel-Chapelon, I. Romieu, K. Straif, et al. 2011. Exposure to environmental tobacco smoke in childhood and incidence of cancer in adulthood in never smokers in the european prospective investigation into cancer and nutrition. *Cancer Causes Control* 22(3):487–494.

Cirigliano, M. 2007. Bioidentical hormone therapy: A review of the evidence. *J Womens Health (Larchmt)* 16(5):600–631.

Clarke, C. A., D. M. Purdie, and S. L. Glaser. 2006. Population attributable risk of breast cancer in white women associated with immediately modifiable risk factors. *BMC Cancer* 6:170.

Cohen, P. A. 2009. American roulette—contaminated dietary supplements. *N Engl J Med* 361(16):1523–1525.

Collaborative Group on Hormonal Factors in Breast Cancer. 1996. Breast cancer and hormonal contraceptives: Collaborative reanalysis of individual data on 53,297 women with breast cancer and 100,239 women without breast cancer from 54 epidemiological studies. *Lancet* 347(9017):1713–1727.

Collaborative Group on Hormonal Factors in Breast Cancer. 1997. Breast cancer and hormone replacement therapy: Collaborative reanalysis of data from 51 epidemiological studies of 52,705 women with breast cancer and 108,411 women without breast cancer. *Lancet* 350(9084):1047–1059.

Collaborative Group on Hormonal Factors in Breast Cancer. 2002. Alcohol, tobacco and breast cancer—collaborative reanalysis of individual data from 53 epidemiological studies, including 58,515 women with breast cancer and 95,067 women without the disease. *Br J Cancer* 87(11):1234–1245.

Collishaw, N. C., N. F. Boyd, K. P. Cantor, S. K. Hammond, K. C. Johnson, J. Millar, A. B. Miller, M. Miller, J. R. Palmer, A. G. Salmon, and F. Turcotte. 2009. *Canadian expert panel on tobacco smoke and breast cancer risk.* OTRU Special Report Series. Toronto, Canada: Ontario Tobacco Research Unit. http://www.otru.org/pdf/special/expert_panel_tobacco_breast_cancer.pdf (accessed October 29, 2011).

Cummings, S. R., J. A. Tice, S. Bauer, W. S. Browner, J. Cuzick, E. Ziv, V. Vogel, J. Shepherd, et al. 2009. Prevention of breast cancer in postmenopausal women: Approaches to estimating and reducing risk. *J Natl Cancer Inst* 101(6):384–398.

Cuzick, J. 2008. Assessing risk for breast cancer. *Breast Cancer Res* 10(Suppl 4):S13.

Cuzick, J., A. DeCensi, B. Arun, P. H. Brown, M. Castiglione, B. Dunn, J. F. Forbes, A. Glaus, et al. 2011. Preventive therapy for breast cancer: A consensus statement. *Lancet Oncol* 12(5):496–503.

Davidson, N. E., and K. J. Helzlsouer. 2002. Good news about oral contraceptives. *N Engl J Med* 346(26):2078–2079.

DeRoo, L. A., P. Cummings, and B. A. Mueller. 2011. Smoking before the first pregnancy and the risk of breast cancer: A meta-analysis. *Am J Epidemiol* 174(4):390-402.

Don, S. 2011. Pediatric digital radiography summit overview: State of confusion. *Pediatr Radiol* 41(5):567–572.

Eliassen, A. H., G. A. Colditz, B. Rosner, W. C. Willett, and S. E. Hankinson. 2006. Adult weight change and risk of postmenopausal breast cancer. *JAMA* 296(2):193–201.

Endocrine Society. 2006. *Position statement: Bioidentical hormones.* http://www.endo-society.org/advocacy/policy/upload/BH_Position_Statement_final_10_25_06_w_Header.pdf (accessed November 8, 2011).

EPA (Environmental Protection Agency). 2005. *Guidelines for carcinogen risk assessment.* EPA/630/P-03/001F. http://www.epa.gov/raf/publications/pdfs/CANCER_GUIDELINES_FINAL_3-25-05.PDF (accessed November 17, 2011).

EPA. 2011. *Integrated risk information system (IRIS).* http://www.epa.gov/ncea/iris/index.html (accessed October 28, 2011).

Fazel, R., H. M. Krumholz, Y. Wang, J. S. Ross, J. Chen, H. H. Ting, N. D. Shah, K. Nasir, et al. 2009. Exposure to low-dose ionizing radiation from medical imaging procedures. *N Engl J Med* 361(9):849–857.

FDA (Food and Drug Administration). 2005. *FDA authority over cosmetics.* http://www.fda.gov/Cosmetics/GuidanceComplianceRegulatoryInformation/ucm074162.htm (accessed November 17, 2011).

FDA. 2009. *Consumer information: Overview of dietary supplements.* http://www.fda.gov/Food/DietarySupplements/ConsumerInformation/ucm110417.htm (accessed November 9, 2011).

FDA. 2010. *Initiative to reduce unnecessary radiation exposure from medical imaging.* http://www.fda.gov/radiation-emittingproducts/radiationsafety/radiationdosereduction/ucm199904.htm (accessed November 17, 2011).

Fennell, T. R., J. P. MacNeela, R. W. Morris, M. Watson, C. L. Thompson, and D. A. Bell. 2000. Hemoglobin adducts from acrylonitrile and ethylene oxide in cigarette smokers: Effects of glutathione S-transferase T1-null and M1-null genotypes. *Cancer Epidemiol Biomarkers Prev* 9(7):705–712.

Fisher, B., J. P. Costantino, D. L. Wickerham, R. S. Cecchini, W. M. Cronin, A. Robidoux, T. B. Bevers, M. T. Kavanah, et al. 2005. Tamoxifen for the prevention of breast cancer: Current status of the National Surgical Adjuvant Breast and Bowel Project P-1 study. *J Natl Cancer Inst* 97(22):1652–1662.

Flegal, K. M., M. D. Carroll, C. L. Ogden, and L. R. Curtin. 2010. Prevalence and trends in obesity among US adults, 1999–2008. *JAMA* 303(3):235–241.

Freedman, A. N., B. Yu, M. H. Gail, J. P. Costantino, B. I. Graubard, V. G. Vogel, G. L. Anderson, and W. McCaskill-Stevens. 2011. Benefit/risk assessment for breast cancer chemoprevention with raloxifene or tamoxifen for women age 50 years or older. *J Clin Oncol* 29(17):2327–2333.

Friedenreich, C. M. 2010. The role of physical activity in breast cancer etiology. *Semin Oncol* 37(3):297–302.

Friedenreich, C. M., C. G. Woolcott, A. McTiernan, R. Ballard-Barbash, R. F. Brant, F. Z. Stanczyk, T. Terry, N. F. Boyd, et al. 2010. Alberta physical activity and breast cancer prevention trial: Sex hormone changes in a year-long exercise intervention among postmenopausal women. *J Clin Oncol* 28(9):1458–1466.

Friedenreich, C. M., C. G. Woolcott, A. McTiernan, T. Terry, R. Brant, R. Ballard-Barbash, M. L. Irwin, C. A. Jones, et al. 2011. Adiposity changes after a 1-year aerobic exercise intervention among postmenopausal women: A randomized controlled trial. *Int J Obes* (Lond) 35(3):427–435.

Gail, M. H., L. A. Brinton, D. P. Byar, D. K. Corle, S. B. Green, C. Schairer, and J. J. Mulvihill. 1989. Projecting individualized probabilities of developing breast cancer for white females who are being examined annually. *J Natl Cancer Inst* 81(24):1879–1886.

GAO (Government Accountability Office). 2005. *Options exist to improve EPA's ability to assess health risks and manage its Chemical Review Program.* GAO-05-458. Washington, DC: GAO. http://www.gao.gov/new.items/d05458.pdf (accessed November 9, 2011).

GAO. 2006. *Actions are needed to improve the effectiveness of EPA's Chemical Review Program.* GAO-06-1032T. Washington, DC: GAO. http://www.gao.gov/new.items/d061032t.pdf (accessed November 9, 2011).

GAO. 2007. *Comparison of U.S. and recently enacted European Union approaches to protect against the risks of toxic chemicals.* GAO-07-825. Washington, DC: GAO. http://www.gao.gov/new.items/d07825.pdf (accessed November 9, 2011).

GAO. 2009a. *Options for enhancing the effectiveness of the Toxic Substances Control Act.* GAO-09-428T. Washington, DC: GAO. http://www.gao.gov/new.items/d09428t.pdf (accessed November 9, 2011).

GAO. 2009b. *Biomonitoring: EPA needs to coordinate its research strategy and clarify its authority to obtain biomonitoring data.* GAO-09-353. Washington, DC: GAO. http://www.gao.gov/new.items/d09353.pdf (accessed October 29, 2011).

Gapstur, S. M., J. D. Potter, T. A. Sellers, and A. R. Folsom. 1992. Increased risk of breast cancer with alcohol consumption in postmenopausal women. *Am J Epidemiol* 136(10):1221–1231.

Gibbons, J., and S. Laha. 1999. Water purification systems: A comparative analysis based on the occurrence of disinfection by-products. *Environ Pollut* 106(3):425–428.

Goss, P. E., J. N. Ingle, J. E. Ales-Martinez, A. M. Cheung, R. T. Chlebowski, J. Wactawski-Wende, A. McTiernan, J. Robbins, et al. 2011. Exemestane for breast-cancer prevention in postmenopausal women. *N Engl J Med* 364(25):2381–2391.

Gray, J. 2010. *State of the evidence: The connection between breast cancer and the environment,* 6th ed. San Francisco, CA: Breast Cancer Fund.

Gronbaek, M. 2009. The positive and negative health effects of alcohol—and the public health implications. *J Intern Med* 265(4):407–420.

Gunzerath, L., V. Faden, S. Zakhari, and K. Warren. 2004. National Institute on Alcohol Abuse and Alcoholism report on moderate drinking. *Alcohol Clin Exp Res* 28(6): 829–847.

Ha, M., K. Mabuchi, A. J. Sigurdson, D. M. Freedman, M. S. Linet, M. M. Doody, and M. Hauptmann. 2007. Smoking cigarettes before first childbirth and risk of breast cancer. *Am J Epidemiol* 166(1):55–61.

Hammond, E. C., I. J. Selikoff, and H. Seidman. 1979. Asbestos exposure, cigarette smoking and death rates. *Ann N Y Acad Sci* 330:473–490.

Hankinson, S. E. 2005–2006. Endogenous hormones and risk of breast cancer in postmenopausal women. *Breast Dis* 24:3–15.

Harvie, M., A. Howell, R. A. Vierkant, N. Kumar, J. R. Cerhan, L. E. Kelemen, A. R. Folsom, and T. A. Sellers. 2005. Association of gain and loss of weight before and after menopause with risk of postmenopausal breast cancer in the Iowa Women's Health Study. *Cancer Epidemiol Biomarkers Prev* 14(3):656–661.

Helzlsouer, K. J. 2011. The numbers game: The risky business of projecting risk. *J Natl Cancer Inst* 103(13):992–993.

HHS (Department of Health and Human Services). 2006. *The health consequences of involuntary exposure to tobacco smoke: A report of the Surgeon General.* Atlanta, GA: Department of Health and Human Services, Centers for Disease Control and Prevention, Coordinating Center for Health Promotion, National Center for Chronic Disease Prevention and Health Promotion, Office on Smoking and Health.

HHS. 2008. *2008 Physical activity guidelines for Americans.* http://www.health.gov/paguidelines/guidelines/default.aspx (accessed August 9, 2011).

Horn-Ross, P. L., A. J. Canchola, D. W. West, S. L. Stewart, L. Bernstein, D. Deapen, R. Pinder, R. K. Ross, et al. 2004. Patterns of alcohol consumption and breast cancer risk in the California Teachers Study cohort. *Cancer Epidemiol Biomarkers Prev* 13(3):405–411.

Huang, Z., S. E. Hankinson, G. A. Colditz, M. J. Stampfer, D. J. Hunter, J. E. Manson, C. H. Hennekens, B. Rosner, et al. 1997. Dual effects of weight and weight gain on breast cancer risk. *JAMA* 278(17):1407–1411.

Hunter, D. J., G. A. Colditz, S. E. Hankinson, S. Malspeis, D. Spiegelman, W. Chen, M. J. Stampfer, and W. C. Willett. 2010. Oral contraceptive use and breast cancer: A prospective study of young women. *Cancer Epidemiol Biomarkers Prev* 19(10):2496–2502.

IARC (International Agency for Research on Cancer). 2010. *Alcohol consumption and ethyl carbamate. IARC monographs on the evaluation of carcinogenic risks to humans.* Vol. 96. Lyon, France: IARC.

IARC. 2011. *IARC monographs on the evaluation of carcinogenic risks to humans.* http://monographs.iarc.fr/ (accessed October 28, 2011).

Key, T., P. Appleby, I. Barnes, and G. Reeves. 2002. Endogenous sex hormones and breast cancer in postmenopausal women: Reanalysis of nine prospective studies. *J Natl Cancer Inst* 94(8):606–616.

Klatsky, A. L. 2009. Alcohol and cardiovascular diseases. *Expert Rev Cardiovasc Ther* 7(5): 499–506.

La Vecchia, C. 2001. Epidemiology of ovarian cancer: A summary review. *Eur J Cancer Prev* 10(2):125–129.

LaCroix, A. Z., R. T. Chlebowski, J. E. Manson, A. K. Aragaki, K. C. Johnson, L. Martin, K. L. Margolis, M. L. Stefanick, et al. 2011. Health outcomes after stopping conjugated equine estrogens among postmenopausal women with prior hysterectomy: A randomized controlled trial. *JAMA* 305(13):1305–1314.

Lee, C. I., A. H. Haims, E. P. Monico, J. A. Brink, and H. P. Forman. 2004. Diagnostic CT scans: Assessment of patient, physician, and radiologist awareness of radiation dose and possible risks. *Radiology* 231(2):393–398.

Luo, J., K. L. Margolis, J. Wactawski-Wende, K. Horn, C. Messina, M. L. Stefanick, H. A. Tindle, E. Tong, et al. 2011. Association of active and passive smoking with risk of breast cancer among postmenopausal women: A prospective cohort study. *BMJ* 342:d1016.

Marchbanks, P. A., J. A. McDonald, H. G. Wilson, S. G. Folger, M. G. Mandel, J. R. Daling, L. Bernstein, K. E. Malone, et al. 2002. Oral contraceptives and the risk of breast cancer. *N Engl J Med* 346(26):2025–2032.

Maskarinec, G., L. Meng, and L. N. Kolonel. 1998. Alcohol intake, body weight, and mortality in a multiethnic prospective cohort. *Epidemiology* 9(6):654–661.

McTiernan, A., S. S. Tworoger, K. B. Rajan, Y. Yasui, B. Sorenson, C. M. Ulrich, J. Chubak, F. Z. Stanczyk, et al. 2004a. Effect of exercise on serum androgens in postmenopausal women: A 12-month randomized clinical trial. *Cancer Epidemiol Biomarkers Prev* 13(7): 1099–1105.

McTiernan, A., S. S. Tworoger, C. M. Ulrich, Y. Yasui, M. L. Irwin, K. B. Rajan, B. Sorensen, R. E. Rudolph, et al. 2004b. Effect of exercise on serum estrogens in postmenopausal women: A 12-month randomized clinical trial. *Cancer Res* 64(8):2923–2928.

Mezzetti, M., C. La Vecchia, A. Decarli, P. Boyle, R. Talamini, and S. Franceschi. 1998. Population attributable risk for breast cancer: Diet, nutrition, and physical exercise. *J Natl Cancer Inst* 90(5):389–394.

Million Women Study Collaborators. 2003. Breast cancer and hormone-replacement therapy in the Million Women Study. *Lancet* 362(9382):419–427.

Monninkhof, E. M., S. G. Elias, F. A. Vlems, I. van der Tweel, A. J. Schuit, D. W. Voskuil, and F. E. van Leeuwen. 2007. Physical activity and breast cancer: A systematic review. *Epidemiology* 18(1):137–157.

Monninkhof, E. M., M. J. Velthuis, P. H. Peeters, J. W. Twisk, and A. J. Schuit. 2009. Effect of exercise on postmenopausal sex hormone levels and role of body fat: A randomized controlled trial. *J Clin Oncol* 27(27):4492–4499.

Moreno, M. A. 2011. Advice for patients: Decreasing unnecessary radiation exposure for children. *Arch Pediatr Adolesc Med* 165(5):480.

NCI (National Cancer Institute). 2010. *SEER cancer statistics review, 1975–2007.* Edited by S. F. Altekruse, C. L. Kosary, M. Krapcho, N. Neyman, R. Aminou, W. Waldron, J. Ruhl, N. Howlader, Z. Tatalovich, H. Cho, A. Mariotto, M. P. Eisner, D. R. Lewis, K. Cronin, H. S. Chen, E. J. Feuer, D. G. Stinchcomb, and B. K. Edwards. Bethesda, MD: NCI. http://seer.cancer.gov/csr/1975_2007/ (accessed January 6, 2011).

NCI. 2011a. *Breast cancer risk assessment tool.* http://www.cancer.gov/bcrisktool/ (accessed December 21, 2011).

NCI. 2011b. *Breast cancer risk assessment tool: About the tool.* http://www.cancer.gov/bcrisktool/about-tool.aspx (accessed December 21, 2011).

Nelson, H. D., R. Fu, J. C. Griffin, P. Nygren, M. E. Smith, and L. Humphrey. 2009. Systematic review: Comparative effectiveness of medications to reduce risk for primary breast cancer. *Ann Intern Med* 151(10):703–715, W-226–W-735.

Neutel, C. I., and H. Morrison. 2010. Could recent decreases in breast cancer incidence really be due to lower HRT use? Trends in attributable risk for modifiable breast cancer risk factors in Canadian women. *Can J Public Health* 101(5):405–409.

NTP (National Toxicology Program). 2007a. *Toxicology and carcinogenesis studies of dibromoacetic acid (CAS No. 631-64-1) in F344/N rats and B6C3F1 mice (drinking water studies)*. NTP TR 537. Research Triangle Park, NC: NTP. http://ntp.niehs.nih.gov/files/537_FINAL_web.pdf (accessed December 21, 2011).

NTP. 2007b. *Toxicology studies of dichloroacetic acid (CAS No. 79-43-6) in genetically modified mice*. NTP GMM 11. Research Triangle Park, NC: NTP. http://www.epa.gov/iris/toxreviews/0654tr.pdf (accessed November 8, 2011).

NTP. 2011. *12th report on carcinogens* (RoC). http://ntp.niehs.nih.gov/index.cfm?objectid=03C9AF75-E1BF-FF40-DBA9EC0928DF8B15 (accessed November 9, 2011).

Palmer, J. R., L. L. Adams-Campbell, D. A. Boggs, L. A. Wise, and L. Rosenberg. 2007. A prospective study of body size and breast cancer in black women. *Cancer Epidemiol Biomarkers Prev* 16(9):1795–1802.

Petracci, E., A. Decarli, C. Schairer, R. M. Pfeiffer, D. Pee, G. Masala, D. Palli, and M. H. Gail. 2011. Risk factor modification and projections of absolute breast cancer risk. *J Natl Cancer Inst* 103(13):1037–1048.

Physical Activity Guidelines Advisory Committee. 2008. *Physical activity guidelines advisory committee report, 2008*. Washington, DC: HHS. http://www.health.gov/paguidelines/committeereport.aspx (accessed August 9, 2011).

Preston, D. L., E. Ron, S. Tokuoka, S. Funamoto, N. Nishi, M. Soda, K. Mabuchi, and K. Kodama. 2007. Solid cancer incidence in atomic bomb survivors: 1958–1998. *Radiat Res* 168(1):1–64.

Ravdin, P. M., K. A. Cronin, N. Howlader, C. D. Berg, R. T. Chlebowski, E. J. Feuer, B. K. Edwards, and D. A. Berry. 2007. The decrease in breast-cancer incidence in 2003 in the United States. *N Engl J Med* 356(16):1670–1674.

Reeves, G. K., V. Beral, J. Green, T. Gathani, and D. Bull. 2006. Hormonal therapy for menopause and breast-cancer risk by histological type: A cohort study and meta-analysis. *Lancet Oncol* 7(11):910–918.

Reynolds, P., S. Hurley, D. E. Goldberg, H. Anton-Culver, L. Bernstein, D. Deapen, P. L. Horn-Ross, D. Peel, et al. 2004. Active smoking, household passive smoking, and breast cancer: Evidence from the California Teachers Study. *J Natl Cancer Inst* 96(1):29–37.

Reynolds, P., D. Goldberg, S. Hurley, D. O. Nelson, J. Largent, K. D. Henderson, and L. Bernstein. 2009. Passive smoking and risk of breast cancer in the California Teachers Study. *Cancer Epidemiol Biomarkers Prev* 18(12):3389–3398.

Richardson, S. D., M. J. Plewa, E. D. Wagner, R. Schoeny, and D. M. Demarini. 2007. Occurrence, genotoxicity, and carcinogenicity of regulated and emerging disinfection by-products in drinking water: A review and roadmap for research. *Mutat Res* 636(1–3):178–242.

Ronksley, P. E., S. E. Brien, B. J. Turner, K. J. Mukamal, and W. A. Ghali. 2011. Association of alcohol consumption with selected cardiovascular disease outcomes: A systematic review and meta-analysis. *BMJ* 342:d671.

Rose, G. 1992. Strategies of prevention: The individual and the population. In *Coronary heart disease epidemiology: From aetiology to public health*. Edited by M. Marmot and P. Elliott. Oxford, UK: Oxford University Press. Pp. 311–324.

Rothman, K. J., and S. Greenland, eds. 1998. *Modern epidemiology*, 2nd ed. Philadelphia, PA: Lippincott–Raven.

Rudel, R. A., K. R. Attfield, J. N. Schifano, and J. G. Brody. 2007. Chemicals causing mammary gland tumors in animals signal new directions for epidemiology, chemicals testing, and risk assessment for breast cancer prevention. *Cancer* 109(12 Suppl):2635–2666.

Rudel, R. A., J. M. Gray, C. L. Engel, T. W. Rawsthorne, R. E. Dodson, J. M. Ackerman, J. Rizzo, J. L. Nudelman, et al. 2011. Food packaging and bisphenol A and bis(2-ethyhexyl) phthalate exposure: Findings from a dietary intervention. *Environ Health Perspect* 119(7):914–920.

Santen, R. J., D. C. Allred, S. P. Ardoin, D. F. Archer, N. Boyd, G. D. Braunstein, H. G. Burger, G. A. Colditz, et al. 2010. Postmenopausal hormone therapy: An Endocrine Society scientific statement. *J Clin Endocrinol Metab* 95(7 Suppl 1):s1–s66.

Schoeny, R. 2010. Disinfection by-products: A question of balance. *Environ Health Perspect* 118(11):A466–A467.

Schwartz, L. M., S. Woloshin, E. L. Dvorin, and H. G. Welch. 2006. Ratio measures in leading medical journals: Structured review of accessibility of underlying absolute risks. *BMJ* 333(7581):1248.

Secretan, B., K. Straif, R. Baan, Y. Grosse, F. El Ghissassi, V. Bouvard, L. Benbrahim-Tallaa, N. Guha, et al. 2009. A review of human carcinogens—Part E: Tobacco, areca nut, alcohol, coal smoke, and salted fish. *Lancet Oncol* 10(11):1033–1034.

Shuryak, I., R. K. Sachs, and D. J. Brenner. 2010. Cancer risks after radiation exposure in middle age. *J Natl Cancer Inst* 102(21):1628–1636.

Smith-Bindman, R. 2010. Is computed tomography safe? *N Engl J Med* 363(1):1–4.

Smith-Bindman, R. 2011. Ionizing radiation exposure to the U.S. population, with a focus on radiation from medical imaging. Paper commissioned by the Committee on Breast Cancer and the Environment (see Appendix F; available at http://www.nap.edu/catalog.php?record_id=13263).

Smith-Warner, S. A., D. Spiegelman, S. S. Yaun, P. A. van den Brandt, A. R. Folsom, R. A. Goldbohm, S. Graham, L. Holmberg, et al. 1998. Alcohol and breast cancer in women: A pooled analysis of cohort studies. *JAMA* 279(7):535–540.

Sovak, M., A. L. Seligson, M. Konas, M. Hajduch, M. Dolezal, M. Machala, and R. Nagourney. 2002. Herbal composition PC-SPES for management of prostate cancer: Identification of active principles. *J Natl Cancer Inst* 94(17):1275–1281.

Taylor, H., and R. Leftman, eds. 2002. Widespread ignorance of regulation and labeling of vitamins, minerals, and food supplements. *Health Care News* 2(23):1–5. http://www.harrisinteractive.com/news/newsletters/healthnews/HI_HealthCareNews2002Vol2_Iss23.pdf (accessed August 9, 2011).

Teras, L. R., M. Goodman, A. V. Patel, W. R. Diver, W. D. Flanders, and H. S. Feigelson. 2011. Weight loss and postmenopausal breast cancer in a prospective cohort of overweight and obese U.S. women. *Cancer Causes Control* 22(4):573–579.

Tseng, M., C. R. Weinberg, D. M. Umbach, and M. P. Longnecker. 1999. Calculation of population attributable risk for alcohol and breast cancer (United States). *Cancer Causes Control* 10(2):119–123.

Tyrer, J., S. W. Duffy, and J. Cuzick. 2004. A breast cancer prediction model incorporating familial and personal risk factors. *Stat Med* 23:1111–1130.

USPSTF (U.S. Preventive Services Task Force). 2002. *Chemoprevention of breast cancer: Recommendations and rationale.* http://www.uspreventiveservicestaskforce.org/uspstf/uspsbrpv.htm (accessed August 9, 2011).

USPSTF. 2005. *Hormone therapy for the prevention of chronic conditions in postmenopausal women: Recommendation statement.* http://www.uspreventiveservicestaskforce.org/uspstf/uspspmho.htm (accessed August 9, 2011).

Vogel, V. G., J. P. Costantino, D. L. Wickerham, W. M. Cronin, R. S. Cecchini, J. N. Atkins, T. B. Bevers, L. Fehrenbacher, et al. 2010. Update of the National Surgical Adjuvant Breast and Bowel Project Study of Tamoxifen and Raloxifene (STAR) P-2 Trial: Preventing breast cancer. *Cancer Prev Res (Phila)* 3(6):696–706.

WCRF/AICR (World Cancer Research Fund/American Institute for Cancer Research). 2007. *Food, nutrition, physical activity, and the prevention of cancer: A global perspective.* Washington DC: AICR.

WCRF/AICR. 2008. The associations between food, nutrition and physical activity and the risk of breast cancer. *WCRF/AICR systematic literature review continuous update report.* http://dietandcancerreport.org/downloads/cu/cu_breast_cancer_report_2008.pdf (accessed October 29, 2011).

WCRF/AICR. 2010. Food, nutrition and physical activity and the prevention of breast cancer. *WCRF/AICR continuous update report summary.* http://dietandcancerreport.org/downloads/cu/cu_breast_cancer_summary_2008.pdf (accessed October 29, 2011).

White, J. 2002. PC-SPES—a lesson for future dietary supplement research. *J Natl Cancer Inst* 94(17):1261–1263.

Winzer, B. M., D. C. Whiteman, M. M. Reeves, and J. D. Paratz. 2011. Physical activity and cancer prevention: A systematic review of clinical trials. *Cancer Causes Control* 22(6):811–826.

Writing Group for the Women's Health Initiative Investigators. 2002. Risks and benefits of estrogen plus progestin in healthy postmenopausal women: Principal results from the Women's Health Initiative randomized controlled trial. *JAMA* 288(3):321–333.

Xue, F., W. C. Willett, B. A. Rosner, S. E. Hankinson, and K. B. Michels. 2011. Cigarette smoking and the incidence of breast cancer. *Arch Intern Med* 171(2):125–133.

Yang, C. Z., S. I. Yaniger, V. C. Jordan, D. J. Klein, and G. D. Bittner. 2011. Most plastic products release estrogenic chemicals: A potential health problem that can be solved. *Environ Health Perspect* 119(7):989–996.

7

Recommendations for Future Research

Although much has been learned about breast cancer and its relation to environmental exposures, much remains unclear. As the preceding chapters have illustrated, this reflects a mixture of circumstances. First, the scientific community is faced with conflicting and inconclusive results from past studies of some risk factors. Second, growing knowledge of the complex biology of breast cancer suggests a need to reframe hypotheses by focusing more on exposures in early life, examining associations with tumors of specific types, and considering mechanistically driven gene–environment interactions. Third, for a wide array of exposures, data are simply inadequate because exposure assessment methodologies have not been developed, informative studies may be nearly impossible to conduct in humans, and/or the existing tools and resources to conduct relevant research in animals or in vitro systems are limited.

With the complexity of breast cancer as a disease and of the combinations of biological and environmental factors that are potential contributors to it, the committee is persuaded that no one perspective will be sufficient to guide the future research that is needed to reduce the toll of this disease. Bringing together the perspectives of many disciplines into a transdisciplinary approach will be needed to generate innovative and cost-effective approaches to framing research questions, designing and conducting studies, developing new tools for data collection and analysis, and translating the results of research on risk factors into interventions that can reduce the risk of breast cancer.

Drawing on the insights developed in the previous chapters, the committee presents in this final chapter recommendations for research that range from further examination of elements of the biology of breast development and carcinogenesis to tests of potential interventions to reduce risk. Important components of the work recommended here provide support for the research necessary to develop better tools for assessing the carcinogenicity of chemicals and pharmaceuticals as well as tools needed to strengthen epidemiologic research. The importance of a life course perspective runs throughout these recommendations.

Many of these recommendations are directed to both researchers and research funders. Researchers will have to conduct the work described here, but they will need the resources that come from a variety of sources. The National Institutes of Health and other federal agencies are major funders of research on breast cancer or they have unique authority or responsibility in certain areas. But the nation's portfolio of research on breast cancer is also shaped in important ways by funders and other organizations in the private sector, such as Susan G. Komen for the Cure, that have the flexibility to pursue research topics and approaches that federal agencies may not. The committee urges effective and innovative collaborations to answer the many unresolved questions about the causes of breast cancer.

APPLYING A LIFE COURSE PERSPECTIVE TO RESEARCH ON BREAST CANCER

Progress has been made in understanding the biology of breast development, molecular mechanisms of carcinogenesis, the influence of the tissue microenvironment on breast cancer development, and some aspects of risk and prevention. But gaps remain in understanding of the etiology of breast cancer and the extent of environmental influences on breast cancer development.

Most epidemiologic studies have been obliged to focus on events in the few years or perhaps one to two decades before a breast cancer diagnosis. As described in Chapter 5, however, growing evidence suggests that events associated with breast carcinogenesis may occur much earlier—in young adulthood, puberty, childhood, and in utero. The effect of radiation, for instance, is greater when exposure occurs around the time of puberty or earlier. Although information about some early life events, such as age when first giving birth or age at menarche, can be reliably retrieved, few studies have collected information on nonreproductive environmental exposures that may influence the occurrence of clinically detectable breast cancer many decades later.

To address gaps in knowledge about the origins of breast cancer, the committee determined that research should increasingly focus on the influence of environmental factors during potential windows of susceptibility over the life course. It is possible that some exposures later in life, after childbearing is complete, have little effect on breast cancer risk whereas similar exposures, if incurred early in life, before completion of breast development, may increase risk for breast cancer. On the other hand, exposures later in life may increase the growth of cancerous cells that have lain dormant for years and that would, without the exposure, have continued to be dormant. Thus the committee recommends that future research address the timing of exposures in relation to a woman's life course and explore vulnerable windows for specific exposures of concern.

Recommendation 1: Breast cancer researchers and research funders should pursue integrated and transdisciplinary studies that provide evidence on etiologic factors and the determinants of breast cancer across the life course, with the goal of developing innovative prevention strategies that can be applied at various times in life.

- Such studies should seek to integrate animal models that capture the whole life course and human epidemiologic cohort studies that follow individuals over long periods of time and allow for investigation of so-called "windows of susceptibility" wherein breast tissue may be especially sensitive to environmental influences (e.g., prenatal, childhood, and adolescent, and childbearing periods). Long-term follow-up of cohorts is critical because new, unexpected evidence frequently arises with extended follow-up.
- Topics warranting attention include (but are not limited to) the biology of breast development; the mechanisms of carcinogenesis early in life, including the role of the tissue microenvironment in tumor suppression and development, and differences that may be related to tumor type; differences in risk by tumor type; the potential contribution of timing of exposure to variation in risk; and analytical tools for investigating the potential for interactions among exposures and the impact of mixtures of environmental agents on biologic processes.

Other work to aid investigation of environmental influences on breast cancer risk includes

- identifying cellular, biochemical, or molecular biomarkers of early events leading to breast cancer and validating their predictive value for future risk for breast cancer;
- determining whether intermediate endpoints, such as indicators of breast development or peak height growth velocity, are valid and predictive biomarkers of risk for breast cancer so that research can

effectively identify predictors of change in risk earlier in life or with shorter study periods;

- investigating the role that environmental factors may have in the origins of breast cancers of different types (e.g., estrogen or progestin receptor positive [ER+, PR+] or receptor negative [ER–, PR–]; HER2/neu positive or negative; or triple negative, meaning being negative for all three types of receptors) to better understand the potential contribution of these factors to disparities in the incidence of types of breast cancers among racial and ethnic groups;

- exploring the value of linking information across cohort studies focused on different stages of life as a way to overcome the challenges of mounting single long-term follow-up studies; and

- ensuring that cohorts established primarily to study genetic determinants of cancer and other diseases improve their capacity to capture information about environmental exposures over the life course.

TARGETING SPECIFIC CONCERNS

Rationale: From its examination of evidence on a selection of environmental factors, the committee sees particular benefit in further research to clarify the mechanisms underlying breast cancer.

Recommendation 2: Breast cancer researchers and research funders should pursue research to increase knowledge of mechanisms of action of environmental factors for which there is provocative, but as yet inconclusive, mechanistic, animal, life course, or human health evidence of a possible association with breast cancer risk.

High-priority topics include the following:

- *Shift work*: There is growing evidence that shift work resulting in the disruption of circadian rhythm is probably associated with increased risk for breast cancer. Currently, there are no known effective interventions other than avoidance of shift work, which will not be an option for many workers. The biological mechanisms and the potential contribution of light exposure during normal sleep periods are poorly understood. More needs to be learned about the biological processes and pathways through which shift work and circadian rhythm disruption, or other factors arising from shift work, relate to breast cancer. This includes investigation of hormonal effects of circadian disruption, the role of "clock genes" and signaling pathways in breast tissue development, how disruption of those signaling pathways may contribute to initiation or progression of breast tumors, developing more detailed and standardized approaches to exposure assessment for use in epidemiologic research, and developing and testing the effectiveness of interventions that could mitigate the carcinogenic effects that may be associated with shift work.

- *Endocrine activity*: Exposure to chemicals with estrogenic or other properties relevant to sex steroid activity, such as bisphenol A (BPA), polybrominated diphenyl ethers (PBDEs), zearalenone, and certain dioxins and dioxin-like compounds, may influence breast cancer risk, especially if those exposures occur at certain life stages or in combination with exposure to other similar chemicals, certain dietary components, or other factors. Although the evidence on the association between breast cancer risk and individual chemicals in this category is not conclusive, current mechanistic hypotheses warrant further research to examine their activity, to investigate additive or greater potency across multiple chemicals, to explore the effects of timing of exposure, and to evaluate interactions with diet, body mass index, and other factors that may influence the relationship of these types of compounds to breast cancer risk.
- *Genotoxicity*: Animal studies have demonstrated that some mutagenic chemicals are capable of inducing malignant mammary tumors, and numerous animal models of breast carcinogenesis routinely use the potent mutagens 7,12-dimethylbenz[a]anthracene (DMBA) and N-methyl-N-nitrosourea (MNU) as reproducible initiators of those tumors. But these studies have shown that the effect is highly sensitive to the timing of the exposures and can be influenced by other factors. More research is needed to understand the degree to which mutagenic chemicals, such as polycyclic aromatic hydrocarbons (PAHs), benzene, and ethylene oxide, acting alone or in combination with other exposures at specific life stages, may contribute to breast cancer risk at current levels of exposure.
- *Epigenetic activity*: Recent studies have demonstrated that some chemicals, including BPA, while not genotoxic per se, can have important influences on gene expression that may be relevant to breast cancer risk. Relatively little is known about the importance for breast cancer risk of such epigenetic modifications by environmental chemicals. More fundamental research on the role of epigenetic modifications in breast cancer risk is needed.
- *Gene–environment interactions*: Although few such interactions have been identified, to some extent this may reflect the small number of discrete exposures for which relevant genes are currently identifiable. Limited evidence indicates, for example, that

genes governing acetylation efficiency may describe a susceptible subset of the population for which exposure to tobacco smoke has substantial influence on breast cancer risk. Likewise, isozymes of different enzymes involved in alcohol metabolism may affect breast cancer risks, particularly among those with high alcohol intake.

EPIDEMIOLOGIC RESEARCH

Studies of Occupational Cohorts and Other Highly Exposed Populations

Rationale: Many known human carcinogens were first identified as a result of studies carried out in occupational settings where workers were subject to chemical and physical exposures that were typically higher than those experienced by the general population. When many of the early occupational studies were carried out, relatively few women were in the workforce. Changes in the typical workplace and the presence of more women in the workforce, both in the United States and internationally, make it appropriate to revisit occupational studies as a possible means to identify some exposures that increase risk for breast cancer. These studies should account for not only comparisons of breast cancer incidence associated with various work assignments or job titles, but also the distribution of known breast cancer risk factors among workers to ensure that the analyses of exposure-related risk are not confounded by differences among types of workers in the prevalence of these other known risk factors.

Outside the workplace, other events such as industrial accidents or contamination episodes can lead to high exposures for specific population groups. Sometimes these events provide opportunities to investigate the impact of specific timing of exposures, as in the case of the survivors of the atomic bombs in Hiroshima and Nagasaki, or the population living in the vicinity of the industrial accident in Seveso, Italy, and exposed to high levels of dioxin. High-dose or long-term medical exposures have also lent themselves to study through the assembly of cohorts from records of patients treated for specific diseases or conditions.

Recommendation 3: Breast cancer researchers and research funders should pursue studies of populations with higher exposures, such as occupational cohorts, persons with event-related high exposures, or patient groups given high-dose or long-term medical treatments. These studies should include collection of information on the prevalence of known breast cancer risk factors among the study population. Support for these studies should include resources for the development of improved exposure assessment methods to quantify chemical and other environmental exposures potentially associated with the development of breast cancer.

New Exposure Assessment Tools

Rationale: A life course perspective on breast cancer suggests that critical periods of vulnerability may exist during in utero development, in childhood, adolescence, and early adulthood, and at older ages. Exposure assessment becomes particularly challenging if the interval between critical exposure events and the point at which breast cancer can be diagnosed extends over decades.

If evidence of exposure is retained in either environmental media or the human body, measurements made long after exposure may provide an adequate basis for estimating an earlier exposure. To be able to do so requires sufficient knowledge of the patterns of persistence of chemical compounds and their metabolites, the determinants of variability in retention, and the variation in exposure levels over time. If evidence of exposure is not retained, one-time measurements are unlikely to be an adequate basis for assessing true exposure unless it is known that an individual's exposure is consistent over long periods.

To effectively study exposures over long time periods, research protocols may need to obtain measurements of exposure at multiple time points. However, because repeated measurements can be prohibitively burdensome, it may be necessary to develop alternative strategies that rely on external indicators of exposure. For instance, if, hypothetically, 50 percent of the body burden of a chemical exposure is from consumption of liquids from plastic bottles, then questionnaires about such behavior patterns may be a more reliable basis for assessing exposure than measurements of urinary metabolites. If, additionally, persons who consume fluids from plastic bottles do so consistently over years or decades, then this approach may be reasonable for establishing past exposures as well.

Recommendation 4: Breast cancer and exposure assessment researchers and research funders should pursue research to improve methodologies for measuring, across the life course, personal exposure to and biologically effective doses of environmental factors that may alter risk for or susceptibility to breast cancer.

Such research should encompass

- improving measurements in the environment and assessing variation over time and space;

- determining routes of exposures and how they vary over time and over the life course;
- evaluating how products are used and the extent to which actual usage deviates from label instructions (e.g., home pesticide applications) as a critical component of exposure assessment, and focusing on the impact on personal exposures;
- incorporating use of advanced environmental dispersion modeling techniques with accurate emissions and air monitoring data to characterize specific population exposures;
- measuring compounds and their metabolites in biospecimens, including specimens obtained by noninvasive means;
- understanding pharmacodynamics and pharmacokinetics and how they vary by lifestage, body weight, nutrition, comorbidity, or other factors;
- developing other biomarkers of exposure through early biologic effects (DNA adducts, methylation, tissue changes, gene expression, etc.);
- using existing and yet-to-be-established human exposure biomonitoring programs (e.g., breast milk repositories) by geographic areas; and
- validating exposure questionnaires through various strategies.

RESEARCH TO ADVANCE PREVENTIVE ACTIONS

Minimizing Exposure to Ionizing Radiation

Rationale: As discussed in Chapters 3 and 6 and Appendix F, some of the strongest evidence reviewed by the committee indicated a strong causal association between breast cancer and ionizing radiation. However, population exposures to ionizing radiation in medical imaging are increasing. Chapter 6 sets forth a series of steps that can be taken by various groups and in various settings to reduce exposures to ionizing radiation and therefore reduce risks for breast and other cancers. However, many unknowns remain about the best ways to achieve these reductions. This work might include investigation of the feasibility of developing cost-effective forms of imaging that do not rely on ionizing radiation. Further research is warranted to clarify the extent of population risks, unnecessary uses of medical radiography, and the best means to maximize its benefits and minimize its harms.

Recommendation 5: The National Institutes of Health, the Food and Drug Administration, and the Agency for Healthcare Research and Quality should support comparative effectiveness research to assess

the relative benefits and harms of imaging procedures and diagnostic/ follow-up algorithms in common practice. This research effort should also assess the most effective ways to fill knowledge gaps among patients, health care providers, hospitals and medical practices, industry, and regulatory authorities regarding practices to minimize exposure to ionizing radiation incurred through medical diagnostic procedures.

Developing and Validating Preventive Measures

Rationale: Some breast cancer risk factors appear to be modifiable, but it is important to determine what modifications of these environmental exposures can be most effective in reducing risk and when during the life course these changes need to occur. For example, overweight and obesity are recognized as increasing risk for postmenopausal breast cancer, but the contribution of weight loss to reducing risk is much less clear.

Recommendation 6: Breast cancer researchers and research funders should pursue prevention research in humans and animal models to develop strategies to alter modifiable risk factors, and to test the effectiveness of these strategies in reducing breast cancer risk, including timing considerations and population subgroups likely to benefit most.

Particular aspects of prevention that require attention include

- when weight loss is most likely to be beneficial in reducing risk for postmenopausal breast cancer;
- effective strategies for achieving and maintaining weight loss in different risk groups;
- effective and sustainable methods to prevent obesity;
- the feasibility of interventions in early life and development that may influence breast cancer risk in adult life such as preventing childhood obesity, increasing physical activity, and minimizing exposures to potentially harmful environmental carcinogens;
- approaches to prevention that respond to the differing breast cancer experience of various racial and ethnic groups; and
- dissemination and adoption of effective prevention strategies.

Chemoprevention—Research on Medications to Reduce Breast Cancer Risk

Rationale: Breast cancer is likely to remain a major source of morbidity for many decades to come. However, if early life events are critical in breast cancer carcinogenesis, then most women may have already had some critical exposures by mid-life, when the incidence of breast cancer increases. Avoiding other exposures later in life, such as hormone therapy, may delay or even prevent breast cancer in some women, but it may be that further reductions in risk later in life are most efficiently achieved through pharmaceutical interventions.

Research has demonstrated that drugs that alter responses to estrogen (e.g., tamoxifen, raloxifene) or production of estrogen (e.g., aromatase inhibitors) can substantially reduce risk of ER+ breast cancer (Cummings et al., 2009; Nelson et al., 2009; Goss et al., 2011). The Food and Drug Administration (FDA) has approved use of tamoxifen and raloxifene for this purpose by women who are considered at increased risk of breast cancer and are not at increased risk for cerebrovascular disease. Other medications, such as bisphosphonates and metformin, are under study to assess their potential role in reducing the risk of either ER+ or ER− breast cancer (Cuzick et al., 2011). But relatively few eligible women have chosen to use tamoxifen and raloxifene, at least in part because they are associated with increased risk for serious adverse health effects, including endometrial cancer and stroke (Fisher et al., 2005; Vogel et al., 2010).

The desirability of drugs that can reduce breast cancer risk must be balanced against any potential dangers associated with the use of those drugs. These dangers are of particular concern for the large numbers of women who would not have developed breast cancer even without medication, as well as for the smaller numbers of women who develop breast cancer despite using them.

Additional research into medications that can reduce risk for breast cancer with minimal added risk of other serious adverse health effects should be fostered and accelerated. Studies should include sufficient follow-up, both during the study when the medications are being used and after what is anticipated to be the typical period of use, to provide an adequate basis for determining the benefits and risks that may be associated with the medication. Furthermore, because the approved drugs only reduce the risk of ER+ breast cancer, research is critically needed to find effective ways to reduce the risk of other forms of breast cancer, including triple negative breast cancer and other hard-to-treat forms of breast cancer that may have a disproportionate impact at younger ages or among African American, Asian, or Hispanic women.

Recommendation 7: Breast cancer researchers and research funders should pursue continued research into new breast cancer chemoprevention agents that have minimal risk for other adverse health effects. This work should include efforts to identify chemopreventive approaches for hormone receptor negative breast cancer.

Adequately sized primary prevention studies will be needed to allow for estimation of both benefits and risks. Research plans should also include long-term follow-up to identify any changes in risk patterns for types of breast cancer or other effects that only become evident beyond the time frame of the initial study and analyses.

TESTING TO IDENTIFY POTENTIAL BREAST CARCINOGENS

In Vivo Testing for Carcinogenicity

Rationale: Testing in animals is currently an established component of the evaluation of the carcinogenicity of chemicals in industry and commerce, but it is unclear which whole-animal test protocols are best suited for screening for possible human breast carcinogens. Human sensitivity to breast cancer has been demonstrated for exposures in utero (e.g., diethylstilbestrol [DES]), before and during puberty (e.g., radiation), and postmenopausally (e.g., combination hormone therapy). Studies in animals have also demonstrated that some exposures early in life that are not themselves carcinogenic may alter susceptibility to carcinogens encountered later in life.

But these age windows are typically not included in standard cancer bioassays such as those used in conjunction with the registration of pesticides and pharmaceuticals. The standard protocols commonly begin exposures when animals are 7 to 8 weeks of age. Thus they miss the rapid mammary ductal growth and branching during pubertal development, a period of heightened sensitivity in the rat to adverse effects from chemical exposures. These protocols also miss gestational exposures and terminate the experiments at 2 years, which omits the older age period, a time of increasing incidence of breast cancer in humans.

Interpretation of rodent bioassays for mammary carcinogenicity is complicated by certain characteristics of the animals typically used for these studies. The mouse strains appear generally insensitive to hormonally induced mammary tumors. Conversely, a commonly used rat strain is overly sensitive to the occurrence of constant estrus and early reproductive senescence. Constant estrus and early reproductive senescence can tend to increase the incidence of mammary tumors, but this phenomenon may not be relevant for humans. Thus results of bioassays of hormonally active agents are confounded when mammary tumors are increased concomitantly with constant estrus in the treated rats. With the insensitivity of mice, negative results from tests in mice are not necessarily a reliable indicator of lack of mammary carcinogenicity.

To increase the ability to detect statistically significant increases in cancer rates in the limited number of animals that can be used in toxicity and carcinogenicity testing, chemicals are typically administered at dose equivalents that are far higher than the exposures humans would normally have. Pharmacokinetic and metabolic differences between high- and low-dose chemical exposures complicate the prediction of risks at lower doses that would be more comparable to human experience.

Finally, standard bioassay protocols for regulatory testing generally test individual chemicals. However, humans are generally exposed throughout life to a myriad of hormonally active and genotoxic chemicals. Some experimental protocols used in cancer research employ mixed exposures (e.g., in utero exposure to one agent and subsequent high-dose exposure to a genotoxic chemical during a period of rapid ductal growth). Other tests look for abnormal development of the mammary gland following in utero or early in life exposure, to identify early predisposing events. In reports from some research studies, it is difficult to assess the level of attention devoted to important design issues such as randomization, blinded assessment of endpoints, and standardization of endpoints.

Recommendation 8: The research and testing communities should pursue a concerted and collaborative effort across a range of relevant disciplines to determine optimal whole-animal bioassay protocols for detection and evaluation of chemicals that potentially increase the risk of human breast cancer.

The development of these protocols should address several issues, including the following:

- potential differences in sensitivity to carcinogenic effects and during different life stages;
- the appropriateness and limitations of the rodent strains and species used for testing, and potential alternatives;
- the frequency, magnitude, and route of dosing, and the possible need for alternative protocols that provide improved relevance for predicting human risk;
- the utility of genetically engineered mouse models, which show promise for studying breast tumor formation and progression and the effectiveness of treatments; and
- standard practices for conducting and reporting results of animal studies.

This work will probably also require targeted mechanistic and pharmacokinetic studies to assess appropriate dosing levels in test protocols to better address human exposure circumstances, including the influence of life stage, genetic variability, and multiple chemical exposures.

New Approaches to Toxicity Testing

Rationale: Most of the thousands of chemicals used in industry and commerce have not been tested for their potential to contribute to breast and other cancers. Screening all chemicals with the standardized approaches used for pharmaceuticals and pesticides is impracticable because of the time and resources (including large numbers of test animals) that would be required (NRC, 2006, 2007). Furthermore, the tests are done chemical-by-chemical, which does not address the potential consequences of exposures to mixtures of chemicals or interactions with other ongoing exposures (e.g., dietary components). The high doses used in testing also introduce uncertainty and limitations for predicting risks at lower doses that are relevant to human exposures.

Under the broad umbrella of the Tox21 (EPA, 2011) and National Toxicology Program initiatives, new toxicity testing approaches are being developed to more rapidly and accurately screen and identify the toxicity of chemicals encountered in human environmental, occupational, and product exposures. This effort relies on the elucidation of key toxicity pathways involved in human disease, and on the development of sensitive, rapid testing approaches to determine a chemical's potential to perturb such pathways and at what concentrations. A variety of tests are being developed and considered: high-throughput in vitro screens that use cell components and engineered cells; toxicogenomic responses following cellular, tissue, and organism exposures; novel animal systems (e.g., the roundworm, *Caenorhabditis elegans*); and limited, targeted testing in laboratory animals to anchor test results and understand mechanisms, new chemistries, and pharmacokinetics (Dix et al., 2007; NRC, 2007).

The new approach also calls for the use of pharmacokinetic evaluations, human biomonitoring data, and epidemiologic results to establish the predictive ability of the tests. Pharmacokinetics will be an important consideration in understanding test results, in studying uptake and distribution to target cells, and in examining the biochemical transformations that make the chemical biologically active or inactive. This aspect of the effort is currently a significant challenge in the development of high-throughput and other in vitro tests.

Because breast cancer is a major contributor to morbidity among women, these tests should address pathways that underlie the basic mechanisms of breast cancer—mutagenesis, estrogen receptor signaling, epigenetic programming, growth promotion via mitogenic cell signaling, and modulation of immune functioning—with particular attention to cell types and environments relevant to breast cancer. They should also take into account alterations at the whole-organ level, and they should be relevant to typical human exposures, which often occur at low doses and as mixtures.

Recommendation 9:

a. The research and testing communities should ensure that new testing approaches developed to serve as alternatives to long-term rodent carcinogenicity studies include components that are relevant for breast cancer.

To be relevant for breast cancer, it will be necessary to be able to assess changes in susceptibility through the life course and mechanisms characteristic of hormonally active agents. The test development should also include exploring the predictive value of in vitro and in vivo experimental testing for site-specific cancer risks for humans.

b. A research initiative should assess the persistence and consequences for mammary carcinogenicity of abnormal mammary development and related intermediate outcomes observed in some toxicological testing.

As useful predictors of increased mammary cancer risk become available, intermediate outcomes may aid in identifying chemicals that may pose increased risk of human breast cancer when exposures occur early in life.

c. Research should be conducted to improve understanding of the potential cumulative effects of multiple, small environmental exposures on risk for breast cancer and the interaction of these exposures with other factors that influence risk for breast cancer.

Improved understanding of both mixed and serial low-dose exposures is critical for the interpretation of in vivo results and is of heightened importance for understanding the results of the emerging in vitro tests. Relevant exposures may come from sources that include food, pharmaceuticals, and the general environment. It is also critical for the understanding of epidemiologic and in vivo and in vitro experimental research results on the health effects of chemical mixtures that are characteristic of human environmental exposures.

Identifying Breast Cancer Risks Associated with Hormonally Active Pharmaceutical Products

The committee sees a need to ensure that mechanisms for detection and assessment of breast cancer risks associated with use of drugs regulated by FDA are adequate. It also recognizes that enhanced methods to detect breast cancer risks represent only one specific dimension of a more general interest in strengthening FDA's ability to ensure the safety and timely availability of prescription and over-the-counter drugs (IOM, 2007a,b) and in strengthening the science to support FDA's regulatory work (e.g., IOM, 2011).

Menopausal hormone therapy was originally developed to control menopausal symptoms. Some health professionals advocated long-term and substantially expanded use in anticipation that it would reduce age-related health problems, including cardiovascular disease and memory disorders, even before clear evidence was in hand. Although these products are effective in reducing menopausal symptoms and osteoporotic fractures while women are taking them, evidence from the Women's Health Initiative examining multiple health outcomes in a randomized trial design showed that use of a combination of estrogen and progestin in postmenopausal hormone preparations increases risk of breast cancer and stroke and does not provide overall benefits for cardiovascular risk or memory disorders (Writing Group for the Women's Health Initiative Investigators, 2002). This experience is an illustration of the dangers of exposing millions of healthy women to pharmacological doses of exogenous hormones without sufficient evidence of net benefit. Decades of study have also confirmed a small excess risk of breast cancer among current users of oral contraceptives (Collaborative Group on Hormonal Factors in Breast Cancer, 1996; Marchbanks et al., 2002; Strom et al., 2004; IARC, 2011). Although the increased risk of breast cancer that is associated with use of combination hormone therapies, including oral contraceptives, declines after treatment stops, women should be aware of the full range of potential harms as well as the benefits when they decide whether to use any form of hormone therapy, including those touted as safe because they are "bioidentical" or "natural."

New Approaches to Testing Hormonally Active Candidate Pharmaceuticals

Rationale: Given the evidence for hormonal influences on the development of breast cancer, the committee is concerned that testing required to gain marketing approval for various hormonally active pharmaceuticals that are already on the market or that are being developed does not adequately address the potential impact on the risk for breast cancer. For example, the 2-year rodent carcinogenicity studies done for Prempro, the combined estrogen–progestin product used in the Women's Health Initiative, showed a reduction in mammary tumors in rats (Ayerst Laboratories, 2003), and premarketing human safety and efficacy studies are generally too small and too brief to detect an effect on the incidence of breast cancer. Given that some hormonal products have been found to increase the risk of breast cancer, it is important that new postmenopausal hormone preparations, including those advertised as bioidentical or natural hormones, have an adequate evidence base to support any claims that they do not cause breast cancer. It is also important to have an adequate understanding of the implications for breast cancer risk of the hormone composition and dosing schedules of new oral contraceptives (e.g., a preparation that causes a woman to have only four menstrual periods per year).

Identifying hormonally active substances is complex, in that various models are used to measure hormonal activity and the activity levels detected for a substance may differ depending on the model and dose used. It is important to assess the effectiveness of current testing protocols for hormonally active products in providing indicators of the potential for increased risk of breast cancer, and to develop and validate new testing practices where needed.

> **Recommendation 10: The pharmaceutical industry and other sponsors of research on new hormonally active pharmaceutical products should support the development and validation of better preclinical screening tests that can be used before such products are brought to market to help evaluate their potential for increasing the risk of breast cancer.**

A suite of in vitro and in vivo tests will likely be needed to address the different mechanisms of action that may be relevant over the life course (in utero, early infancy, pre- and postpuberty, pregnancy, and pre- and postmenopause). If such tests can be developed and validated, FDA should require submission of the results as part of the process for approving the introduction of new hormonal preparations for prescription or over-the-counter use. These tests may also prove useful in testing environmental chemicals.

Postmarketing Studies of Hormonally Active Products

Rationale: With the demonstration that use of certain hormonally active prescription drugs is associated with an increased risk of breast cancer and other adverse health effects, it is important to investigate whether use of other hormonally active drugs is also associated with increased risk. The Food and Drug Administration Amendments Act of 2007 gave FDA the authority to require postmarketing studies or clinical trials for approved drugs when adverse event reporting would not be sufficient to assess a known or suspected serious risk (FDA, 2011). Because adverse event reporting systems are generally better suited to the detection of adverse events that occur soon after use of a drug than to events such as breast cancer that take years to develop, formally conducted studies appear necessary to assess the potential breast cancer risk.

Recommendation 11: FDA should use its authority under the Food and Drug Adminisration Amendments Act of 2007 to engage the pharmaceutical industry and scientific community in postmarketing studies or clinical trials for hormonally active prescription drugs for which the potential impact on breast cancer risk has not been well characterized.

Study oversight should be designed to mitigate against bias and conflict of interest of study sponsors. Special attention should be accorded to those products that represent a substantial change in pharmacologic composition or dosage schedule from products currently on the market. The studies should be adequately powered to quantitatively explore the possible contribution of the products to breast cancer risk, as well as other risks that have been associated with these classes of drugs (e.g., cardiovascular effects).

UNDERSTANDING BREAST CANCER RISKS

Researchers, health care providers, and the public all have an incomplete picture of the components of breast cancer risk. Further work is needed to clarify the contribution of recognized risk factors to differences and changes in the incidence of breast cancer and to determine the most effective ways to convey information about breast cancer risk.

Risk Modeling

Rationale: Public health messages about ways to reduce risk should rest on strong science on the attribution of risks to various causal factors. Systematic modeling approaches are needed to refine the estimates of the proportion of breast cancer in the United States and other countries that can be attributed to known factors, especially modifiable factors. Substantial proportions of the increase in breast cancer incidence rates in the United States over the past century, and of the differences in rates of breast cancer between less developed countries and more affluent countries, are probably due in large part to differences over time and between countries in the prevalence of established breast cancer risk factors (e.g., age at menarche, age at first birth, parity, use of menopausal hormone therapy, physical activity, weight and weight change). Few reliable estimates of these temporal and international differences in risk factor prevalence exist.

Developing data on changes in the prevalence of known risk factors, along with changes in breast cancer incidence, should permit statistical modeling of the size of these proportions associated with individual risk factors and combinations of these risk factors. This information will also help in determining the magnitude of risk associated with other unidentified factors, which may include other environmental exposures. Of particular interest are the modifiable risk factors.

Risk modeling on both the individual and population levels will benefit greatly from improved understanding of the etiology of breast cancer. As the science improves, risk models can also help guide future research investments and policy decisions for population-level interventions. A collaborative approach, such as that used by the Cancer Intervention and Surveillance Modeling Network (CISNET) consortium, may be a cost-effective way to pursue some of this work.

> **Recommendation 12: Breast cancer researchers and research funders should support efforts to (1) develop statistical methodology for the estimation of risk of breast cancer for given sets of risk factors and that takes the life course perspective into account, (2) determine the proportion of the total temporal and geographic differences in breast cancer rates that can be plausibly attributed to established risk factors, and (3) develop modeling tools that allow for calculation of breast cancer risk, in both absolute and relative terms, with the goal of assessing potential risk reduction strategies, at both personal and public health levels.**

Communicating About Breast Cancer Risks

Rationale: Accurate and effective communication of breast cancer risks is important for individuals, the public at large, and policy makers and public health officials. Individuals need to be able to make informed choices regarding risk factors, prevention opportunities, and health care appropriate to their risk circumstances. Research indicates that women may have a poor understanding of their risk of breast cancer, with both over- and underestimates of risk observed (Lipkus et al., 2001; Apicella et al., 2009; Waters et al., 2011). A systematic review under the auspices of the Cochrane Collaborative found that both health care providers and consumers understood risks of health outcomes better when those risks were presented as frequencies rather than as probabilities (Akl et al., 2011). Both thought the risks were lower when presented as absolute risk reduction than as relative risk reduction, and both were more persuaded by relative than absolute risks in terms of potential behavioral change. To allow a fair comparison of risks and benefits, supplementing presentation of relative measures with absolute ones is useful because other disease endpoints may be more or less common than breast cancer.

From a public health policy and practice perspective, it is important to determine where risks lie and the potential for benefit and risk at a population level. Uncertainty is inherent in risk prediction, and it can be difficult or impossible to establish that an exposure is not associated with cancer risk. However, moderate or large risks can be ruled out with reasonable confidence when studies with robust and appropriate research designs and analyses have been conducted in populations with relevant exposures. Meaningful differences in risk need to be effectively communicated to the public, health care providers, and policy makers so that limited funds can be invested in the most promising research and intervention strategies.

Recommendation 13: Breast cancer researchers and research funders should pursue research to identify the most effective ways of communicating accurate breast cancer risk information and statistics to the general public, health care professionals, and policy makers.

Because people differ in their health literacy, their numeracy (ability to understand numerical information), and in their preferred modes of learning, multiple communication strategies, modes, and messaging tactics will be needed to reach diverse communities and stakeholders. Among the topics that should receive attention in this research are

- perception and comprehension of different ways to present messages (numbers, graphs, text), modalities of communication (audio, video, print, face-to-face, and multiple modalities, etc.), as well as the content of the messages themselves;
- ways that personal experiences (e.g., family history) affect the ability to absorb messages;
- determination of the similarities and differences in how individuals from diverse racial, ethnic, educational, and occupational groups understand and respond to breast cancer risk information that is presented various ways;
- comprehension of terms such as relative risks, absolute risks, and hazards;
- ways to improve translation of research results into messages that can effectively convey the implications of the results for women in different risk categories, women from diverse racial and ethnic groups, health care providers, and public health decision makers; and
- ways to convey information about chemicals for which there is suggestive evidence of risk from experimental studies.

CONCLUDING OBSERVATIONS

Breast cancer is a leading cause of cancer morbidity among women in the United States and many other countries. Major advances have been made in understanding its biology and diversity, but more needs to be learned about the causes of breast cancer and how to prevent it. Familiar advice about healthful lifestyles appears relevant, but it remains difficult to discern what contribution a diverse array of other environmental factors may be making. Important targets for research are the biologic significance of life stages at which environmental risk factors are encountered, what steps may counter their effects, when preventive actions can be most effective, and whether opportunities for prevention can be found for the variety of forms of breast cancer.

REFERENCES

Akl, E. A., A. D. Oxman, J. Herrin, G. E. Vist, I. Terrenato, F. Sperati, C. Costiniuk, D. Blank, et al. 2011. Using alternative statistical formats for presenting risks and risk reductions. *Cochrane Database Syst Rev* (3):CD006776.

Apicella, C., S. J. Peacock, L. Andrews, K. Tucker, M. B. Daly, and J. L. Hopper. 2009. Measuring and identifying predictors of women's perceptions of three types of breast cancer risk: Population risk, absolute risk and comparative risk. *Br J Cancer* 100(4):583–589.

Ayerst Laboratories. 2003. *Prempro™*. [product labeling] http://www.accessdata.fda.gov/drugsatfda_docs/label/2003/20527s30bl.pdf (accessed October 18, 2011).

Collaborative Group on Hormonal Factors in Breast Cancer. 1996. Breast cancer and hormonal contraceptives: Collaborative reanalysis of individual data on 53,297 women with breast cancer and 100,239 women without breast cancer from 54 epidemiological studies. *Lancet* 347(9017):1713–1727.

Cummings, S. R., J. A. Tice, S. Bauer, W. S. Browner, J. Cuzick, E. Ziv, V. Vogel, J. Shepherd, et al. 2009. Prevention of breast cancer in postmenopausal women: Approaches to estimating and reducing risk. *J Natl Cancer Inst* 101(6):384–398.

Cuzick, J., A. DeCensi, B. Arun, P. H. Brown, M. Castiglione, B. Dunn, J. F. Forbes, A. Glaus, et al. 2011. Preventive therapy for breast cancer: A consensus statement. *Lancet Oncol* 12(5):496–503.

Dix, D. J., K. A. Houck, M. T. Martin, A. M. Richard, R. W. Setzer, and R. J. Kavlock. 2007. The ToxCast program for prioritizing toxicity testing of environmental chemicals. *Toxicol Sci* 95(1):5–12.

EPA (Environmental Protection Agency). 2011. *Tox21*. http://www.epa.gov/ncct/Tox21/ (accessed July 29, 2011).

FDA (Food and Drug Administration). 2011. *Guidance for industry. Postmarketing studies and clinical trials—implementation of Section 505(o)(3) of the Federal Food, Drug, and Cosmetic Act.* http://www.fda.gov/downloads/Drugs/GuidanceComplianceRegulatory Information/Guidances/UCM172001.pdf (accessed April 26, 2011).

Fisher, B., J. P. Costantino, D. L. Wickerham, R. S. Cecchini, W. M. Cronin, A. Robidoux, T. B. Bevers, M. T. Kavanah, et al. 2005. Tamoxifen for the prevention of breast cancer: Current status of the National Surgical Adjuvant Breast and Bowel Project P-1 study. *J Natl Cancer Inst* 97(22):1652–1662.

Goss, P. E., J. N. Ingle, J. E. Ales-Martinez, A. M. Cheung, R. T. Chlebowski, J. Wactawski-Wende, A. McTiernan, J. Robbins, et al. 2011. Exemestane for breast-cancer prevention in postmenopausal women. *N Engl J Med* 364(25):2381–2391.

IARC (International Agency for Research on Cancer). 2011. *IARC monographs on the evaluation of carcinogenic risks to humans.* Vol. 100, Part A: Pharmaceuticals. Lyon, France: IARC.

IOM (Institute of Medicine). 2007a. *Challenges for the FDA: The future of drug safety, workshop summary.* Washington, DC: The National Academies Press.

IOM. 2007b. *The future of drug safety: Promoting and protecting the health of the public.* Washington, DC: The National Academies Press.

IOM. 2011. *Building a national framework for the establishment of regulatory science for drug development: Workshop summary.* Washington, DC: The National Academies Press.

Lipkus, I. M., W. M. Klein, and B. K. Rimer. 2001. Communicating breast cancer risks to women using different formats. *Cancer Epidemiol Biomarkers Prev* 10(8):895–898.

Marchbanks, P. A., J. A. McDonald, H. G. Wilson, S. G. Folger, M. G. Mandel, J. R. Daling, L. Bernstein, K. E. Malone, et al. 2002. Oral contraceptives and the risk of breast cancer. *N Engl J Med* 346(26):2025–2032.

Nelson, H. D., R. Fu, J. C. Griffin, P. Nygren, M. E. Smith, and L. Humphrey. 2009. Systematic review: Comparative effectiveness of medications to reduce risk for primary breast cancer. *Ann Intern Med* 151(10):703–715, W-226–W-735.

NRC (National Research Council). 2006. *Toxicity testing for assessment of environmental agents: Interim report.* Washington, DC: The National Academies Press.

NRC. 2007. *Toxicity testing in the 21st century: A vision and a strategy.* Washington, DC: The National Academies Press.

Strom, B. L., J. A. Berlin, A. L. Weber, S. A. Norman, L. Bernstein, R. T. Burkman, J. R. Daling, D. Deapen, et al. 2004. Absence of an effect of injectable and implantable progestin-only contraceptives on subsequent risk of breast cancer. *Contraception* 69(5):353–360.

Vogel, V. G., J. P. Costantino, D. L. Wickerham, W. M. Cronin, R. S. Cecchini, J. N. Atkins, T. B. Bevers, L. Fehrenbacher, et al. 2010. Update of the National Surgical Adjuvant Breast and Bowel Project Study of Tamoxifen and Raloxifene (STAR) P-2 Trial: Preventing breast cancer. *Cancer Prev Res (Phila)* 3(6):696–706.

Waters, E. A., W. M. Klein, R. P. Moser, M. Yu, W. R. Waldron, T. S. McNeel, and A. N. Freedman. 2011. Correlates of unrealistic risk beliefs in a nationally representative sample. *J Behav Med* 34(3):225–235.

Writing Group for the Women's Health Initiative Investigators. 2002. Risks and benefits of estrogen plus progestin in healthy postmenopausal women: Principal results from the Women's Health Initiative randomized controlled trial. *JAMA* 288(3):321–333.

Appendix A

Agendas for Public Meetings

MEETING 1
Committee on Breast Cancer and the Environment:
The Scientific Evidence, Research Methodology, and Future Directions

The Keck Center of the National Academies
Washington, DC

Wednesday, April 14, 2010

10:45 a.m.	**Introductory Remarks** *Irva Hertz-Picciotto, Ph.D.* *Chair, Committee on Breast Cancer and the* *Environment: The Scientific Evidence, Research* *Methodology, and Future Directions* Introductions by committee members and meeting attendees

11:00 a.m.   **Study Context and Goals, Sponsor Perspective**
                    Amelie Ramirez, Dr.P.H.
                       Member, Susan G. Komen for the Cure Scientific
                       Advisory Board
                       Director, Institute for Health Promotion Research
                       University of Texas Health Science Center at San
                       Antonio

                    Questions and discussion with the committee

12:00 p.m.   **Lunch**

12:45–2:30 p.m.   **Comments from Breast Cancer Research and**
                    Advocacy Organizations
                    • What are their concerns and priorities regarding
                      environmental risk factors for breast cancer?
                    • What do they want to make sure the committee is
                      aware of?
                    • What do they hope the study will contribute?

                    Organizations Planning to Present
                    Avon Foundation: Marc Hurlbert, Ph.D.
                    Breast Cancer Fund: Janet Gray, Ph.D.
                    Breast Cancer Research Foundation: Mary Beth Terry,
                       Ph.D. (phone)
                    National Breast Cancer Coalition: Fran Visco
                    National Institutes of Health Breast Cancer and
                       Environment Activities: Gwen Collman, Ph.D.
                    Young Survival Coalition: Marcia Stein

2:45 p.m.   **Break**

3:00–4:30 p.m.   **Introduction to Issues in Studying Breast Cancer and**
                    the Environment
                    Presentations by Committee Members

4:30 p.m.   **Opportunity for Individual Public Comment**

                    Kathleen Burns, Ph.D. (phone)
                       Director, Sciencecorps
                       Lexington, MA

William Mimiaga, Major, USMC (RET) (phone)
California

Michael Partain (phone)
The Few, The Proud, The Forgotten

James Fontella (phone)
Shelby Township, MI

5:00 p.m.   **Adjourn Open Session**

MEETING 2
Committee on Breast Cancer and the Environment:
The Scientific Evidence, Research Methodology, and Future Directions

The Hyatt Regency San Francisco
San Francisco, CA

Tuesday, July 6, 2010

2:00 p.m.   **Introductory Remarks**
                   Irva Hertz-Picciotto, Ph.D.
                        Chair, Committee on Breast Cancer and the
                        Environment: The Scientific Evidence, Research
                        Methodology, and Future Directions

                   Introductions by committee members and meeting
                   attendees

                   Opportunity for Sponsor Comment
                   Amelie Ramirez, Dr.P.H.
                        Member, Susan G. Komen for the Cure Scientific
                        Advisory Board
                        Director, Institute for Health Promotion Research
                        University of Texas Health Science Center at San
                        Antonio

2:15 p.m.   **Role of Animal Models in Studying Environmental Factors for Breast Cancer**
                     Helmut Zarbl, Ph.D.
                         Professor, University of Medicine and Dentistry of New Jersey and Robert Wood Johnson Medical School
                     Questions and discussion with the committee

3:00 p.m.   **Role of Stem Cells in Environmental Risks for Breast Cancer**
                     Zena Werb, Ph.D.
                         Professor of Anatomy
                         University of California, San Francisco
                     Questions and discussion with the committee

3:45 p.m.   **Linking Prenatal and Neonatal Exposures to Breast Cancer Risks**
                     Dimitrios Trichopoulos, M.D.
                         Professor of Cancer Prevention and
                         Professor of Epidemiology
                         Harvard School of Public Health
                     Questions and discussion with the committee

4:30 p.m.   **General Discussion as Needed**

4:45–5:25 p.m.   **Comments from Breast Cancer Research and Advocacy Organizations**
                     • What are their concerns and priorities regarding environmental risk factors for breast cancer?
                     • What do they want to make sure the committee is aware of?
                     • What do they hope the study will contribute?

                     Organizations Planning to Present
                     American Cancer Society: Michael Thun, M.D.
                     Breast Cancer Action: Kim Irish, J.D.
                     Breast Cancer Fund: Nancy Buermeyer
                     Zero Breast Cancer: Janice Barlow

5:25 p.m.   **Adjourn Open Session**

Wednesday, July 7, 2010

8:30 a.m.   **Introductory Remarks**
                    Irva Hertz-Picciotto, Ph.D.
                    *Chair, Committee on Breast Cancer and the
                    Environment: The Scientific Evidence, Research
                    Methodology, and Future Directions*

                    Introductions by committee members and meeting
                    attendees

8:45 a.m.–   **Reaching Conclusions About Carcinogenicity**
12:00 p.m.

8:45 a.m.   Weighing Evidence from NTP Bioassays
                    John Bucher, Ph.D.
                    Associate Director, National Toxicology Program
                    Questions and discussion

9:30 a.m.   Evaluation of Human and Experimental Evidence to
                    Identify Cancer Hazard and Estimate Risk
                    Kathryn Z. Guyton, Ph.D., D.A.B.T.
                    *Toxicologist, National Center for Environmental
                    Assessment*
                    Office of Research and Development, U.S. EPA
                    Questions and discussion

10:15 a.m.   Identifying Breast Carcinogens at IARC
                    Vincent James Cogliano, Ph.D.
                    Head, IARC Monographs Program
                    Questions and discussion

11:00 a.m.   How Toxicology Can Advance Breast Cancer
                    Prevention by Informing Chemicals Policies and
                    Epidemiologic Study Design
                    Ruthann Rudel, M.S.
                    Director of Research, Silent Spring Institute
                    Questions and discussion

11:45 a.m.   General discussion as needed

12:00 p.m.   **Lunch**

| 1:00–2:00 p.m. | **Introduction to Issues in Studying Breast Cancer and the Environment, Committee Presentations Continued** |

1:00 p.m.   An Epidemiologic Perspective on Environmental Risk Factors
Peggy Reynolds, Ph.D.
  Senior Research Scientist, Cancer Prevention Institute of California

1:30 p.m.   Breast Cancer and Environment Research Centers: Experience and Plans
Robert Hiatt, M.D.
  Deputy Director, Helen Diller Family Comprehensive Cancer Center

2:00–3:00 p.m.   Initiatives on Breast Cancer and the Environment: Research Gaps and Policy Proposals

2:00 p.m.   The California Breast Cancer Research Program's Special Research Initiatives on Environment and Disparities
Marion (Mhel) Kavanaugh-Lynch, M.D., M.P.H.
  Director, California Breast Cancer Research Program
Questions and discussion

2:30 p.m.   Breast Cancer and Chemicals Policy Project
Sarah Janssen, M.D., Ph.D., M.P.H.
  Senior Scientist, Natural Resources Defense Council
  and
Megan Schwarzman, M.D., M.P.H.
  Research Scientist, Berkeley Center for Green Chemistry
  School of Public Health, University of California, Berkeley
Questions and discussion

3:00 p.m.   Opportunity for Individual Public Comment

Nancy Bellen, Santa Rosa, CA
Marika Holmgren, Breast Cancer Survivor
Susan Braun, Executive Director, Commonweal

3:30 p.m.   Adjourn Open Session

MEETING 3
Committee on Breast Cancer and the Environment:
The Scientific Evidence, Research Methodology, and Future Directions

The Keck Center of the National Academies
Washington, DC

Wednesday, September 15, 2010

2:15 p.m.	**Introductory Remarks**
	Irva Hertz-Picciotto, Ph.D.
	    Chair, Committee on Breast Cancer and the
	    Environment: The Scientific Evidence, Research
	    Methodology, and Future Directions

	Introductions by committee members and meeting attendees

	Opportunity for Sponsor Comment
	Amelie Ramirez, Dr.P.H.
	    Member, Susan G. Komen for the Cure Scientific
	    Advisory Board
	    Director, Institute for Health Promotion Research
	    University of Texas Health Science Center at San
	    Antonio

2:30 p.m.	**Environmental Pollution and Breast Cancer: Adding Epidemiological Studies to Biological Evidence**
	Julia G. Brody, Ph.D.
	    Executive Director
	    Silent Spring Institute

	Questions and discussion with the committee

3:15 p.m.	**Early Life Exposures and Breast Cancer Risk**
	Michele R. Forman, Ph.D., M.S.
	    Professor, Department of Epidemiology
	    University of Texas MD Anderson Cancer Center

	Questions and discussion with the committee

4:00 p.m.	**Windows of Susceptibility to Breast Cancer and Environmental Exposures**

Windows of Susceptibility to Breast Cancer and Environmental Exposures
Jose Russo, M.D.
> *Director, Breast Cancer Research Laboratory and NCI-NIEHS Breast Cancer and the Environment Research Center*
> *Fox Chase Cancer Center*

Questions and discussion with the committee

4:45 p.m.   **General Discussion as Needed**

5:00 p.m.   **Opportunity for Individual Public Comment**

Rebecca Shaloff, Washington, DC
Victoria Pavelko, Reston, VA
Heather Rogers, Alexandria, VA

5:30 p.m.   **Adjourn Open Session**

Appendix B

Biographical Sketches of Committee Members

Irva Hertz-Picciotto (*Chair*), is a professor in the Department of Public Health Sciences, School of Medicine, and at the Medical Investigations of Neurodevelopmental Disorders (MIND) Institute, University of California, Davis, and is chief of the Division of Environmental and Occupational Health. She also is deputy director of the Center for Children's Environmental Health at UC Davis and director of the Northern California Center for the National Children's Study. She has published widely on environmental exposures, including metals, pesticides, PCBs, and air pollution, and their effects on pregnancy, the neonate, and early child development, as well as on methods in epidemiologic research. In 2002, she turned her attention to identifying causes of autism, and launched the CHARGE Study (Childhood Autism Risk from Genetics and the Environment) and subsequently the MARBLES Study (Markers of Autism Risk in Babies–Learning Early Signs). She has served or currently sits on editorial boards for the *American Journal of Epidemiology*, *Environmental Health Perspectives*, *Epidemiology*, and *Autism Research*. Dr. Hertz-Picciotto has served as president of the Society for Epidemiologic Research and of the International Society for Environmental Epidemiology. She has held appointments on the Governor's Carcinogen Identification Committee for the State of California, the scientific advisory boards/panels for the Environmental Protection Agency, National Institute for Occupational Safety and Health, National Toxicology Program, and National Institutes of Health Interagency Coordinating Committee on Autism Research. Dr. Hertz-Picciotto has also chaired two previous Institute of Medicine committees. She received a Ph.D. and an M.P.H. in epidemiology and an M.A. in biostatistics from UC Berkeley. Before joining the faculty at UC Davis, Dr. Hertz-Picciotto was a professor in the Department of Epidemiology at the School of Public Health, University of North Carolina, Chapel Hill.

Lucile Adams-Campbell joined the Georgetown University Lombardi Comprehensive Cancer Center in 2008 as associate dean for Community Health and Outreach. Previously, she had served as director of the Howard University Cancer Center for 13 years. She also serves as the co-principal investigator for the Black Women's Health Study. She focuses on community outreach and community-based participatory research, particularly cancer-related health disparities in minority populations, with an emphasis on cancer prevention. Her research interests include understanding the biological basis of health disparities in those cancers that disproportionately affect minority and underserved populations via clinical trials; cancer epidemiology using minority cohorts; and behavioral epidemiology as it relates to physical activity and nutrition interventions. She aims to export prevention-based clinical trials and behavioral interventions targeting nutrition and exercise strategies to address obesity from the laboratory setting into the community. Dr. Adams-Campbell was elected to the Institute of Medicine in 2008. She currently serves on the editorial board of or as a reviewer for eight journals. Dr. Adams-Campbell received an M.S. in biomedical science from Drexel University and a Ph.D. in epidemiology from the University of Pittsburgh. She completed a National Institutes of Health-funded postdoctoral fellowship at the University of Pittsburgh.

Peggy Devine is the founder and president of Cancer Information and Support Network, an organization that seeks to increase public awareness on all aspects of cancer, including the importance of cancer research. She served as the multisite advocate coordinator for the American College of Radiology Imaging Network (ACRIN) MRI/CALGB Correlative Science clinical trial (I SPY 1); is an advocate in the University of California, San Francisco Breast Specialized Program of Research Excellence (SPORE); and is a research advocate consultant for many groups, including Los Alamos Laboratory, where she serves on a Department of Defense-funded team award grant, Breast Cancer: Catch It with Ultrasound, 2011–2015. Ms. Devine has served as a trainer for advocacy and professional associations, including the American Society of Clinical Oncology Gastrointestinal Cancer, Summit Series on Clinical Trials Advocate Training, American College of Surgeons Oncology Group, ACRIN, and Coalition of Cancer Cooperative Groups. Ms. Devine also sits on many advisory boards, including the National Cancer Institute's Office of Biorepositories and Biospecimen Research Steering Committee for a grant entitled "Research Studies in Cancer and Normal Pre-Analytical Variables and Their Effects on Molecular Integrity," 2010–2014. Ms. Devine has also reviewed grants for the Department of Defense, the National Cancer Institute, Avon, Susan G. Komen for the Cure, and the California Breast Cancer Research Program. Ms. Devine has a B.S. in chemistry and biological science from Michigan State University. She then completed a year of training in clinical laboratory science at Huntington Memorial Hospital and holds federal and state licensure in that field.

David Eaton is associate vice provost for research and a professor in the Department of Environmental and Occupational Health Sciences at the School of Public Health of the University of Washington. He is also the director of the Center for Ecogenetics and Environmental Health at the University of Washington. He joined the faculty of the University of Washington in 1979. Dr. Eaton's research and teaching focus on the molecular basis for environmental causes of cancer, and how human genetic differences in biotransformation enzymes may increase or decrease individual susceptibility to chemicals found in the environment. He has served as president of the Society of Toxicology and as a member of the board of trustees of the Academy of Toxicological Sciences. He is an elected fellow of the American Association for the Advancement of Science and the Academy of Toxicological Sciences. Dr. Eaton has served on several committees for the National Academy of Sciences, most recently chairing the Committee for Review of the Federal Strategy to Address Environmental, Health, and Safety Research Needs for Engineered Nanoscale Materials. He received his Ph.D. in pharmacology and toxicology and completed a postdoctoral fellowship in toxicology at the University of Kansas Medical Center.

S. Katharine Hammond is a professor in the Division of Environmental Health Sciences in the School of Public Health of the University of California, Berkeley. Her research interests focus on health effects of exposure to airborne materials, including responses of asthmatic children to short-term fluctuations in particulate air pollution, neurologic and reproductive effects of hexane on workers, secondhand smoke, and the effects of exposure to polycyclic aromatic hydrocarbons on asthmatic children and users of coal in China. She has received the National Institute for Occupational Safety and Health's Alice Hamilton Award for Excellence in Occupational Safety and Health and the American Industrial Hygiene Association's Rachel Carson Environmental Award. Dr. Hammond currently serves on the Scientific Review Panel on Toxic Air Contaminants for the California Environmental Protection Agency and has served as a consultant to the Science Advisory Board of the Environmental Protection Agency. She is also a member of the World Health Organization's Tobacco Product Regulation Study Group. She has served on numerous committees for the Institute of Medicine and the National Research Council. Dr. Hammond received a

Ph.D. in chemistry from Brandeis University and an M.S. in environmental health sciences from the Harvard School of Public Health.

Kathy J. Helzlsouer is the director of the Prevention and Research Center at Mercy Medical Center in Baltimore, Maryland. She is also an adjunct professor of epidemiology at the Johns Hopkins University Bloomberg School of Public Health. Dr. Helzlsouer's work focuses on clinical epidemiology, cancer epidemiology, and cancer prevention. In 2008, Dr. Helzlsouer was named chair of the Maryland State Council on Cancer Control. She also is chair-elect of the Molecular Epidemiology Group (MEG) of the American Association for Cancer Research. She serves on the Physician Data Query (PDQ) Cancer Screening and Prevention Committee of the National Cancer Institute, and she is a member of the advisory board for the AARP Cohort Study. Dr. Helzlsouer holds an M.D. from the University of Pittsburgh School of Medicine and an M.H.S. in epidemiology from the Johns Hopkins University School of Hygiene and Public Health. She is board certified in medical oncology.

Robert A. Hiatt is professor and chair of Epidemiology and Biostatistics at University of California, San Francisco (UCSF). He is the director of Population Sciences and associate director of the UCSF Helen Diller Family Comprehensive Cancer Center. Dr. Hiatt holds adjunct appointments at the UC Berkeley School of Public Health and the Division of Research at Kaiser Permanente Northern California. He has been the principal investigator for the Bay Area Breast Cancer and the Environment Research Center for the past 7 years and now leads the Coordinating Center for the national program that continues to explore the influence of environmental factors on pubertal maturation as a window to understanding the causes of breast cancer. From 1998 to early 2003, Dr. Hiatt was the first deputy director of the Division of Cancer Control and Population Sciences at the National Cancer Institute (NCI), where he oversaw cancer research in epidemiology and genetics, surveillance, and health services research. Before then he was the director of Prevention Sciences at the Northern California Cancer Center and also assistant director for epidemiology at the Division of Research, Kaiser Permanente Medical Care Program in Northern California. He is board certified in preventive medicine and, until taking his NCI position, practiced general internal medicine. He is a past president of the American College of Epidemiology and the American Society for Preventive Oncology. Dr. Hiatt received an M.D. from the University of Michigan and a Ph.D. in epidemiology from the University of California, Berkeley.

Chanita Hughes Halbert is an associate professor in the Department of Psychiatry at the University of Pennsylvania and director of the Center for Community-Based Research and Health Disparities. She is also director of the Community Diversity Initiative at the Abramson Cancer Center and associate director for Community Engagement in the Robert Wood Johnson Clinical Scholars Program. Dr. Hughes Halbert's research focuses on understanding the sociocultural underpinnings of cancer prevention and control behaviors among ethnically diverse populations and translating this knowledge into interventions designed to reduce ethnic and racial differences in cancer morbidity and mortality. She is principal investigator (PI) of an academic–community partnership funded by the National Institute on Minority Health and Health Disparities and the National Cancer Institute to develop and evaluate interventions for cancer prevention and control in community settings. She is also PI of grants funded by the National Human Genome Research Institute to identify barriers and facilitators of African American participation in cancer genetics research and to understand the long-term psychological and behavioral impact of genetic testing for *BRCA1* and *BRCA2* mutations. She earned her Ph.D. in personality psychology from Howard University. In addition to her doctoral training, Dr. Hughes Halbert completed pre- and postdoctoral training at the Lombardi Cancer Center at Georgetown University.

David J. Hunter is the dean for Academic Affairs and Vincent L. Gregory Professor in Cancer Prevention in the Departments of Epidemiology and Nutrition in the Harvard School of Public Health. His principal research interests are the etiology of cancer—particularly breast, prostate, and skin cancer. He was an investigator on the Nurses' Health Study, a long-running cohort of 121,000 U.S. women, and a project director for the Nurses' Health Study II, a newer cohort of 116,000 women. He also analyzes inherited susceptibility to cancer and other chronic diseases using molecular techniques and studies molecular markers of environmental exposures. Dr. Hunter was the director for the Harvard Center for Cancer Prevention and was co-director of the National Cancer Institute (NCI) Cancer Genetic Markers of Susceptibility (CGEMS) Special Initiative. He is the director of the Program in Molecular and Genetic Epidemiology at the Harvard School of Public Health, and co-chair of the NCI Breast and Prostate Cancer Cohort Consortium. Dr. Hunter received his M.P.H. and Sc.D. from Harvard University. He earned his M.B., B.S. from the University of Sydney.

Barnett (Barry) Kramer is editor-in-chief of the *Journal of the National Cancer Institute* (*JNCI*) and of the Screening and Prevention Editorial Board of the National Cancer Institute (NCI) Physician Data Query (PDQ). He was the associate director for disease prevention and head of the Office of Disease Prevention in the Office of the Director at the National Institutes of Health from 2001 to 2010. Previously, he served as the director of the

Office of Medical Applications of Research (OMAR), a component of the
Office of Disease Prevention. He has also previously served as deputy director of NCI's Division of Cancer Prevention. Dr. Kramer received his M.D.
from the University of Maryland Medical School and is board certified in
internal medicine and medical oncology. He received an M.P.H. from the
Johns Hopkins University School of Hygiene and Public Health.

Bryan M. Langholz is a professor of research in the Division of Biostatistics and a visiting professor in the department of preventive medicine
at the Keck School of Medicine of the University of Southern California
(USC). He also is a research adjunct professor in the USC Department of
Mathematics and a member of the Children's Oncology Group. Dr. Langholz's research interests are cohort sampling methods, statistical methods
in epidemiology and occupational health, and statistical methods in genetic
epidemiology. Additionally, he has done biostatistics research in pesticides
and cancer, electromagnetic fields and cancer, radiation and cancer, and
genetic and environmental factors in type 1 diabetes. Dr. Langholz is currently a co-principal investigator on a National Cancer Institute-funded
study examining the link between ultraviolet exposure and melanogenesis.
Dr. Langholz is a member of the American Statistical Association, the
International Biometrics Society, the International Genetic Epidemiology
Society, and the International Society of Clinical Biostatistics. He has served
as a reviewer for 28 journals and was an associate editor for *Biometrics*.
Dr. Langholz previously served on the Institute of Medicine Committee to
Review the Health Effects in Vietnam Veterans of Exposure to Herbicides:
First Biennial Update. Dr. Langholz received an M.S. and a Ph.D. in bio-
mathematics from the University of Washington.

Peggy Reynolds is a senior research scientist at the Cancer Prevention
Institute of California, a consulting professor for the Department of Health
Research and Policy in the Stanford University School of Medicine, and a
member of the Stanford Cancer Center. Dr. Reynolds spent several years as
an epidemiologist for the California Tumor Registry and San Francisco Bay
Area SEER (Surveillance, Epidemiology, and End Results) program, and
previously served as the chief of the Environmental Epidemiology Section in
the California Department of Health Services. Over the years, she has conducted a number of cancer epidemiology studies, with a concentration on
environmental risk factors. Her research currently focuses on female breast
cancer and cancers in children. She was a founding member of the California Teachers Study (CTS), an ongoing prospective study of 133,479 women
established in 1995. Dr. Reynolds was a co-investigator for an influential
multicenter study on the risk of lung cancer from secondhand smoke. She
and her research team further pursued the role of secondhand smoke and

breast cancer in a more detailed assessment of reported lifetime exposures in the CTS. In addition, Dr. Reynolds has served as principal investigator for a study of regional variations in breast cancer in California, a study of body burden levels of endocrine disruptors in breast cancer patients, a study of breast cancer in young women, studies of breast cancer incidence in flight attendants and cosmetologists, a study of malignant melanoma among Lawrence Livermore Laboratory employees, and a statewide study of patterns of childhood cancer. Dr. Reynolds earned a Ph.D. in epidemiology from the University of California, Berkeley.

Joyce S. Tsuji is a principal scientist within the Center for Toxicology and Mechanistic Biology of the Health Sciences practice of Exponent. She is a board-certified toxicologist and a fellow of the Academy of Toxicological Sciences. She has conducted risk assessment and toxicology studies in the United States and internationally for industry, trade associations, the Environmental Protection Agency (EPA) and state agencies, the Department of Justice, the Australian EPA, municipalities, and private citizens. Dr. Tsuji's experience includes human health and environmental toxicology related to metals and a wide variety of other chemicals in the environment. She has designed and directed dietary and environmental exposure studies and community programs involving health education and biomonitoring for populations potentially exposed to chemicals in the environment, including soil, water, and food-chain exposures. She has also assessed exposure and health risks associated with chemicals in air, foods, and a variety of consumer products. She has served on committees for several National Research Council (NRC) studies and is currently a member of the NRC Board on Environmental Studies and Toxicology and the NRC Committee on Toxicology. Dr. Tsuji received a Ph.D. in environmental physiology and completed a postdoctoral fellowship in quantitative genetics at the Department of Zoology of the University of Washington.

Cheryl Lyn Walker recently became a Welch Professor and director of the Institute of Biosciences and Technology of Texas A&M Health Science Center. Prior to this appointment, she was the Ruth and Walter Sterling Professor of Carcinogenesis at the University of Texas MD Anderson Cancer Center, where she also directed the Center for Environmental and Molecular Carcinogenesis of the Institute for Basic Science. Dr. Walker's research interests include the genetic basis of susceptibility to cancer, specifically the interaction of environmental agents with genes during tumor development; the effects of endocrine disruptors on human health; and animal models for human disease. She also studies the molecular mechanisms of kidney, breast, and uterine cancers and the mechanisms by which environmental hormones reprogram the epigenome to increase susceptibility to these cancers. She has served on the Board of Scientific Counselors of the National Cancer Institute and the National Institute of Environmental Health Sciences National Toxicology Program and is a past president of the Society of Toxicology. Dr. Walker received her Ph.D. from the Department of Cell Biology at the University of Texas Health Science Center, Southwestern Medical School (Dallas), and completed postdoctoral training at the National Institutes of Health.

Lauren Zeise is chief of the Reproductive and Cancer Hazard Assessment Branch of the California Environmental Protection Agency's Office of Environmental Health Hazard Assessment. She oversees or is involved in a variety of California's risk assessment activities, including cancer and reproductive toxicant assessments; development of frameworks and methodologies for assessing toxicity, cumulative impact, nanotechnology, green chemistry/safer alternatives, and susceptible populations; the California Environmental Contaminant Biomonitoring Program; and health risk characterizations for environmental media, food, fuels and consumer products. Dr. Zeise's research focuses on human interindividual variability, dose response, uncertainty, and risk. She was the 2008 recipient of the Society of Risk Analysis's Outstanding Practitioners Award. She has served on advisory boards and committees of the Environmental Protection Agency (EPA), Office of Technology Assessment, World Health Organization, and National Institute of Environmental Health Sciences. Dr. Zeise has also served on numerous National Research Council and Institute of Medicine committees and boards. Most recently, she was a member of the Committee to Review EPA's Title 42 Hiring Authority for Highly Qualified Scientists and Engineers, and the Committee on Use of Emerging Science for Environmental Health Decisions. Dr. Zeise received a Ph.D. from Harvard University.

Appendix C

Classification Systems Used in Evidence Reviews

TABLE C-1 Compilation of Evidence Categories Used by Selected
Organizations

International Agency for Research on Cancer (IARC)	Environmental Protection Agency (EPA)	National Toxicology Program (NTP)
Overall evaluation		
Finally, the body of evidence is considered as a whole, in order to reach an overall evaluation of the carcinogenicity of the agent to humans.		
An evaluation may be made for a group of agents that have been evaluated by the Working Group. In addition, when supporting data indicate that other related agents, for which there is no direct evidence of their capacity to induce cancer in humans or in animals, may also be carcinogenic, a statement describing the rationale for this conclusion is added to the evaluation narrative; an additional evaluation may be made for this broader group of agents if the strength of the evidence warrants it.		
The agent is described according to the wording of one of the following categories, and the designated group is given. The categorization of an agent is a matter of scientific judgement that reflects the strength of the evidence derived from studies in humans and in experimental animals and from mechanistic and other relevant data.		

World Cancer Research Fund/ American Institute for Cancer Research (WCRF/AICR)	Institute of Medicine (IOM)

Special upgrading factors

These are factors that form part of the assessment of the evidence that, when present, can upgrade the judgement reached. So an exposure that might be deemed a "limited—suggestive" causal factor in the absence, say, of a biological gradient, might be upgraded to "probable" in its presence. The application of these factors (listed below) requires judgement, and the way in which these judgements affect the final conclusion in the matrix are stated.

- Presence of a plausible biological gradient ("dose response") in the association. Such a gradient need not be linear or even in the same direction across the different levels of exposure, so long as this can be explained plausibly.
- A particularly large summary effect size (an odds ratio or relative risk of 2.0 or more, depending on the unit of exposure) after appropriate control for confounders.
- Evidence from randomised trials in humans.
- Evidence from appropriately controlled experiments demonstrating one or more plausible and specific mechanisms actually operating in humans.
- Robust and reproducible evidence from experimental studies in appropriate animal models showing that typical human exposures can lead to relevant cancer outcomes.

The committee relied entirely on clinical and human epidemiologic studies to draw its conclusions about the strength of evidence regarding associations between deployment to the Gulf War and health outcomes seen in Gulf War veterans. The committee acknowledges, however, that animal studies might prove helpful in providing biologic understanding of many of the effects seen in humans from specific exposures, such as pesticides, solvents, and nerve agents, which have been reported by troops deployed in the Gulf War. Furthermore, information from molecular and cellular biology, neuroimaging, and other types of human studies can be used to understand the biological mechanisms and identification of biomarkers for clinical outcomes. Such studies, however, are not, in general, included in this review.

continued

TABLE C-1 Continued

International Agency for Research on Cancer (IARC)	Environmental Protection Agency (EPA)	National Toxicology Program (NTP)
Group 1: The agent is *carcinogenic to humans.* This category is used when there is sufficient evidence of carcinogenicity in humans. Exceptionally, an agent may be placed in this category when evidence of carcinogenicity in humans is less than *sufficient* but there is *sufficient evidence of carcinogenicity* in experimental animals and strong evidence in exposed humans that the agent acts through a relevant mechanism of carcinogenicity.	**Carcinogenic to humans** This descriptor indicates strong evidence of human carcinogenicity. It covers different combinations of evidence. 1. This descriptor is appropriate when there is convincing epidemiologic evidence of a *causal* association between human exposure and cancer. 2. Exceptionally, this descriptor may be equally appropriate with a lesser weight of epidemiologic evidence that is strengthened by other lines of evidence. It can be used when all of the following conditions are met: (a) there is strong evidence of an association between human exposure and either cancer or the key precursor events of the agent's mode of action but not enough for a causal association, and (b) there is extensive evidence of carcinogenicity in animals, and (c) the mode(s) of carcinogenic action and associated key precursor events have been identified in animals, and (d) there is strong evidence that the key precursor events that precede the cancer response in animals are anticipated to occur in humans and progress to tumors, based on available biological information. In this case, the narrative includes a summary of both the experimental and epidemiologic information on mode of action and also an indication of the relative weight that each source of information carries, e.g., based on human information, based on limited human and extensive animal experiments.	**Known to be human carcinogen** There is sufficient evidence of carcinogenicity from studies in humans,* which indicates a causal relationship between exposure to the agent, substance, or mixture, and human cancer.

World Cancer Research Fund/ American Institute for Cancer Research (WCRF/AICR)	Institute of Medicine (IOM)
Convincing These criteria are for evidence strong enough to support a judgement of a convincing causal relationship, which justifies goals and recommendations designed to reduce the incidence of cancer. A convincing relationship should be robust enough to be highly unlikely to be modified in the foreseeable future as new evidence accumulates. All of the following were generally required: • Evidence from more than one study type. • Evidence from at least two independent cohort studies. • No substantial unexplained heterogeneity within or between study types or in different populations relating to the presence or absence of an association, or direction of effect. • Good-quality studies to exclude with confidence the possibility that the observed association results from random or systematic error, including confounding, measurement error, and selection bias. • Presence of a plausible biological gradient ("dose response") in the association. Such a gradient need not be linear or even in the same direction across the different levels of exposure, so long as this can be explained plausibly. • Strong and plausible experimental evidence, either from human studies or relevant animal models, that typical human exposures can lead to relevant cancer outcomes.	**Sufficient evidence of a causal relationship** Evidence is sufficient to conclude that a causal relationship exists between being deployed to the Gulf War and a health outcome. The evidence fulfills the criteria for sufficient evidence of a causal association in which chance, bias, and confounding can be ruled out with reasonable confidence. The association is supported by several of the other considerations used to assess causality: strength of association, dose–response relationship, consistency of association, temporal relationship, specificity of association, and biologic plausibility.

continued

TABLE C-1 Continued

International Agency for Research on Cancer (IARC)	Environmental Protection Agency (EPA)	National Toxicology Program (NTP)

Group 2:

This category includes agents for which, at one extreme, the degree of evidence of carcinogenicity in humans is almost *sufficient*, as well as those for which, at the other extreme, there are no human data but for which there is evidence of carcinogenicity in experimental animals. Agents are assigned to either Group 2A (*probably carcinogenic to humans*) or Group 2B (*possibly carcinogenic to humans*) on the basis of epidemiological and experimental evidence of carcinogenicity and mechanistic and other relevant data. The terms *probably carcinogenic* and *possibly carcinogenic* have no quantitative significance and are used simply as descriptors of different levels of evidence of human carcinogenicity, with *probably carcinogenic* signifying a higher level of evidence than *possibly carcinogenic*.

World Cancer Research Fund/ American Institute for Cancer Research (WCRF/AICR)	Institute of Medicine (IOM)

continued

TABLE C-1 Continued

International Agency for Research on Cancer (IARC)	Environmental Protection Agency (EPA)	National Toxicology Program (NTP)
Group 2A: The agent is *probably carcinogenic to* **humans.** This category is used when there is *limited evidence of carcinogenicity* in humans and *sufficient evidence of carcinogenicity* in experimental animals. In some cases, an agent may be classified in this category when there is inadequate evidence of carcinogenicity in humans and sufficient evidence of carcinogenicity in experimental animals and strong evidence that the carcinogenesis is mediated by a mechanism that also operates in humans. Exceptionally, an agent may be classified in this category solely on the basis of limited evidence of carcinogenicity in humans. An agent may be assigned to this category if it clearly belongs, based on mechanistic considerations, to a class of agents for which one or more members have been classified in Group 1 or Group 2A.	**Likely to be carcinogenic to humans** This descriptor is appropriate when the weight of the evidence is adequate to demonstrate carcinogenic potential to humans but does not reach the weight of evidence for the descriptor "Carcinogenic to Humans." Adequate evidence consistent with this descriptor covers a broad spectrum. As stated previously, the use of the term "likely" as a weight of evidence descriptor does not correspond to a quantifiable probability. The examples below are meant to represent the broad range of data combinations that are covered by this descriptor; they are illustrative and provide neither a checklist nor a limitation for the data that might support use of this descriptor. Moreover, additional information, e.g., on mode of action, might change the choice of descriptor for the illustrated examples. Supporting data for this descriptor may include: • an agent demonstrating a plausible (but not definitively causal) association between human exposure and cancer, in most cases with some supporting biological, experimental evidence, though not necessarily carcinogenicity data from animal experiments; • an agent that has tested positive in animal experiments in more than one species, sex, strain, site, or exposure route, with or without evidence of carcinogenicity in humans; • a positive tumor study that raises additional biological concerns beyond that of a statistically significant result, for example, a high degree of malignancy, or an early age at onset; • a rare animal tumor response in a single experiment that is assumed to be relevant to humans; or	**Reasonably anticipated to be human carcinogen:** There is limited evidence of carcinogenicity from studies in humans,* which indicates that causal interpretation is credible, but that alternative explanations, such as chance, bias, or confounding factors, could not adequately be excluded, or there is sufficient evidence of carcinogenicity from studies in experimental animals, which indicates there is an increased incidence of malignant and/or a combination of malignant and benign tumors (1) in multiple species or at multiple tissue sites, or (2) by multiple routes of exposure, or (3) to an unusual degree with regard to incidence, site, or type of tumor, or age at onset, or there is less than sufficient evidence of carcinogenicity in humans or laboratory animals; however, the agent, substance, or mixture belongs to a well-defined, structurally related class of substances whose members are listed in a previous *Report on Carcinogens* as either known to be a human carcinogen or reasonably anticipated to be a human carcinogen, or there is convincing relevant information that the agent acts through mechanisms indicating it would likely cause cancer in humans.

World Cancer Research Fund/ American Institute for Cancer Research (WCRF/AICR)	Institute of Medicine (IOM)
Probable	**Sufficient evidence of an association**

Probable

These criteria are for evidence strong enough to support a judgement of a probable causal relationship, which would generally justify goals and recommendations designed to reduce the incidence of cancer.

All the following were generally required:

- Evidence from at least two independent cohort studies, or at least five case–control studies.
- No substantial unexplained heterogeneity between or within study types in the presence or absence of any association, or direction of effect.
- Good quality studies to exclude with confidence the possibility that the observed association results from random or systematic error, including confounding, measurement error, and selection bias.
- Evidence for biological plausibility.

Sufficient evidence of an association

Evidence suggests an association, in that a positive association has been observed between deployment to the Gulf War and a health outcome in humans; however, there is some doubt as to the influence of chance, bias, and confounding.

continued

TABLE C-1 Continued

International Agency for Research on Cancer (IARC)	Environmental Protection Agency (EPA)	National Toxicology Program (NTP)
	• a positive tumor study that is strengthened by other lines of evidence, for example, either plausible (but not definitively causal) association between human exposure and cancer or evidence that the agent or an important metabolite causes events generally known to be associated with tumor formation (such as DNA reactivity or effects on cell growth control) likely to be related to the tumor response in this case.	Conclusions regarding carcinogenicity in humans or experimental animals are based on scientific judgment, with consideration given to all relevant information. Relevant information includes, but is not limited to, dose response, route of exposure, chemical structure, metabolism, pharmacokinetics, sensitive sub-populations, genetic effects, or other data relating to mechanism of action or factors that may be unique to a given substance. For example, there may be substances for which there is evidence of carcinogenicity in laboratory animals, but there are compelling data indicating that the agent acts through mechanisms which do not operate in humans and would therefore not reasonably be anticipated to cause cancer in humans.

World Cancer Research Fund/ American Institute for Cancer Research (WCRF/AICR)	Institute of Medicine (IOM)

continued

TABLE C-1 Continued

International Agency for Research on Cancer (IARC)	Environmental Protection Agency (EPA)	National Toxicology Program (NTP)

Group 2B: The agent is *possibly carcinogenic to humans.*

This category is used for agents for which there is *limited evidence of carcinogenicity* in humans and less than *sufficient evidence of carcinogenicity* in experimental animals. It may also be used when there is *inadequate evidence of carcinogenicity* in humans but there is *sufficient evidence of carcinogenicity* in experimental animals. In some instances, an agent for which there is *inadequate evidence of carcinogenicity* in humans and less than *sufficient evidence of carcinogenicity* in experimental animals together with supporting evidence from mechanistic and other relevant data may be placed in this group. An agent may be classified in this category solely on the basis of strong evidence from mechanistic and other relevant data.

Suggestive evidence of carcinogenic potential

This descriptor of the database is appropriate when the weight of evidence is suggestive of carcinogenicity; a concern for potential carcinogenic effects in humans is raised, but the data are judged not sufficient for a stronger conclusion. This descriptor covers a spectrum of evidence associated with varying levels of concern for carcinogenicity, ranging from a positive cancer result in the only study on an agent to a single positive cancer result in an extensive database that includes negative studies in other species. Depending on the extent of the database, additional studies may or may not provide further insights. Some examples include:

- a small, and possibly not statistically significant, increase in tumor incidence observed in a single animal or human study that does not reach the weight of evidence for the descriptor "Likely to Be Carcinogenic to Humans." The study generally would not be contradicted by other studies of equal quality in the same population group or experimental system (see discussions of *conflicting evidence* and *differing results*, below);
- a small increase in a tumor with a high background rate in that sex and strain, when there is some but insufficient evidence that the observed tumors may be due to intrinsic factors that cause background tumors and not due to the agent being assessed. (When there is a high background rate of a specific tumor in animals of a particular sex and strain, then there may be biological factors operating independently of the agent being assessed that could be responsible for the development of the observed tumors.) In this case, the reasons for determining that the tumors are not due to the agent are explained;

World Cancer Research Fund/ American Institute for Cancer Research (WCRF/AICR)	Institute of Medicine (IOM)
Limited—suggestive These criteria are for evidence that is too limited to permit a probable or convincing causal judgement, but where there is evidence suggestive of a direction of effect. The evidence may have methodological flaws, or be limited in amount, but shows a generally consistent direction of effect. This almost always does not justify recommendations designed to reduce the incidence of cancer. Any exceptions to this require special explicit justification. All the following were generally required:	**Limited/suggestive evidence of an association** Some evidence of an association between deployment to the Gulf War and a health outcome in humans exists, but this is limited by the presence of substantial doubt regarding chance, bias, and confounding.

- Evidence from at least two independent cohort studies or at least five case–control studies.
- The direction of effect is generally consistent though some unexplained heterogeneity may be present.
- Evidence for biological plausibility.

continued

TABLE C-1 Continued

International Agency for Research on Cancer (IARC)	Environmental Protection Agency (EPA)	National Toxicology Program (NTP)
	• evidence of a positive response in a study whose power, design, or conduct limits the ability to draw a confident conclusion (but does not make the study fatally flawed), but where the carcinogenic potential is strengthened by other lines of evidence (such as structure-activity relationships); **or** • a statistically significant increase at one dose only, but no significant response at the other doses and no overall trend.	
Group 3: The agent is *not classifiable as to its carcinogenicity to humans.* This category is used most commonly for agents for which the evidence of carcinogenicity is *inadequate* in humans and *inadequate* or *limited* in experimental animals. Exceptionally, agents for which the evidence of carcinogenicity is *inadequate* in humans but *sufficient* in experimental animals may be placed in this category when there is strong evidence that the mechanism of carcinogenicity in experimental animals does not operate in humans. Agents that do not fall into any other group are also placed in this category. An evaluation in Group 3 is not a determination of non-carcinogenicity or overall safety. It often means that further research is needed, especially when exposures are widespread or the cancer data are consistent with differing interpretations.	**Inadequate information to assess carcinogenic potential** This descriptor of the database is appropriate when available data are judged inadequate for applying one of the other descriptors. Additional studies generally would be expected to provide further insights. Some examples include: – little or no pertinent information; – conflicting evidence, that is, some studies provide evidence of carcinogenicity but other studies of equal quality in the same sex and strain are negative. *Differing results*, that is, positive results in some studies and negative results in one or more different experimental systems, do not constitute *conflicting evidence,* as the term is used here. Depending on the overall weight of evidence, differing results can be considered either suggestive evidence or likely evidence; **or** – negative results that are not sufficiently robust for the descriptor, "Not Likely to Be Carcinogenic to Humans."	

World Cancer Research Fund/ American Institute for Cancer Research (WCRF/AICR)	Institute of Medicine (IOM)

Limited—no conclusion

Evidence is so limited that no firm conclusion can be made. This category represents an entry level, and is intended to allow any exposure for which there are sufficient data to warrant Panel consideration, but where insufficient evidence exists to permit a more definitive grading. This does not necessarily mean a limited quantity of evidence. A body of evidence for a particular exposure might be graded "limited—no conclusion" for a number of reasons. The evidence might be limited by the amount of evidence in terms of the number of studies available, by inconsistency of direction of effect, by poor quality of studies (for example, lack of adjustment for known confounders), or by any combination of these factors. Exposures that are graded "limited—no conclusion" do not appear in the matrices presented in Chapters 4–6, but do appear in Chapters 7 and 8.

When an exposure is graded "limited—no conclusion", this does not necessarily indicate that the Panel has judged that there is evidence of no relationship. With further good quality research, any exposure graded in this way might in the future be shown to increase or decrease the risk of cancer. Where there is sufficient evidence to give confidence that an exposure is unlikely to have an effect on cancer risk, this exposure will be judged "substantial effect on risk unlikely."

There are also many exposures for which there is such limited evidence that no judgement is possible. In these cases, evidence is recorded in the full SLR reports contained on the CD included with this Report. However, such evidence is usually not included in the summaries and is not included in the matrices in this printed Report.

Inadequate/insufficient evidence to determine whether an association exists

The available studies are of insufficient quality, validity, consistency, or statistical power to permit a conclusion regarding the presence or absence of an association between deployment to the Gulf War and a health outcome in humans.

continued

TABLE C-1 Continued

International Agency for Research on Cancer (IARC)	Environmental Protection Agency (EPA)	National Toxicology Program (NTP)
Group 4: *The agent is probably not carcinogenic to humans.* This category is used for agents for which there is *evidence suggesting lack of carcinogenicity* in humans and in experimental animals. In some instances, agents for which there is *inadequate evidence of carcinogenicity* in humans but *evidence suggesting lack of carcinogenicity* in experimental animals, consistently and strongly supported by a broad range of mechanistic and other relevant data, may be classified in this group.	**Not likely to be carcinogenic to humans** This descriptor is appropriate when the available data are considered robust for deciding that there is no basis for human hazard concern. In some instances, there can be positive results in experimental animals when there is strong, consistent evidence that each mode of action in experimental animals does not operate in humans. In other cases, there can be convincing evidence in both humans and animals that the agent is not carcinogenic. The judgment may be based on data such as: • animal evidence that demonstrates lack of carcinogenic effect in both sexes in well-designed and well-conducted studies in at least two appropriate animal species (in the absence of other animal or human data suggesting a potential for cancer effects), • convincing and extensive experimental evidence showing that the only carcinogenic effects observed in animals are not relevant to humans, • convincing evidence that carcinogenic effects are not likely by a particular exposure route (see Section 2.3), or • convincing evidence that carcinogenic effects are not likely below a defined dose range. A descriptor of "not likely" applies only to the circumstances supported by the data. For example, an agent may be "Not Likely to Be Carcinogenic" by one route but not necessarily by another. In those cases that have positive animal experiment(s) but the results are judged to be not relevant to humans, the narrative discusses why the results are not relevant.	

| World Cancer Research Fund/ American Institute for Cancer Research (WCRF/AICR) | Institute of Medicine (IOM) |

Substantial effect on risk unlikely

Evidence is strong enough to support a judgement that a particular food, nutrition, or physical activity exposure is unlikely to have a substantial causal relation to a cancer outcome. The evidence should be robust enough to be unlikely to be modified in the foreseeable future as new evidence accumulates.

All of the following were generally required:

- Evidence from more than one study type.
- Evidence from at least two independent cohort studies.
- Summary estimate of effect close to 1.0 for comparison of high versus low exposure categories.
- No substantial unexplained heterogeneity within or between study types or in different populations.
- Good quality studies to exclude, with absence of an observed association results from random or systematic error, including inadequate power, imprecision or error in exposure measurement, inadequate range of exposure, confounding, and selection bias.
- Absence of a demonstrable biological gradient ("dose response").
- Absence of strong and plausible experimental evidence, either from human studies or relevant animal models, that typical human exposures lead to relevant cancer outcomes.

Factors that might misleadingly imply an absence of effect include imprecision of the exposure assessment, an insufficient range of exposure in the study population, and inadequate statistical power. Defects in these and other study design attributes might lead to a false conclusion of no effect.

The presence of a plausible, relevant biological mechanism does not necessarily rule out a judgement of "substantial effect on risk unlikely." But the presence of robust evidence from appropriate animal models or in humans that a specific mechanism exists, or that typical exposures can lead to cancer outcomes, argues against such a judgement.

Limited/suggestive evidence of no association

There are several adequate studies, covering the full range of levels of exposure that humans are known to encounter, that are consistent in not showing an association between deployment to the Gulf War and a health outcome. A conclusion of no association is inevitably limited to the conditions, levels of exposure, and length of observation covered by the available studies. In addition, the possibility of a very small increase in risk at the levels of exposure studied can never be excluded.

continued

TABLE C-1 Continued

International Agency for Research on Cancer (IARC)	Environmental Protection Agency (EPA)	National Toxicology Program (NTP)

*This evidence can include traditional cancer epidemiology studies, data from clinical studies, and/or data derived from the study of tissues or cells from humans exposed to the substance in question that can be useful for evaluating whether a relevant cancer mechanism is operating in people.

SOURCES: EPA (2005); NTP (2005); IARC (2006, pp. 22–23), used with permission: From *Monographs on the evaluation of carcinogenic risks to humans: Preamble.* Lyon, France: IARC. http://monographs.iarc.fr/ENG/Preamble/CurrentPreamble.pdf; WCRF/AICR (2007, pp. 60–61), used with permission; IOM (2010).

World Cancer Research Fund/ American Institute for Cancer Research (WCRF/AICR)	Institute of Medicine (IOM)
Because of the uncertainty inherent in concluding that an exposure has no effect on risk, the criteria used to judge an exposure "substantial effect on risk unlikely" are roughly equivalent to the criteria used with at least a "probable" level of confidence. Conclusions of "substantial effect on risk unlikely" with a lower confidence than this would not be helpful, and could overlap with judgements of "limited—suggestive" or "limited—no conclusion."	

TABLE C-2 Supplemental Criteria Used by IARC and WCRF/AICR in Evaluation of Evidence

IARC	
Human	Animal
Sufficient evidence of carcinogenicity: The Working Group considers that a causal relationship has been established between exposure to the agent and human cancer. That is, a positive relationship has been observed between the exposure and cancer in studies in which chance, bias and confounding could be ruled out with reasonable confidence. A statement that there is *sufficient evidence* is followed by a separate sentence that identifies the target organ(s) or tissue(s) where an increased risk of cancer was observed in humans. Identification of a specific target organ or tissue does not preclude the possibility that the agent may cause cancer at other sites.	*Sufficient evidence of carcinogenicity:* The Working Group considers that a causal relationship has been established between the agent and an increased incidence of malignant neoplasms or of an appropriate combination of benign and malignant neoplasms in (a) two or more species of animals or (b) two or more independent studies in one species carried out at different times or in different laboratories or under different protocols. An increased incidence of tumours in both sexes of a single species in a well-conducted study, ideally conducted under Good Laboratory Practices, can also provide sufficient evidence. A single study in one species and sex might be considered to provide sufficient evidence of carcinogenicity when malignant neoplasms occur to an unusual degree with regard to incidence, site, type of tumour or age at onset, or when there are strong findings of tumours at multiple sites.
Limited evidence of carcinogenicity: A positive association has been observed between exposure to the agent and cancer for which a causal interpretation is considered by the Working Group to be credible, but chance, bias or confounding could not be ruled out with reasonable confidence.	*Limited evidence of carcinogenicity:* The data suggest a carcinogenic effect but are limited for making a definitive evaluation because, e.g., (a) the evidence of carcinogenicity is restricted to a single experiment; (b) there are unresolved questions regarding the adequacy of the design, conduct or interpretation of the studies; (c) the agent increases the incidence only of benign neoplasms or lesions of uncertain neoplastic potential; or (d) the evidence of carcinogenicity is restricted to studies that demonstrate only promoting activity in a narrow range of tissues or organs.
Inadequate evidence of carcinogenicity: The available studies are of insufficient quality, consistency or statistical power to permit a conclusion regarding the presence or absence of a causal association between exposure and cancer, or no data on cancer in humans are available.	*Inadequate evidence of carcinogenicity:* The studies cannot be interpreted as showing either the presence or absence of a carcinogenic effect because of major qualitative or quantitative limitations, or no data on cancer in experimental animals are available.

Mechanism

The strongest indications that a particular mechanism operates in humans derive from data on humans or biological specimens obtained from exposed humans. The data may be considered to be especially relevant if they show that the agent in question has caused changes in exposed humans that are on the causal pathway to carcinogenesis. Such data may, however, never become available, because it is at least conceivable that certain compounds may be kept from human use solely on the basis of evidence of their toxicity and/or carcinogenicity in experimental systems.

The conclusion that a mechanism operates in experimental animals is strengthened by findings of consistent results in different experimental systems, by the demonstration of biological plausibility and by coherence of the overall database. Strong support can be obtained from studies that challenge the hypothesized mechanism experimentally, by demonstrating that the suppression of key mechanistic processes leads to the suppression of tumour development. The Working Group considers whether multiple mechanisms might contribute to tumour development, whether different mechanisms might operate in different dose ranges, whether separate mechanisms might operate in humans and experimental animals and whether a unique mechanism might operate in a susceptible group. The possible contribution of alternative mechanisms must be considered before concluding that tumours observed in experimental animals are not relevant to humans. An uneven level of experimental support for different mechanisms may reflect that disproportionate resources have been focused on investigating a favoured mechanism.

continued

TABLE C-2 Continued

IARC

Evidence suggesting lack of carcinogenicity: There are several adequate studies covering the full range of levels of exposure that humans are known to encounter, which are mutually consistent in not showing a positive association between exposure to the agent and any studied cancer at any observed level of exposure. The results from these studies alone or combined should have narrow confidence intervals with an upper limit close to the null value (e.g., a relative risk of 1.0). Bias and confounding should be ruled out with reasonable confidence, and the studies should have an adequate length of follow-up. A conclusion of *evidence suggesting lack of carcinogenicity* is inevitably limited to the cancer sites, conditions and levels of exposure, and length of observation covered by the available studies. In addition, the possibility of a very small risk at the levels of exposure studied can never be excluded.

Evidence suggesting lack of carcinogenicity: Adequate studies involving at least two species are available which show that, within the limits of the tests used, the agent is not carcinogenic. A conclusion of evidence suggesting lack of carcinogenicity is inevitably limited to the species, tumour sites, age at exposure, and conditions and levels of exposure studied.

WCRF/AICR

Class 1
- In vivo data from studies in human volunteers (controlled human feeding studies).
- In vivo data from studies using genetically modified animal models related to human cancer (such as gene knockout or transgenic mouse models).
- In vivo data from studies using rodent cancer models designed to investigate modifiers of the cancer process.

Class 2
- In vitro data from studies using human cells validated with an in vivo model; for example, a transgenic model.
- In vitro data from studies using primary human cells.
- In vitro data from studies using human cell lines.

SOURCES: IARC (2006, pp. 19–21), used with permission: From *Monographs on the evaluation of carcinogenic risks to humans: Preamble.* Lyon, France: IARC. http://monographs.iarc. fr/ENG/Preamble/CurrentPreamble.pdf; WCRF/AICR (2007, p. 55), used with permission.

Class 3
- In vitro data from studies on animal cells.
- Data from mechanistic test systems; for example, isolated enzymes or genes.

REFERENCES

EPA (Environmental Protection Agency). 2005. *Guidelines for carcinogen risk assessment.* Washington, DC: EPA.

IARC (International Agency for Research on Cancer). 2006. *Monographs on the evaluation of carcinogenic risks to humans: Preamble.* Lyon, France: IARC. http://monographs.iarc. fr/ENG/Preamble/CurrentPreamble.pdf (accessed November 18, 2011).

IOM (Institute of Medicine). 2010. *Gulf War and health: Update of health effects of serving in the Gulf War.* Washington, DC: The National Academies Press.

NTP (National Toxicology Program). 2005. *Report on carcinogens,* 11th ed. Washington, DC: U.S. Department of Health and Human Services.

WCRF/AICR (World Cancer Research Fund/American Institute for Cancer Research). 2007. *Food, nutrition, physical activity, and the prevention of cancer: A global perspective.* Washington, DC: AICR.

Appendix D

Summary Table on Estimates of Population Attributable Risk

TABLE D-1 Summary of Estimates of Population Attributable Risk for
Risk Factors for Breast Cancer

Source	Study Location and Type	Study Population Characteristics
United States		
Madigan et al. (1995)	U.S. Cohort	NHANES I Epidemiologic Follow-up Study (NHEFS) (1971–1987) Women, ages 25–74 at initial examination 7,508 in analytic cohort 193 breast cancer cases
Rockhill et al. (1998)	U.S. Unmatched case–control with randomized recruitment	Carolina Breast Cancer Study (1993–1996) Women, ages 20–74 (white women for PAR analysis) 513 cases 445 controls
Tseng et al. (1999)	U.S. Survey data for risk factor prevalence Adjusted RR from meta-analysis Incidence rates from SEER data (1990–1994)	NHANES III (1988–1994) No age restrictions U.S. women

Risk Factors and Relative Risk Estimates[a]	Population Attributable Risk in Percent[b] (95% CI, where available)	Impact on Absolute Risk
History of breast cancer in 1st degree relative = 2.6 Income in upper two thirds of U.S. population = 1.7 Age at first birth > than 29 yrs = 1.9 Nulliparity = 1.8	*Estimates for U.S.* All factors: 41 (1.6–80.0) Family history: 9 (3.0–15.2) Higher income: 19 (-4.3–42.1) Later age at first birth or nulliparity: 30 (5.6–53.3) *Estimates for NHEFS* All factors: 47 (16.7–76.7) Family history: 8 (2.3–13.9) Higher income: 23 (5.4–39.9) Later age at first birth or nulliparity: 30 (8.9–51.4)	NHEFS age adjusted baseline incidence per 100,000 (standardized to 1970 U.S. population) History of breast cancer in 1st degree relative = 470 Income in upper two thirds of U.S. population = 259 Age at first birth > than 29 = 260 Nulliparity = 259
Early menarche (< age 12 = 1.24; age 12–13 = 1.08) Later age of first full-term pregnancy (≥ age 20) 1.08 to 1.53, depending on age group Breast cancer in mother or sister = 1.36 History of benign breast biopsy = 1.06	All: 25 (6–48) With menarche at < age 14 and first birth at ≥ age 20, or nulliparity All: 15 (~5–20) With menarche at < age 12 and first birth at ≥ age 30, or nulliparity	Not reported
Alcohol consumption None = 1.0 Light (0.1–6.4 g/day) = 1.0 Moderate (6.5–25.9 g/day) = 1.1 Heavy (≥ 26 g/day) = 1.3	Alcohol consumption: 2 (~1.2–2.9)	Not reported

continued

TABLE D-1 Continued

Source	Study Location and Type	Study Population Characteristics
Clarke et al. (2006)	U.S.	California Health Interview Survey, 2001
	Survey data for risk factor prevalence	White, non-Hispanic women, ages 40–79
	Published literature for RRs	Analysis for counties and hypothetical populations
	Incidence from California Cancer Registry (1998–2002)	3,781,621 women 13,019 breast cancer cases
Sprague et al. (2008)	U.S.	Collaborative Breast Cancer Study— Wisconsin, Massachusetts, New Hampshire (1996–2000)
	Unmatched case–control	
	Cases from cancer registries; population-based controls	Women, ages 20–69 (95% white, non-Hispanic)
		3,499 cases 4,213 controls

Risk Factors and Relative Risk Estimates[a]	Population Attributable Risk in Percent[b] (95% CI, where available)	Impact on Absolute Risk
Alcohol (2+ drinks/day on 10+ days/mo) = 1.4 HT (E+P) = 1.26 Physical inactivity (no vigorous/moderate activity in past month) = 1.3	Alcohol: 3.5 HT (E+P): 4.4 Physical inactivity: 7.5	Alcohol: 450 cases/yr HT (E+P): 567 cases/yr Physical inactivity: 1,422 cases/yr
Not modifiable Age at menarche (< age 15 yrs) = 1.20–1.37 Age at menopause (≥ age 45 yrs) = 1.22–1.40 Age at first full-term pregnancy (< age 20 yrs, parous women only) = 1.02–1.42 Parity (< 4 births) = 1.13–1.35 1st degree family history = 1.66 History of benign breast disease = 1.53 Height at age 25 (≥ 1.6 m) = 1.11–1.27 Modifiable Alcohol (≥ 1 drink/wk) = 1.12–1.43 HT, current use (E, E+P, or other) = 0.96–1.31 Physical inactivity (≤ 5 hrs/wk) = 1.17–1.26 Weight gain (since age 18, > 5 kg) = 1.27–1.57	Not modifiable 57.3 (47.5–65.4) Age at menarche: 18.8 (7.9–29.0) Age at menopause: 13.7 (6.6–19.6) Age at first full-term pregnancy: 5.2 (–3.2–13.9) Parity: 13.3 (6.9–19.8) 1st degree family history: 8.5 (6.5–10.5) History of benign breast disease: 9.7 (7.3–12.0) Height at age 25: 11.0 (3.5–18.5) Modifiable 40.7 (23.0–55.1) Alcohol: 6.1 (2.1–10.3), HT, current use: 4.6 (–3.5–11.9) Physical activity: 15.7 (–6.5–33.7) Weight gain: 21.3 (13.1–29.3)	Not reported

continued

TABLE D-1 Continued

Source	Study Location and Type	Study Population Characteristics
Europe and Canada		
Mezzetti et al. (1998)	Italy Case–control	Unmatched hospital cases (ages 23–74, median 55 yrs) and controls (ages 20–74, median 56 yrs) (1991–1994) 2,569 cases 2,588 controls
Bakken et al. (2004)	Norway Cohort Cases identified from national cancer registry	Norwegian Women and Cancer study Postmenopausal women, ages 45–64, recruited in 1991–1992 or 1996–1997 31,451 in cohort
Neutel and Morrison (2010)	Canada Survey data for risk factor prevalence Published literature for RRs National age-adjusted cancer incidence rates	National Population Health Survey (1994–2006) Canadian women ages 50–69

Risk Factors and Relative Risk Estimates[a]	Population Attributable Risk in Percent[b] (95% CI, where available)	Impact on Absolute Risk
Low levels of physical activity = 1.5 Alcohol consumption (> 20 g/day) = 1.25	Aggregate: 19.2 (1.5–36.8) Physical activity: 11.6 (–0.1–23.3) Alcohol: 10.7 (4.4–17.0)	Not reported
HT, ever used = 1.9 HT, current use = 2.1 (HT formulations different from those in U.S.)	Current use of HT: 27	300 cases/yr
Modifiable risk factors: Alcohol (> 9 drinks/wk) = 1.4 HT use = 1.4 Obesity (BMI ≥ 30 kg/m(2)) = 1.4 Physical inactivity = 1.15 Smoking, current = 1.25	*2000* All: 28.9 Alcohol: 1.8 HT use: 11.5 Obesity: 7.6 Physical inactivity: 8.0 Smoking: 3.8 *2006* All: 23.6 Alcohol: 2.6 HT use: 5.2 Obesity: 8.8 Physical inactivity: 6.4 Smoking: 3.1	Not reported

continued

TABLE D-1 Continued

Source	Study Location and Type	Study Population Characteristics
Barnes et al. (2010)	Germany Case–control	Mammary carcinoma Risk factor Investigation (MARIE) (2001–2005) Postmenopausal women, ages 50–74 3,074 cases 6,386 controls
Friedenreich et al. (2010)	Europe (15 countries) Survey data for risk factor prevalence Review of published literature for RR Incidence from estimates by IARC	Eurobarometer, Wave 58.2 (2002) Incidence (2008)

Risk Factors and Relative Risk Estimates[a]	Population Attributable Risk in Percent[b] (95% CI, where available)	Impact on Absolute Risk
Not modifiable Age at menarche (< age 15 yrs) = 1.11–1.16 Age at menopause (≥ age 45 yrs) = 1.12–1.36 Parity (< 3 births) = 1.08–1.30 1st degree family history = 1.49 History of benign breast disease = 1.24	Not modifiable: 37.2 (27.1–47.2) Age at menarche: 7.7 (0.2–14.1) Age at menopause: 12.0 (3.9–20.2) Parity: 10.9 (1.3–18.8) 1st degree family history: 5.7 (4.1–7.5) History of benign breast disease: 7.9 (4.4–11.6)	Not reported
Modifiable Alcohol (≥ 1 g/day) = 0.93–0.93 BMI (> 22.4 kg/m(2)) = 0.93–1.06 HT, current use (E, E+P, or other) = 1.19–2.25 Physical activity (< 76.5 MET hrs/wk of recreational activities since age 50 yrs) = 1.16–1.23	Modifiable: 26.3 (13.7–37.5) Alcohol: –7.6 (-21.1–3.6) BMI: 2.4 (-2.8–7.4) HT: 19.4 (15.9–23.2) Physical activity: 12.8 (5.5–20.8)	
Physical activity = 0.75 Sufficiently active (3,000 MET-minutes in 7 days or 1,500 MET-minutes of vigorous activity over 3 or more days) Not sedentary (≥ 600 MET-minutes over 7 days)	Insufficiently active: 20 Sedentary: 10	Insufficiently active: 83,353 cases/yr Sedentary: 42,837 cases/yr

continued

TABLE D-1 Continued

Source	Study Location and Type	Study Population Characteristics
Petracci et al. (2011)	Italy Model to predict absolute risk; tested with independent data	Absolute risks: Florence cancer registry (1989–1993) For model development: Case–control subjects (1991–1994), ages 20–74 For validation: Florence-EPIC cohort (1998–2004), ages 35–64

NOTES: Variation across studies in estimated PAR values reflects differences in the prevalence of exposure, in overlap among multiple risk factors, in susceptibility to the risk factor, and in the degree of control for confounding. For these and other reasons, the PARs should be viewed as ballpark estimates based on current science and the assumption that measured associations for these factors are primarily causal.

REFERENCES

Bakken, K., E. Alsaker, A. E. Eggen, and E. Lund. 2004. Hormone replacement therapy and incidence of hormone-dependent cancers in the Norwegian Women and Cancer study. *Int J Cancer* 112(1):130–134.

Barnes, B. B., K. Steindorf, R. Hein, D. Flesch-Janys, and J. Chang-Claude. 2010. Population attributable risk of invasive postmenopausal breast cancer and breast cancer subtypes for modifiable and non-modifiable risk factors. *Cancer Epidemiol* 35(4):345–352.

Clarke, C., D. Purdie, and S. Glaser. 2006. Population attributable risk of breast cancer in white women associated with immediately modifiable risk factors. *BMC Cancer* 6(1):170.

Friedenreich, C. M., H. K. Neilson, and B. M. Lynch. 2010. State of the epidemiological evidence on physical activity and cancer prevention. *Eur J Cancer* 46(14):2593–2604.

Madigan, M. P., R. G. Ziegler, J. Benichou, C. Byrne, and R. N. Hoover. 1995. Proportion of breast cancer cases in the United States explained by well-established risk factors. *J Natl Cancer Inst* 87(22):1681–1685.

Mezzetti, M., C. La Vecchia, A. Decarli, P. Boyle, R. Talamini, and S. Franceschi. 1998. Population attributable risk for breast cancer: Diet, nutrition, and physical exercise. *J Natl Cancer Inst* 90(5):389–394.

Risk Factors and Relative Risk Estimates[a]	Population Attributable Risk in Percent[b] (95% CI, where available)	Impact on Absolute Risk
Aggregate contribution of elimination of risk factors (no alcohol consumption, physical activity of ≥ 2 hrs per week, and BMI at age ≥ 50 yrs of < 25 kg/m²)	Entire population 10-yr period Age 45: 20.5 (11.0–29.2) Age 55: 24.5 (14.7–34.2) Age 65: 24.5 (14.6–34.1) 20-yr period Age 45: 20.9 (11.6–29.6) Age 55: 24.0 (14.4–33.7) Age 65: 24.0 (14.3–33.6)	Absolute risk reduction from elimination of risk factors, reported in percentage point change (95% CI) Entire population 10-yr period Age 45: 0.6 (0.3–1.0) Age 55: 0.8 (0.5–1.1) Age 65: 0.9 (0.5–1.3) 20-yr period Age 45: 1.4 (0.7–2.0) Age 55: 1.6 (0.9–2.3) Age 65: 1.6 (0.9–2.3)

Abbreviations: BMI, body mass index; CI, confidence interval; E, estrogen-only hormone therapy; E+P, estrogen–progestin hormone therapy; EPIC, European Prospective Investigation into Cancer; HT, hormone therapy; IARC, International Agency for Research on Cancer; MET, metabolic equivalent; NHANES, National Health and Nutrition Examination Survey; PAR, population attributable risk; RR, relative risk.

[a]Includes relative risks or odds ratios, when reported; range of relative risks provided when multiple risk categories were used in the original report.

[b]Population attributable risk is the fraction of all cases of breast cancer in the studied population in which the factor of interest appears to play a role.

Neutel, C. I., and H. Morrison. 2010. Could recent decreases in breast cancer incidence really be due to lower HRT use? Trends in attributable risk for modifiable breast cancer risk factors in Canadian women. *Can J Public Health* 101(5):405–409.

Petracci, E., A. Decarli, C. Schairer, R. M. Pfeiffer, D. Pee, G. Masala, D. Palli, M. H. Gail. 2011. Risk factor modification and projections of absolute breast cancer risk. *J Natl Cancer Inst* 103(13):1037–1048.

Rockhill, B., C. R. Weinberg, and B. Newman. 1998. Population attributable fraction estimation for established breast cancer risk factors: Considering the issues of high prevalence and unmodifiability. *Am J Epidemiol* 147(9):826–833.

Sprague, B. L., A. Trentham-Dietz, K. M. Egan, L. Titus-Ernstoff, J. M. Hampton, and P. A. Newcomb. 2008. Proportion of invasive breast cancer attributable to risk factors modifiable after menopause. *Am J Epidemiol* 168(4):404–411.

Tseng, M., C. R. Weinberg, D. M. Umbach, and M. P. Longnecker. 1999. Calculation of population attributable risk for alcohol and breast cancer (United States). *Cancer Causes Control* 10(2):119–123.

Appendix E

Glossary

Absolute risk: Risk of developing a disease over a set time period. Similar in concept to the incidence rate, calculated by dividing the number of new cases of a given disease by the number of people at risk for the disease over a defined period of time.

Adaptive immunity: A second line of defense, distinct from the more generalized innate immune response, in which the body reacts in a manner specific to invading pathogens and any toxic molecules they produce. Adaptive immune responses often confer long-lasting protection from a specific pathogen (adapted from Alberts et al., 2002).

Adipose tissue: Specialized connective tissue composed of adipocytes, or fat cells.

Allele: One of two or more versions of a gene. An individual inherits two alleles for each gene, one from each parent. If the two alleles are the same, the individual is homozygous for that gene; if the two alleles are different, the individual is said to be heterozygous (NHGRI, 2011).

Androgen: A sex hormone responsible for the development and maintenance of male sex characteristics.

Apoptosis: A complex program of cellular self-destruction and death (adapted from Weinberg, 2007).

Aromatase: An enzyme that catalyzes the conversion of testosterone to estradiol.

Association: Statistical relationship between two or more events, characteristics, or other variables (CDC, 2010).

Attributable risk: The percentage of cases that occur in the exposed group that is in excess of the cases in the comparison group.

Benign: Describing a growth that is confined to a specific site within a tissue and gives no evidence of invading adjacent tissue (Weinberg, 2007). Benign tumors are typically not life threatening.

Biomarker: A characteristic that is objectively measured and evaluated as an indicator of normal biological processes, pathogenic processes, or pharmacologic responses to an intervention (IOM, 2010).

Body mass index (BMI): A measure of adiposity calculated from an individual's height and weight as an alternative for direct measures of body fat (CDC, 2011).

BRCA1/BRCA2: Human genes that belong to a class of genes known as tumor suppressors. Mutation of these genes has been linked to hereditary breast and ovarian cancers (NCI, 2011).

Carcinogen: An agent that contributes to the formation of a tumor (Weinberg, 2007).

Case–control study: A type of observational analytic study. Enrollment into the study is based on presence ("case") or absence ("control") of disease. Characteristics such as previous exposure are then compared between cases and controls (CDC, 2010).

Chromatin: Complex of DNA, histones, and non-histone proteins found in the nucleus of a eukaryotic cell. The material of which chromosomes are made (Alberts et al., 2002).

Circadian rhythm: Physical, mental and behavioral changes that follow a roughly 24-hour cycle, responding primarily to light and darkness in an organism's environment (NIGMS, 2011).

Cluster: An aggregation of cases of a disease or other health-related condition, particularly cancer and birth defects, which are closely grouped in time and place. The number of cases may or may not exceed the expected number; frequently the expected number is not known (CDC, 2010).

Cohort: A well-defined group of people who are followed up for the incidence of new diseases or events, as in a cohort or prospective study. A group of people born during a particular period or year is called a birth cohort (CDC, 2010).

Confidence interval: A range of values for a statistic of interest, such as a rate, constructed so that this range has a specified probability of including the true value of the variable. The specified probability is called the confidence level, and the endpoints of the confidence interval are called the confidence limits (CDC, 2010). It may be thought of as the range of values that are consistent at a given level of confidence with a quantitative observation or measurement.

Confidence limit: The minimum or maximum value of a confidence interval (CDC, 2010).

Confounding factor: A variable or characteristic that is causally related to the outcome of interest, and that is also related to the exposure, but is not a consequence of the exposure.

Control: In a case–control study, comparison group of persons without disease (CDC, 2010).

Correlation: Indicative of a relationship between two measurements.

Cytochrome p450 (CYP) enzyme system: A superfamily of hundreds of closely related hemeproteins found throughout the phylogenetic spectrum, from animals, plants, fungi, to bacteria. They include numerous complex monooxygenases (mixed-function oxygenases). In animals, these CYP enzymes serve two major functions: (1) biosynthesis of steroids, fatty acids, and bile acids; (2) metabolism of endogenous and a wide variety of exogenous substrates, such as toxins and drugs (biotransformation) (National Library of Medicine, 2011).

DMBA: 7,12-Dimethylbenz[a]anthracene, a polycyclic aromatic hydrocarbon that is a potent carcinogen often used as a tumor initiator in laboratory studies.

DNA: Deoxyribonucleic acid or the double-stranded helix of nucleotides carrying the genetic information of the cell. Encodes information for proteins and is able to self-replicate.

DNA adduct: Covalent adducts between chemical mutagens and DNA. Such couplings activate DNA repair processes and, unless repaired prior to DNA replication, may lead to nucleotide substitutions, deletions, and chromosome rearrangements (Rieger et al., 2007).

DNA methylation: Addition of a methyl group to DNA. Extensive methylation of the cytosine base in CG sequences is used in vertebrates to keep genes in an inactive state; one mechanism for "epigenetic" regulation of gene expression.

Ductal carcinoma in situ (DCIS): A noninvasive breast lesion in which abnormal cells multiply and form a growth within the milk duct of the breast. Some of these lesions can progress to become invasive cancers, but the risk of progression is poorly quantified.

Ecological fallacy: The thinking that relationships observed for groups necessarily hold for individuals (Freedman, 1999).

Endocrine: Referring to a gland that secretes fluids into the general circulation, or the signal pathway of a hormone or factor.

Endocrine disrupting compound (EDC): An exogenous agent that interferes with the production, release, transport, metabolism, binding, action, or elimination of the natural hormones in the body responsible for the maintenance of homeostasis and the regulation of developmental processes (EPA, 2011).

Endogenous: Developing or originating within; produced normally by the body.

Environment: Factors not directly heritable through DNA.

Epigenetics: An epigenetic trait is a stably heritable phenotype resulting from changes in a chromosome without alterations in the DNA sequence (Berger et al., 2009)

Epigenome: A set of changes to the genome passed down from one generation to the next that does not include alterations in genetic material.

Epoxide: An organic compound containing a three-membered ring of an oxygen atom and two carbon atoms.

Estrogen: Female reproductive hormone produced by the ovaries.

Estrogen response element (ERE): Recognition site in the regulatory portion of a gene to which the estrogen receptor binds when complexed with a "ligand" (estrogen or estrogen-like chemicals).

Etiology: Cause or origin (of a disease).

Exogenous: Originating from external factors; not produced internally by the body.

Extracellular matrix: A component of tissues largely filled by an intricate network of macromolecules and composed of a variety of proteins and polysaccharides that are secreted locally and assembled into an organized meshwork in close association with the surface of the cell that produced them (Alberts et al., 2002).

G protein coupled receptor (GPCR): A 7 transmembrane domain receptor that activates a protein upon ligand binding, resulting in a second messenger-initiated pathway involving a series of intracellular events.

Genome: The totality of genetic information belonging to a cell or an organism; in particular, the DNA that carries this information (Alberts et al., 2002).

Hazard: Potential to cause harm; differs from risk.

Heritable: Able to be passed from one generation to the next.

High-risk group: A group in the community with an elevated risk of disease (CDC, 2010).

Histone: One of a group of small abundant proteins, rich in arginine and lysine, four of which form the nucleosome on the DNA in eukaryotic chromosomes (Alberts et al., 2002).

Immunosurveillance: A process in which the immune system of a host recognizes antigens of newly arising tumors and eliminates these tumors before they become clinically evident (adapted from Ochsenbein, 2002).

in situ: Occurring at the site of origin (Weinberg, 2007).

in utero: Occurring in the uterus during embryonic or fetal development (Weinberg, 2007).

in vitro: Occurring in tissue culture, in cell lysates, or in purified reaction systems in the test tube (Weinberg, 2007).

in vivo: Occurring in a living organism (Weinberg, 2007).

Incidence rate: A measure of the frequency with which an event, such as a new case of illness, occurs in a population over a period of time. The denominator is the population at risk; the numerator is the number of new cases occurring during a given time period (CDC, 2010). For example, an incidence rate of ovarian cancer is only calculated in women, and the incidence of prostate cancer is only calculated in men.

Lactation: Period following birth in which milk is secreted.

Latency: Elapsed time between exposure and response.

Malignant: Describing a growth that shows evidence of being locally invasive and possibly even metastatic; malignant tumors are generally considered life threatening.

Mammary gland: The milk-producing glands of female mammals.

Melatonin: A hormone secreted by the pineal gland, which helps regulate other hormones and maintain the body's circadian rhythm.

Menarche: First occurrence of menstruation in a woman.

Menopause: The point at which menstruation ceases in a woman's life.

Messenger RNA: See mRNA.

Meta-analysis: Statistical technique for combining the findings from independent studies (Crombie and Davies, 2009).

Metastasis: The spread of cancer from its site of origin to another part of the body.

Methylation: See DNA methylation.

Mitosis: The process by which a single cell separates its complement of chromosomes into two equal sets in preparation for the division into two daughter cells (Weinberg, 2007).

Morbidity: Any departure, subjective or objective, from a state of physiological or psychological well-being (CDC, 2010).

Mortality rate: A measure of the frequency of occurrence of death in a defined population during a specified interval of time (CDC, 2010).

mRNA: RNA sequences that serve as templates for protein synthesis (Reiger et al., 2007).

Multistage carcinogenesis: A conceptual division of carcinogenesis into four steps: tumor initiation, tumor promotion, malignant conversion, and tumor progression (Alberts et al., 2002).

Mutagen: An agent that induces a mutation.

Mutagenesis: The process by which a change in nucleotide sequence, or mutation, occurs.

Mutation: Heritable change in the nucleotide sequence of a chromosome (Alberts et al., 2002).

Neoplasm: An abnormal new tumor mass. Neoplasms may be benign or malignant.

Observational study: Epidemiologic study in situations where nature is allowed to take its course. Changes or differences in one characteristic are studied in relation to changes or differences in others, without the intervention of the investigator (CDC, 2010).

Odds ratio: A measure of association that quantifies the relationship between an exposure and health outcome from a comparative study. For low-frequency outcomes, such as cancer, the odds ratio generally provides a good estimate of relative risk (see Relative risk) (CDC, 2010).

Oncogene: A cancer-inducing gene (Weinberg, 2007).

p53: Tumor suppressor gene found mutated in about half of human cancers. It encodes a gene regulatory protein that is activated by damage to DNA and is involved in blocking further progression through the cell cycle (Alberts et al., 2002).

Paracrine: Referring to the signaling path of a hormone or factor that is released by one cell and acts on a nearby cell (Weinberg, 2007).

Parity: The status of having given birth (parous), or not (nulliparous); the number of children a woman has borne (Weinberg, 2007).

Phytoestrogen: A chemical produced by plants that can mimic the hormone estrogen (Sprecher Institute for Comparative Research, 2001).

Polymorphism: Genetic polymorphism (literally means "many forms") is defined as the occurrence of two or more relatively common normal alleles for a single locus. The difference between a polymorphism and a mutation is that a polymorphism occurs commonly and it is associated with a normal phenotype (Kufe et al., 2003).

Population: The total number of inhabitants of a given area or country. In sampling, the population may refer to the units from which the sample is drawn, not necessarily the total population of people (CDC, 2010).

Population attributable risk (PAR): Population-based measure of the percentage of excess cases associated with the exposure of interest (assuming that the relationship between the exposure and the disease outcome is causal) that also takes into account the distribution of that exposure within the population.

Prevalence: The number or proportion of cases or events or conditions in a given population at a given time (CDC, 2010).

Prospective cohort study: A type of observational study that follows over a period of time a group of individuals, or cohort, who differ with respect to certain factors under study, in order to determine how these factors affect rates of an outcome.

Randomized controlled trial: A study that randomly assigns individuals to an intervention group or to a control group, in order to measure the effects of the intervention (U.S. Department of Education, 2004). When practical and ethical, this study design provides the most direct evidence of a causal association between an exposure and an outcome.

Reactive oxygen species: Molecules and ions of oxygen that have an unpaired electron, thus rendering them extremely reactive. Superoxide anion and hydroxyl radicals are the most common examples.

Receptor: A protein that is capable of specifically binding a signal molecule.

Relative risk: A ratio of the absolute risk (incidence) of disease in an exposed group (or groups with different levels of exposure) to the absolute risk (incidence) of disease in an unexposed group (or some other designated comparison group).

Risk: The probability that an event will occur or the likelihood of harm, for example, that an individual will become ill or die within a stated period of time or by a certain age (CDC, 2010).

RNA: Ribonucleic acid produced through DNA transcription, exists in ribosomal (rRNA), messenger (mRNA), and transfer (tRNA) forms, and provides the template for protein translation.

Statistical power: Power of a statistical test is the probability that the test will reject the null hypothesis when the alternative hypothesis is true.

Stem cell: Cell type within a tissue that is capable of self-renewal and is also capable of generating daughter cells that develop new phenotypes, including those that are more differentiated than the phenotype of the cell (Weinberg, 2007).

Stroma: The mesenchymal components of the epithelial and hematopoetic tissues and tumors, which may include fibroblasts, adipocytes, endothelial cells, and various immunocytes as well as associated extracellular matrix (Weinberg, 2007).

Target cell: A cell that is acted on by an outside agent; a particular cell type that is specifically or uniquely altered by a xenobiotic, hormone, or other stimulus that is external to the cell.

Thelarche: Onset of breast development in prepubertal girls.

Toxicology: The study of the adverse effects of chemical or physical agents on living organisms (Klaassen, 2001).

Transcription factor: Protein that binds to specific DNA sequences, facilitating the transfer of information from DNA to RNA.

Tumor: An abnormal growth of body tissue.

Tumor suppressor gene: A gene whose partial or complete inactivation, occurring in either the germ line or the genome of a somatic cell, leads to an increased likelihood of cancer progression (Weinberg, 2007).

Validity: The degree to which a measurement actually measures or detects what it is supposed to measure (CDC, 2010).

Variable: Any characteristic or attribute that can be measured (CDC, 2010).

Variance: A measure of the dispersion shown by a set of observations, defined by the sum of the squares of deviations from the mean, divided by the number of degrees of freedom in the set of observations (CDC, 2010).

Xenoestrogens: Compounds that mimic estrogens, although they may differ chemically from the forms of estrogen produced normally in the body.

REFERENCES

Alberts, B., A. Johnson, J. Lewis, M. Raff, K. Roberts, and P. Walter. 2002. *Molecular biology of the cell*, 4th ed. New York: Garland Science. http://www.ncbi.nlm.nih.gov/books/NBK21070 (accessed November 18, 2011).

Berger, S. L., T. Kouzarides, R. Shiekhattar, and A. Shilatifard. 2009. An operational definition of epigenetics. *Genes Dev* 23(7):781–783.

CDC (Centers for Disease Control and Prevention). 2010. *Epi glossary*. http://www.cdc.gov/reproductivehealth/Data_Stats/Glossary.htm (accessed December 21, 2011).

CDC. 2011. *About BMI for adults*. http://www.cdc.gov/healthyweight/assessing/bmi/adult_bmi/index.html (accessed November 18, 2011).

Crombie, I. K., and H. T. O. Davies. 2009. *What is meta-analysis?* http://www.whatisseries.co.uk/whatis/ (accessed November 18, 2011).

EPA (Environmental Protection Agency). 2011. *Endocrine disruptors research*. http://www.epa.gov/ncer/science/endocrine/index.html (accessed November 18, 2011).

Freedman, D. A. 1999. *Ecological inference and the ecological fallacy*. http://www.stanford.edu/class/ed260/freedman549.pdf (accessed November 18, 2011).

IOM (Institute of Medicine). 2010. *Evaluation of biomarkers and surrogate endpoints in chronic disease*. Washington, DC: The National Academies Press.

Klaassen, C. D., ed. 2001. *Casarett and Doull's Toxicology: The basic science of poisons*, 6th ed. New York: McGraw Hill.

Kufe, D. W., R. E. Pollock, R. R. Weichselbaum, R. C. Bast, T. S. Gansler, J. F. Holland, and E. Frei (eds). 2003. *Holland-Frei cancer medicine*, 6th ed. Hamilton, Ontario, Canada: BC Decker. http://www.ncbi.nlm.nih.gov/books/NBK12354/ (accessed November 18, 2011).

National Library of Medicine. 2011. *Genetics home reference: Your guide to understanding genetic conditions*. http://ghr.nlm.nih.gov/glossary=cytochromep450 (accessed November 18, 2011).

NCI (National Cancer Institute). 2011. http://www.cancer.gov (accessed November 18, 2011).

NHGRI (National Human Genome Research Institute). 2011. *Talking glossary of genetic terms.* http://www.genome.gov/glossary/?id=4 (accessed November 18, 2011).

NIGMS (National Institute of General Medical Sciences). 2011. *Circadian rhythms fact sheet.* http://www.nigms.nih.gov/Education/Factsheet_CircadianRhythms.htm (accessed November 18, 2011).

Ochsenbein, A. F. 2002. Principles of tumor immunosurveillance and implications for immunotherapy. *Cancer Gene Ther* 9(12):1043–1055.

Rieger, R., A. Michaelis, and M. M. Green. 1991. *Glossary of genetics: Classical and molecular,* 5th ed. New York: Springer-Verlag. http://www.reference.md/files/D018/mD018736.html (accessed November 18, 2011).

Sprecher Institute for Comparative Research. 2001. *Phytoestrogens and breast cancer.* http://envirocancer.cornell.edu/factsheet/diet/fs1.phyto.cfm (accessed November 18, 2011).

U.S. Department of Education. 2004. *Identifying and implementing educational practices supported by rigorous evidence: A user friendly guide.* http://www2.ed.gov/rschstat/research/pubs/rigorousevid/guide_pg5.html (accessed November 18, 2011).

Weinberg, R. A. 2007. *The biology of cancer.* New York: Garland Science.

Appendix F

Ionizing Radiation Exposure to the U.S. Population, with a Focus on Radiation from Medical Imaging

Appendix F is available online at

http://www.nap.edu/catalog.php?record_id=13263